Applications of EPR and NMR Spectroscopy in Homogeneous Catalysis

Applications of EPR and NMR Spectroscopy in Homogeneous Catalysis

Evgenii Talsi
Konstantin Bryliakov

CRC Press
Taylor & Francis Group
Boca Raton London New York

CRC Press is an imprint of the
Taylor & Francis Group, an **informa** business

CRC Press
Taylor & Francis Group
6000 Broken Sound Parkway NW, Suite 300
Boca Raton, FL 33487-2742

© 2017 Taylor & Francis Group, LLC
CRC Press is an imprint of Taylor & Francis Group, an Informa business

No claim to original U.S. Government works

Printed on acid-free paper

International Standard Book Number-13: 978-1-4987-4263-4 (Hardback)

This book contains information obtained from authentic and highly regarded sources. Reasonable efforts have been made to publish reliable data and information, but the author and publisher cannot assume responsibility for the validity of all materials or the consequences of their use. The authors and publishers have attempted to trace the copyright holders of all material reproduced in this publication and apologize to copyright holders if permission to publish in this form has not been obtained. If any copyright material has not been acknowledged please write and let us know so we may rectify in any future reprint.

Except as permitted under U.S. Copyright Law, no part of this book may be reprinted, reproduced, transmitted, or utilized in any form by any electronic, mechanical, or other means, now known or hereafter invented, including photocopying, microfilming, and recording, or in any information storage or retrieval system, without written permission from the publishers.

For permission to photocopy or use material electronically from this work, please access www. copyright.com (http://www.copyright.com/) or contact the Copyright Clearance Center, Inc. (CCC), 222 Rosewood Drive, Danvers, MA 01923, 978-750-8400. CCC is a not-for-profit organization that provides licenses and registration for a variety of users. For organizations that have been granted a photocopy license by the CCC, a separate system of payment has been arranged.

Trademark Notice: Product or corporate names may be trademarks or registered trademarks, and are used only for identification and explanation without intent to infringe.

Visit the Taylor & Francis Web site at
http://www.taylorandfrancis.com

and the CRC Press Web site at
http://www.crcpress.com

Contents

Preface...ix
Authors...xi
Abbreviations.. xiii

Chapter 1   Basic Principles of EPR and NMR Spectroscopy...............................1

    1.1   Electron and Nuclear Magnetic Moments in the Magnetic
         Field...1
    1.2   Instrumentation for CW EPR Spectroscopy...............................4
    1.3   Main Parameters Characterizing EPR Spectra...........................4
    1.4   EPR Spectra of Frozen Solutions...8
    1.5   Instrumentation for Pulsed FT NMR Spectroscopy..............12
    1.6   Main Characteristics of NMR Spectra...................................12
         1.6.1   Chemical Shift...12
         1.6.2   Spin–Spin Coupling..14
         1.6.3   NMR Line Width..14
    1.7   Bloch Equations..16
    1.8   Pulsed FT NMR Spectroscopy...21
    References...24

Chapter 2   Some NMR Spectroscopic Techniques Used in Homogeneous
Catalysis...27

    2.1   Measuring ^1H and ^{13}C NMR Spectra of a Sample
         Compound...29
         2.1.1   One- and Two-Dimensional ^1H and ^{13}C NMR
                Spectra..29
         2.1.2 ^1H COSY versus ^1H TOCSY Spectra.......................37
    2.2 ^1H Spectra of Paramagnetic Molecules.................................39
         2.2.1   Probing the Structure of an Unknown Ni(II)
                Complex by Multinuclear NMR..................................40
         2.2.2   Temperature Dependence of the Paramagnetic
                Shift..42
              2.2.2.1   Temperature Dependence
                      of the Paramagnetic Shift
                      of Monomeric Compounds.........................43
             2.2.2.2   Temperature Dependence
                      of Paramagnetic Shift of
                      Antiferromagnetic Dimers..........................45
              2.2.2.3   Measuring Magnetic Susceptibility
                      (Evans Method) for Studying Spin
                      Equilibrium...47
    References...51

v

vi   Contents

Chapter 3 NMR and EPR Spectroscopy as a Tool for the Studies of
Intermediates of Transition Metal–Catalyzed Oxidations 55

   3.1   Superoxo Complexes 56
       3.1.1   Superoxo Complexes of Co(III) 56
       3.1.2   Superoxo Complexes of Pd(II) 59
       3.1.3   Superoxo Complexes of Ni(II) 61
       3.1.4   Superoxo Complexes of Copper(II) 63
       3.1.5   Superoxo Complexes of Iron(III) 65
   3.2   Alkylperoxo Complexes 67
       3.2.1   Alkylperoxo Complexes of Molybdenum 68
       3.2.2   Alkylperoxo Complexes of Titanium 72
       3.2.3   Alkylperoxo Complexes of Vanadium 76
   3.3   Peroxo Complexes 80
       3.3.1   Peroxo Complexes of Molybdenum 80
       3.3.2   Peroxo Complexes of Vanadium 83
       3.3.3   Peroxo Complexes of Titanium 87
   3.4   Oxo Complexes 88
       3.4.1   Oxocomplexes $[Cr^V = O(Salen)]^+$ 88
       3.4.2   Oxocomplexes $[Mn^V – O(Salen)]^+$ 91
       3.4.3   Oxocomplexes $[(L)Fe^V{=}O]^{3+}$ (L = Tetradentate
N-Donor Ligand) as Proposed Active Species of
Selective Epoxidation of Olefins 95
       3.4.4   EPR Spectroscopic Detection of the Elusive
$Fe^V{=}O$ Intermediates in Selective Catalytic
Epoxidation of Olefins Mediated by Ferric
Complexes with Substituted Aminopyridine
Ligands 104
   3.5   Structure of Co(III) Acetate in Solution 113
   References 116

Chapter 4 NMR and EPR Spectroscopy in the Study of the Mechanisms
of Metallocene and Post-Metallocene Polymerization and
Oligomerization of α-Olefins 127

   4.1   Metallocene Catalysts 127
       4.1.1   Introduction 127
       4.1.2   Size of MAO Oligomers 128
       4.1.3   On the Active Centers of MAO 130
       4.1.4   Structure of Ion Pairs Formed upon the
Interaction of Cp_2ZrMe_2 with MAO 132
       4.1.5   Detection of Ion Pairs Formed upon Activation
of $(Cp-R)_2ZrCl_2$ (R = nBu, tBu) with MAO 134
       4.1.6   Detection of Ion Pairs Formed in the Catalyst
Systems $(Cp-R)_2ZrCl_2$/MAO (R = Me, 1,2-Me_2,
1,2,3-Me_3, 1,2,4-Me_3, Me_4) 137

Contents

vii

4.1.7   Ion Pairs Formed upon Activation of *Ansa*-Zirconocenes with MAO 139

4.1.8   Ion Pairs Formed upon Interaction of Cp_2TiCl_2 and *Rac*-$C_2H_4(Ind)_2TiCl_2$ with MAO 143

4.1.9   Ion Pairs Formed upon Activation of (C_5Me_5) $TiCl_3$ and $[(Me_4C_5)SiMe_2N^tBu]TiCl_2$ with MAO 148

4.1.10 Observation of Ion Pairs Formed in the Catalyst Systems Zirconocene/MMAO 149

4.1.11 Ion Pairs Formed in the Catalyst Systems Metallocene/AliBu$_3$/[Ph$_3$C][B(C$_6$F$_5$)$_4$] 152

4.1.12 Ion Pairs Operating in the Catalyst Systems Zirconocene/Activator/α-Olefin 156

4.2   Post-Metallocene Catalysts 159

4.2.1   Bis(imino)pyridine Iron Ethylene Polymerization Catalysts 159

      4.2.1.1   Activation of $L^{2iPr}FeCl_2$ with MAO 160

      4.2.1.2   Activation of $L^{2iPr}FeCl_2$ with $AlMe_3$ 161

4.2.2   Bis(imino)pyridine Cobalt Ethylene Polymerization Catalysts 166

      4.2.2.1   Activation of $L^{2iPr}Co^{II}Cl_2$ with MAO 166

      4.2.2.2   Activation of $L^{2iPr}Co^{II}Cl_2$ with $AlMe_3$ 169

4.2.3 α-Diimine Vanadium(III) Ethylene Polymerization Catalysts 171

      4.2.3.1   System $L^{2Me}VCl_3/AlMe_3/[Ph_3C]$ $[B(C_6F_5)_4]$ 172

      4.2.3.2   System $L^{2Me}VCl_3/MAO$ 175

4.2.4   Ethylene Polymerization Precatalyst Based on Calix[4]arene Vanadium(V) Complex 176

      4.2.4.1   Reaction of Calix[4]arene Vanadium(V) Complex with $AlEt_2Cl$ 176

      4.2.4.2   Reaction of Calix[4]arene Vanadium(V) Complex with $AlMe_2Cl$ 178

      4.2.4.3   Reaction of Calix[4]arene Vanadium(V) Complex with $AlEt_3$ 181

4.2.5   Neutral $Ni^{II}\kappa^2$-(N,O)-salicylaldiminato Olefin Polymerization Catalysts 182

      4.2.5.1   Chain-Propagating Species Formed upon Ethylene Polymerization with Neutral Salicylaldiminato Nickel(II) Catalysts 182

      4.2.5.2   Evaluation of the Size of Ni–Polymeryl Species by PFG NMR Spectroscopy 187

      4.2.5.3   Catalyst Deactivation 188

4.2.6   Formation of Cationic Intermediates upon the Activation of Bis(imino)pyridine Nickel Catalysts 190

viii   Contents

        4.2.7 Cationic Intermediates Formed upon the
               Activation of Ni(II) Catalysts with $AlMe_2Cl$
               and $AlEt_2Cl$ 194
  4.3 On the Origin of Living Polymerization over
       o-Fluorinated Post-Titanocene Catalysts 198
  4.4 Selective Ethylene Trimerization by Titanium Complex
       Bearing Phenoxy Imine Ligand 202
  References 206

Index 219

Preface

The high quality of everyday life of the modern society is tightly connected with the progress of all branches of industrial chemistry, which is, in turn, entirely dependent on the permanent upgrading of the existing catalysts and catalytic technologies, as well as designing novel advanced ones. Although the trial-and-error approach continues to be extensively exploited in engineering research directed at designing new catalysts and processes, it is obvious that achieving deep insight into the reaction mechanisms can rationalize the enhancement of existing catalyst systems and provide the keys to the construction of novel catalyzed processes on the basis of well-grounded understanding of the mechanisms of catalytic action of both mature and emerging catalyst systems. The use of powerful spectroscopic methods is often crucial (and in some cases indispensable) for gathering the mechanistic information on the working catalysts.

This book is focused on the applications of nuclear magnetic resonance (NMR) and electron paramagnetic resonance (EPR) spectroscopy to the identification and reactivity studies of active intermediates of industrially attractive catalytic processes. Initially, we aimed at advertising the rich and versatile capabilities of the magnetic resonance spectroscopy for a broad audience of homogeneous catalytic and organic chemists. When elaborating the book proposal, it became clear that the book should also serve to uncover the basic principles of the NMR and EPR phenomena, and advantages and drawbacks of spectroscopic techniques founded thereupon (essentially keeping in mind the needs of graduate students and their supervisors). That is why the book begins with a short introduction to the basic principles of NMR and EPR spectroscopy (Chapter 1), which could be helpful for the readers in following the interpretation of experimental spectra discussed in the book. Chapter 2 is aimed at demonstrating the utility of various ^1H and ^{13}C NMR spectroscopic techniques, and at getting insight into the information it can provide. Also, some aspects of the NMR of paramagnetic molecules are discussed in Chapter 2.

The rest of the book is an overview of advancements in the NMR and EPR spectroscopic investigations of the mechanisms of two types of homogeneous catalytic processes: (1) chemo- and stereoselective oxidation of organic substrates, promoted by transition metal complexes (Chapter 3), and (2) coordination–insertion polymerization and oligomerization of olefins over single-site (metallocene and post-metallocene) catalysts (Chapter 4). The scope of the review is purely the authors' choice; we mostly adhered to those research areas that have attracted much interest from industry and academia in the last 20–30 years.

Evgenii Talsi
Konstantin Bryliakov

Authors

Evgenii Talsi obtained a *Cand. Chem. Sci.* degree (PhD, 1983) in chemical physics from the Institute of Chemical Kinetics and Combustion (Novosibirsk). In 1991, he obtained a *Dr. Chem. Sci.* degree in catalysis from the Boreskov Institute of Catalysis (Novosibirsk), where he currently leads the Laboratory of the Mechanistic Studies of Catalytic Reactions. His research interests encompass NMR and EPR spectroscopic characterization of key intermediates of homogeneous transition metal–catalyzed oxidations and polymerizations.

Konstantin Bryliakov is a leading research scientist at the Boreskov Institute of Catalysis (Novosibirsk). He obtained a *Cand. Chem. Sci.* degree (PhD, 2001) in chemical physics from the Institute of Chemical Kinetics and Combustion (Novosibirsk). In 2008, Konstantin Bryliakov obtained a *Dr. Chem. Sci.* degree in catalysis from the Boreskov Institute of Catalysis (Novosibirsk). His research interests include selective transition metal–catalyzed oxidative transformations and single-site olefin polymerizations, and mechanistic aspects of those reactions.

Abbreviations

CSA	chemical shift anisotropy
CW	continuous wave
EPR	electronic paramagnetic resonance
ETA	ethyl trichloroacetate
FID	free induction decay
FT	Fourier transform
HF	hyperfine
hfs	hyperfine splitting
HMWPE	high-molecular-weight polyethylene
ICT	intervalence charge transfer
MAO	methylalumoxane
MMAO	modified methylalumoxane
MW	microwave
MWD	molecular weight distribution
NB	norbornene
NMR	nuclear magnetic resonance
NOE	nuclear Overhauser effect
PE	polyethylene
PFGSE	pulsed field-gradient spin echo
RF	radio frequency
SANS	small-angle neutron scattering
shf	superhyperfine
TEMPO	2,2′,6,6′-tetramethylpiperidine-N-oxyl radical
TIBA	triisobutylaluminum
TMA	trimethylaluminum

1 Basic Principles of EPR and NMR Spectroscopy

1.1 ELECTRON AND NUCLEAR MAGNETIC MOMENTS IN THE MAGNETIC FIELD

The basic physical principles of electron paramagnetic resonance (EPR) spectroscopy and nuclear magnetic resonance (NMR) spectroscopy are similar [1–6]. Both methods allow probing with electromagnetic irradiation the energy levels emerging in the molecules when the latter are subjected to an external stationary magnetic field $\vec{B}_0$. In the case of EPR, these levels typically arise from splitting the otherwise degenerate spin states due to the interaction of the magnetic moment of the unpaired electron(s) $\vec{\mu}_e$ with $\vec{B}_0$. In the case of NMR, the magnetic moment of nuclei $\vec{\mu}_N$ (if it is nonzero) interacts with $\vec{B}_0$, giving rise to two (or more, for $I > 1/2$) nondegenerate states, differing in the mutual orientation of $\vec{\mu}_N$ and $\vec{B}_0$ (parallel and antiparallel). The magnetic moment of the unpaired electron or nucleus is proportional to the corresponding angular momentum (spin).

The external magnetic field $\vec{B}_0$ is commonly taken as directed along the z axis. For the simplest (and most widespread) case of $S = 1/2$ compounds, two projections μ_z^e of the electron spin magnetic moment $\vec{\mu}_e$ along the magnetic field $\vec{B}_0$ are possible:

$$\mu_z^e = -g_e \beta_e m_S \tag{1.1}$$

where $g_e = 2.002319$ is the electron spin g-factor, $\beta_e = |e|\hbar/2m_e$ is the *Bohr magneton*, $m_S = \pm 1/2$, and $\hbar$ is the reduced Planck constant.

Similarly, the projections of the proton magnetic moment $\vec{\mu}_P$ (originating from the proton spin) on the z-aligned external magnetic field $\vec{B}_0$ are

$$\mu_z^P = g_P \beta_N m_I \tag{1.2}$$

where $g_P = 5.585695$ is the proton g-factor, $\beta_N = |e|\hbar/2m_P$ is the *nuclear magneton*, and $m_I = \pm 1/2$ (for the case of $I = 1/2$ nuclei). The value of g_e can be calculated theoretically and measured experimentally with very high precision [1], whereas the value of g_P (and g-factors g_N of other nuclei) is less precisely measured [1]. The value of μ_z^e exceeds that of μ_z^P by a factor of ca. 658.2.

The g-factors for nuclei other than proton are rarely used in the NMR literature. Typically, μ_z^N is presented as

$$\mu_z^N = \gamma_N \hbar m_I \tag{1.3}$$

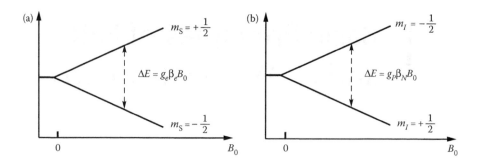

FIGURE 1.1 Zeeman splitting of electron spin (a) and proton spin (b) levels in the external magnetic field.

where γ_N is the so-called *gyromagnetic ratio*; $\gamma_N = g_N \mu_N / \hbar$, where g_N stands for g-factor for nucleus N. The gyromagnetic ratios for various magnetic nuclei can be found in the NMR literature [2].

Taking into account the expression for the energy of interaction of a magnetic moment $\vec{\mu}$ with external magnetic field $\vec{B}$, $E = -\vec{\mu} \cdot \vec{B}$, the splitting of energy levels of magnetic moments of an electron $\vec{\mu}_e$ or a proton $\vec{\mu}_P$ in the external magnetic field $\vec{B}_0$ are expressed as

$$\Delta E^e = g_e \beta_e B_0 \tag{1.4}$$

$$\Delta E^P = |\gamma_P| \hbar B_0 = |g_P| \beta_P B_0 \tag{1.5}$$

Graphically, these splittings are presented in Figure 1.1.

EPR or NMR spectroscopists deal with large ensembles of spins, giving rise to the macroscopic magnetization $\vec{M}$. The equations describing the net magnetization for ensembles of electron and nuclear spins are similar. The relative populations of the two electron spin states $|\alpha\rangle$ ($m_S = +1/2$) and $|\beta\rangle$ ($m_S = -1/2$) are given by the Boltzmann distribution:

$$\frac{n_\alpha}{n_\beta} = \exp\left(-\frac{\Delta E}{kT}\right) \tag{1.6}$$

In a high-temperature approximation ($\Delta E \ll kT$), Equation 1.6 is simplified to Equation 1.7

$$\exp\left(-\frac{\Delta E}{kT}\right) \approx 1 - \frac{\Delta E}{kT} \tag{1.7}$$

Basic Principles of EPR and NMR Spectroscopy

Denoting $N = n_\alpha + n_\beta$ and $\Delta n = n_\beta - n_\alpha$, where n_α and n_β are the number of spins in the $|\alpha\rangle$ and $|\beta\rangle$ states, respectively, one can evaluate the population difference as

$$\Delta n \approx \frac{g_e \beta_e B_0 N}{2kT} \tag{1.8}$$

and the equilibrium magnetization will be defined as

$$\vec{M}_0 = -g_e \beta_e m_S \Delta n \approx \frac{N g_e^2 \beta_e^2 B_0}{4kT} \tag{1.9}$$

The integral intensities of the signals measured by the magnetic resonance methods will be proportional to the macroscopic magnetization. According to Equation 1.9, the latter is proportional to N (the total number of spins in the system). This important result demonstrates the possibility of the quantitative measurement of concentrations by means of EPR and NMR spectroscopy.

The magnetic resonance phenomenon is based on the fact that transitions between the $|\alpha\rangle$ and $|\beta\rangle$ states are possible if the system is irradiated with oscillating electromagnetic field $\vec{B}_1$, which is radio frequency (RF) for NMR (10–1000 MHz) and microwave (MW) frequency for EPR (9–10 GHz for the most widely used X-band EPR spectrometers). The condition for the transitions induced by the oscillating electromagnetic field is fulfilled when the frequency of the oscillating field ν obeys the resonance condition $h\nu = \Delta E$, where ΔE is defined by Equation 1.4 or 1.5. The oscillating magnetic field induces the $|\alpha\rangle \rightarrow |\beta\rangle$ and $|\beta\rangle \rightarrow |\alpha\rangle$ transitions with equal probability. That is why the *resonance absorption* will only be detected if the population difference Δn is nonzero.

The scope of objects that can be studied by EPR spectroscopy is limited to paramagnetic species (organic radicals, some transition metal complexes). At the same time, virtually all chemical substances contain nuclei with nonzero spin (magnetic nuclei) and are thus potentially NMR active [3]. The important advantage of EPR spectroscopy is its higher sensitivity (stemming eventually from higher electron magnetic moment as compared to the magnetic moments of nonzero-spin nuclei [4]). That is why EPR spectroscopy often appears more fruitful for the investigations of the highly reactive transient paramagnetic intermediates of catalytic reactions, whose steady-state concentrations are too low to be observed by NMR. Another advantage of EPR may be the absence of background signals from diamagnetic compounds present in the reaction solution, which enables one to selectively observe the target paramagnetic species.

There are alternative approaches to the detection of magnetic resonance absorption: *continuous-wave* (CW) and *pulsed* magnetic resonance spectroscopy. Accordingly, all NMR and EPR spectrometers belong to one of the two types: CW or pulsed spectrometers. In CW spectrometers, the electromagnetic irradiation is permanently applied to the sample, and the absorption spectrum is obtained by slowly changing the frequency ν ($\Delta E \approx h\nu$) or magnetic field B. In pulsed spectrometers, a short pulse

4   Applications of EPR and NMR Spectroscopy in Homogeneous Catalysis

of electromagnetic radiation containing a range of frequencies centered about the frequency ν (typically, this range is wide enough to excite the whole spectrum of a given molecule) is applied to the sample. The energy absorbed by the sample during this RF pulse is then returned to the receiver coil in the form of oscillating decaying electric signal, the so-called *free induction decay* (FID). Subsequent Fourier transform (FT) converts the time-domain signal (FID) to the frequency-domain signal (spectrum). The latter appears identical to those obtained by using CW spectrometers. The success of FT NMR spectroscopy has been closely connected with the progress of computer technology, which resulted in the emergence of fast and compact computers that could be used together with the spectrometer for performing the FT procedure. Nowadays, pulsed FT NMR spectrometers have completely replaced their CW predecessors, ensuring much higher sensitivity (due to the possibility of repeated application of many RF pulses, with accumulation of data collected after each pulse) and providing a broadest collection of advanced spectroscopic tools for structural characterizations of molecules. On the contrary, the sensitivity restriction rarely occurs in EPR spectroscopic studies, such that good signal-to-noise ratios can be achieved after measuring a single spectrum, without data accumulation. That is why X-band CW EPR spectroscopy remains the method of choice for chemists, both for routine sample measurements and for advanced tasks, such as studies of transient species in catalytic reactions.

1.2   INSTRUMENTATION FOR CW EPR SPECTROSCOPY

A block diagram for a typical CW EPR spectrometer is presented in Figure 1.2. The source of MW radiation is a vacuum tube called a *klystron*. The sample is placed in the resonator cavity, which is connected to the MW source by the wave conductor equipped with an attenuator. In X-band EPR spectrometers, the resonant cavity, placed in the middle of an electromagnet, is a rectangular metal box with characteristic dimensions of ~3 cm. The sample is located in the antinode of the MW field. The waveguides (rectangular open-ended metallic tubes) are used for transferring microwaves from the MW source to the sample and from the sample to the detector (a microwave diode). The majority of EPR spectrometers are *reflection* spectrometers, measuring the amount of radiation that is reflected back out of the resonant cavity. The changes in the level of reflected microwave energy at various values of B_0 represent the EPR spectrum. Technically, the detector is sensitive to a broad frequency range. To reduce the noise, the magnetic field is sinusoidally modulated with an oscillating field (typically 100 kHz frequency), and only the modulated part of the diode output voltage is detected. This approach ensures a drastically enhanced signal-to-noise ratio and assumes the detection of the first derivative of the absorption spectrum rather than the absorption line itself (Figure 1.3). The maximum of the absorbance line corresponds to a zero point of the first derivative.

1.3   MAIN PARAMETERS CHARACTERIZING EPR SPECTRA

As was mentioned, the majority of routine EPR measurements are performed on CW instruments. In such spectrometers, the sample is permanently irradiated at constant

Basic Principles of EPR and NMR Spectroscopy

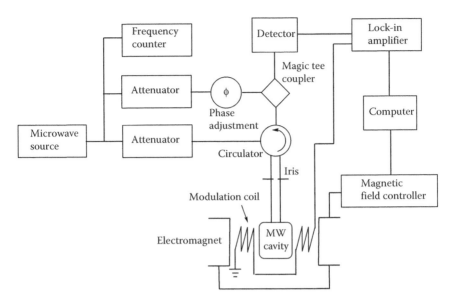

FIGURE 1.2 Block diagram of a CW EPR spectrometer.

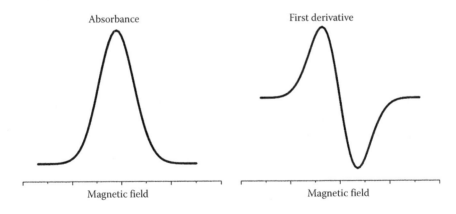

FIGURE 1.3 Absorbance (left) and first derivative (right) line shapes.

MW frequency (for the most widespread X-band, it is 9–10 GHz). The magnetic field is slowly changed in the available range (ca. 0–7000 G), and the EPR spectrum appears as absorption intensity versus magnetic field. For technical reasons (see above), the first derivative of the absorption signal is usually detected.

As an example, the EPR spectrum of stable nitroxyl radical (2,2,6,6-tetramethylpiperidin-1-yl)oxyl (TEMPO) in liquid solution is presented in Figure 1.4. This spectrum appears as a triplet (1:1:1), centered at the field position corresponding to the isotropic g-factor (g_0). The triplet structure illustrates a typical case when the EPR signal is splitted on magnetic nucleus or nuclei present in the structure of the free radical. This *hyperfine* (HF) splitting into a triplet is due to the interaction of the

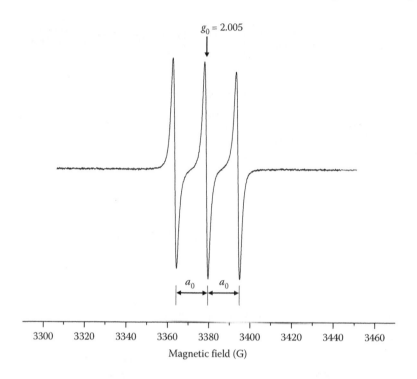

FIGURE 1.4 EPR spectrum (toluene, +20°C) of stable nitroxyl radical (2,2,6,6-tetramethylpiperidin-1-yl)oxyl.

magnetic moment of the unpaired electron $\vec{\mu}_e$ with the magnetic moment of nitrogen nucleus $\vec{\mu}_N$ ($I = 1$). This so-called *Fermi contact interaction* is analogous to the *scalar coupling* (*J*-coupling) of nuclear spins (Section 1.6.2). The energy of the HF interaction is given by

$$E = a_0 \vec{\mu}_e \cdot \vec{\mu}_N \qquad (1.10)$$

where a_0 is the *HF splitting constant*.

The deviation of experimentally observed g from g_e, $((g - g_e)/g_e)$, is an analog of the *chemical shift* in NMR. This deviation is much larger than chemical shifts (see Section 1.6.1).

The electron spin angular momentum is associated with isotropic g-factor $g_e = 2.0023$. In the general case, the magnetic moment of an atom or ion is expressed as

$$\vec{\mu} = -\beta_e(g_e\vec{S} + \vec{L}) \qquad (1.11)$$

where $\vec{S}$ is the *electron angular momentum* and $\vec{L}$ is the *orbital angular momentum* for the ground-state configuration of the atom or ion considered; the g-factor for the pure orbital angular momentum is unity.

Basic Principles of EPR and NMR Spectroscopy

In the ground state, the majority of molecules (including radicals) have zero orbital angular momentum. Therefore, one might expect that the g-factor of the molecule containing one unpaired electron would have the "spin-only" value of 2.0023. However, the electron spin is involved in spin–orbit coupling (between $\vec{S}$ and $\vec{L}$, $\hat{H}_{S-O} = \lambda \hat{L}\hat{S}$) that admixes the "pure spin" ground state with certain excited states and adds a small amount of orbital angular momentum to the pure spin angular momentum. Introducing effective spin $\tilde{S}$, which includes the electron spin $\bar{S}$ and the contribution owing to the spin–orbit coupling, one can eventually write the energy of the $S = 1/2$ atom (or ion) with $I = 0$ nondegenerate electronic ground state as

$$E = \beta_e \cdot \vec{B}\hat{g}\tilde{S} \tag{1.12}$$

where $\hat{g}$ is a second-rank tensor (a symmetric (3×3) matrix). In the arbitrary coordinate system, this tensor has the following form:

$$\begin{bmatrix} g'_{xx} & g'_{xy} & g'_{xz} \\ g'_{yx} & g'_{yy} & g'_{yz} \\ g'_{zx} & g'_{zy} & g'_{zz} \end{bmatrix} \tag{1.13}$$

In the principal-axis system, the above matrix is diagonalized:

$$\begin{bmatrix} g_{xx} & & \\ & g_{yy} & \\ & & g_{zz} \end{bmatrix} \tag{1.14}$$

where g_{xx}, g_{yy}, and g_{zz} are the principal values of g-tensor. The principal-axis system often coincides with the symmetry axes system of the molecule. For the axially symmetric paramagnetic centers, the following notations of principal values of g-tensor are used: $g_{zz} = g_{\parallel}$, $g_{xx} = g_{yy} = g_{\perp}$.

The admixture of orbital angular momentum to the electron spin angular momentum leads to g-factor anisotropy. Using perturbation theory, the following expression for the given elements of g-tensor can be obtained [6,7]:

$$g_{ij} = g_e \delta_{ij} - 2\lambda \sum_{n=0}^{\infty} \frac{\langle 0|L_i|n\rangle\langle n|L_j|0\rangle}{E_n - E_0} \tag{1.15}$$

where δ_{ij} is the Kronecker symbol ($\delta = 0$ for $i \neq j$ and $\delta = 1$ for $i = j$); L_i and L_j are the components of the orbital angular momentum operator; 0 represents the ground state and, n is the different excited states; and E_0 and E_n are the energies of the ground state and the excited states, respectively. Equation 1.15 shows that the deviation of the g-factor from g_e is proportional to λ. The contribution of excited states

8   Applications of EPR and NMR Spectroscopy in Homogeneous Catalysis

is inversely proportional to $E_n - E_0$; in effect, the low-lying excited states make the major contribution to the g-factor deviation from g_e.

Typically, the spin–orbit coupling constant λ is higher for heavier elements, and the values of $E_n - E_0$ are smaller for transition metal complexes, as compared to organic radicals. Therefore, according to Equation 1.15, transition metal complexes could be expected to exhibit higher g-factor anisotropy than organic radicals. Indeed, while principal g-tensor values mostly fit within the range of 1.99–2.015 for organic radicals, for $S = 1/2$ transition metal complexes, they may spread over the range of 1.4–3. Transition metal complexes with $S > 1/2$ can exhibit more complicated spectra, owing to the interaction between unpaired electrons.

In the case of transition metal complexes, theoretical calculations of g-tensor are still rather approximate. Therefore, only tentative information on the composition and structure of paramagnetic center can be derived from the analysis of g-tensor. At the same time, the observed magnitude and anisotropy of HF coupling can provide more detailed information on the electronic and spatial structure of the paramagnetic center.

EPR spectroscopy of both liquid and frozen solutions is widely used for mechanistic studies in homogeneous catalysis. In the case of frozen solutions, the EPR spectra of the systems with one unpaired electron ($S = 1/2$) are characterized by principal values of g- and A-tensors ($g_{xx}, g_{yy}, g_{zz}, A_{xx}, A_{yy}, A_{zz}$). Like the g-factor, the HF coupling is anisotropic, resulting in the (3×3) A-matrix with principal values A_{xx}, A_{yy}, and A_{zz}. Typically, a_0 and A_{ij} values are expressed in frequency units (MHz) or in magnetic field units (mT or G). In liquid solutions with unrestricted motion of molecules, isotropic g-factor and HF coupling values, $g_0 = (g_{xx} + g_{yy} + g_{zz})/3$ and $a_0 = (A_{xx} + A_{yy} + A_{zz})/3$, respectively, are observed.

1.4   EPR SPECTRA OF FROZEN SOLUTIONS

The EPR spectra of frozen solutions are similar to the EPR spectra of polycrystalline materials (powder spectra). Aggregation of dissolved paramagnetic compound on freezing can cause line broadening due to dipole–dipole interaction between magnetic moments of the unpaired electrons. The use of glass-forming solvents allows avoiding the undesired aggregation.

In the principal-axis coordinate system, the *spin Hamiltonian* (Hamiltonian operator incorporating spin operators) of paramagnetic center ($S = 1/2$) with HF splitting from one nucleus can be presented as follows:

$$\hat{H} = \mu_B \left(g_{xx} B_x \hat{S}_x + g_{yy} B_y \hat{S}_y + g_{zz} B_z \hat{S}_z \right) + A_{xx} \hat{I}_x \hat{S}_x + A_{yy} \hat{I}_y \hat{S}_y + A_{zz} \hat{I}_z \hat{S}_z \qquad (1.16)$$

Frozen solutions incorporate a large number of single crystals randomly oriented in space. The resulting spectrum from these crystals is spread over resonant field range determined by the principal g-values. Fortunately, the intensity of the spectrum is not uniform, and characteristic peaks at the positions corresponding to the principal values of g-tensor are observed. The examples of simulated axial and

Basic Principles of EPR and NMR Spectroscopy

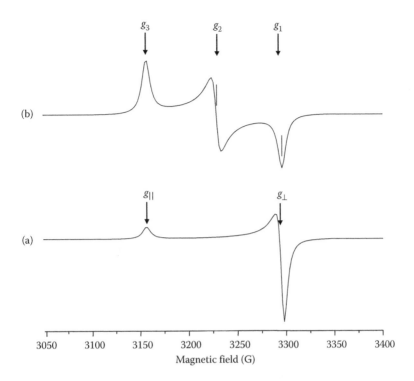

FIGURE 1.5 Simulated EPR spectra of $S = 1/2$ system with axial (a) and rhombic (b) g-tensor anisotropy. $g_\parallel = 2.148$, $g_\perp = 2.100$; $g_1 = 2.057$, $g_2 = 2.100$, and $g_3 = 2.148$.

rhombic EPR spectra (apart from the HF interactions), corresponding to the spin Hamiltonians (1.17) and (1.18), respectively,

$$\hat{H} = \mu_B \left[g_\perp \left(B_x \hat{S}_x + B_y \hat{S}_y \right) + g_\parallel B_z \hat{S}_z \right] \quad (1.17)$$

$$\hat{H} = \mu_B \left[g_{xx} B_x \hat{S}_x + g_{yy} B_y \hat{S}_y + g_{zz} B_z \hat{S}_z \right] \quad (1.18)$$

are presented in Figure 1.5a and b. For the axially anisotropic spectrum (Figure 1.5a), peaks corresponding to $g_\perp$ and $g_\parallel$ can be readily distinguished, since the former is more intense, than the latter. In the case of rhombically anisotropic spectrum (Figure 1.5b), one cannot unequivocally assign the observed peaks to the corresponding molecular axis (x, y, z). Therefore, these peaks are commonly denoted as g_1, g_2, and g_3. The corresponding HF splitting constants (if available) are denoted as A_1, A_2, and A_3.

The simulated EPR spectrum of the $S = 1/2$ and $I = 1/2$ system is presented in Figure 1.6. It is seen that the principal values of g- and A-tensors can be readily derived from this spectrum.

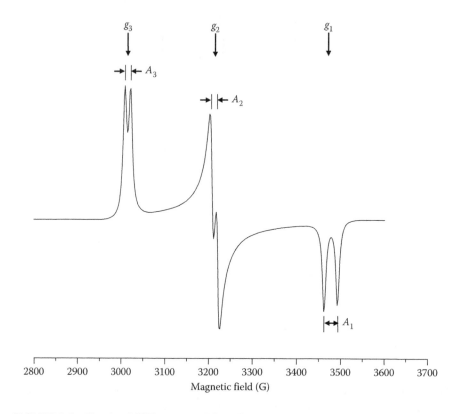

FIGURE 1.6 Simulated EPR spectra of $S = 1/2$ system with $g_1 = 1.95$, $g_2 = 2.11$, $g_3 = 2.25$, $A_1 = 30$ G, $A_2 = 15$ G, and $A_3 = 15$ G. Line widths $d_1 = d_2 = d_3 = 5$ G.

However, the interpretation of the EPR spectrum of frozen solutions may be complicated by the presence of the so-called *extra absorption peaks* arising from the interplay of g- and A-anisotropies. For axially symmetric complexes, these peaks appears at $\theta = \theta_e \neq 0, \pi/2$, where θ is the angle between the magnetic symmetry axis z of the complex and the direction of the magnetic field [8]. As an example, experimental and simulated EPR spectra of frozen solutions of bis(acetylacetonato)Cu(II) in a toluene/chloroform 1:1 mixture are presented (Figure 1.7). The simulated spectrum parameters are almost axial: $g_1 = g_2$, and the difference between A_1 and A_2 is very small compared to the difference between A_1 and A_3 and A_2 and A_3 (see the caption of Figure 1.7). The values of g_3 and A_3 can be easily retrieved from the experimental spectrum. Contrariwise, the intense extra adsorption peaks (denoted with asterisks in Figure 1.8) complicate the evaluation of g_1, g_2 and A_1, A_2. The simulation of powder or frozen solution spectra is essentially equal to numeric summing-up EPR spectra of an ensemble of single crystals, with random orientations of the molecular axis with respect to the laboratory frame of axis. In effect, the simulated spectrum represents an averaged curve depending on $g_x, g_y, g_z, A_x, A_y, A_z$, and d_x, d_y, d_z, where the latter three parameters correspond to the principal values of the line width tensor of single crystals. Quite often, the experimental EPR spectra of frozen solutions

Basic Principles of EPR and NMR Spectroscopy

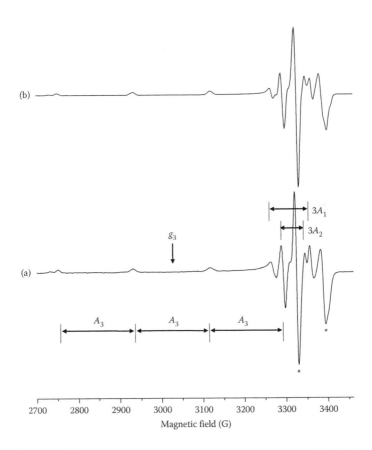

FIGURE 1.7 Experimental (a) and simulated (b) EPR spectra of bis(acetylacetonato) Cu(II). Experimental spectrum recorded in a chloroform/toluene (1:1) mixture at −196°C, [Cu(acac)$_2$] = 5·10^{-3} M. The following parameters were used for simulation: g_1 = 2.05146, g_2 = 2.04822, and g_3 = 2.24850. Hyperfine splitting from two magnetic isotopes of Cu was taken into account, ^{63}Cu (69%, I = $^3/_2$): A_1 = 33.90 G, A_2 = 22.62 G, and A_3 = 220.67 G; ^{65}Cu (31%, I = $^3/_2$): A_1 = 31.65 G, A_2 = 21.12 G, and A_3 = 206 G. Asterisks mark the *extra absorption peaks*.

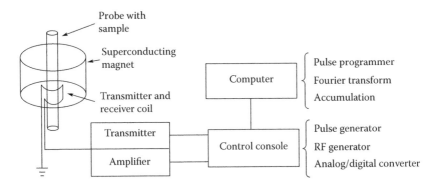

FIGURE 1.8 Block diagram of an FT NMR spectrometer.

12 Applications of EPR and NMR Spectroscopy in Homogeneous Catalysis

can be interpreted only with the aid of computer simulation. The latter may also have some restrictions; for example, if the major details of the experimental spectrum are masked by line broadening, accurate g and A parameters cannot be derived from this spectrum even by computer simulation.

1.5 INSTRUMENTATION FOR PULSED FT NMR SPECTROSCOPY

The block diagram of an FT NMR spectrometer is schematically presented in Figure 1.8. The strong and highly homogeneous magnetic field B_0 is typically generated by a superconducting magnet. The probe, which contains a coil, serving to transmit the RF pulse to the sample and subsequently detect the FID, is inserted into the magnet. The coil is connected to two capacitors, "tune" and "match"; adjusting the "tune" capacitor brings the circuit in resonance to the desired nucleus, and adjusting the "match" brings the circuit impedance for the maximum energy transfer (to efficiently transfer the FID to the receiver). Tuning and matching the probe are technically similar to tuning a radio receiver to a particular radio station. After applying a pulse of RF irradiation, the sample's response (FID) is detected, which is a very weak electric signal, usually in the μV range. This signal is amplified and then digitized by the analog-to-digital converter (ADC). The ADC yields the time-domain signal as an array of data points. The digital data are transferred to the computer, which can collect as many FIDs (*scans*) from the same sample as necessary for attaining appropriate signal-to-noise ratio. After the *acquisition* is complete, the accumulated time-domain FID is Fourier transformed to the frequency-domain *spectrum*.

1.6 MAIN CHARACTERISTICS OF NMR SPECTRA

Solution NMR spectroscopy usually provides the most detailed information on the structure of molecules as compared to other spectroscopic methods. As an example, Figure 1.9 shows the ^1H NMR spectrum of a chemical substance containing three groups of protons. The x axis of the spectrum is called the delta scale (δ) with units of ppm (see below); the intensity is expressed in dimensionless units. The spectrum displays three separate multiplets originating from the protons of OH (triplet, t), CH_2 (doublet of quartets, dq), and CH_3 (triplet, t) groups. The area under each multiplet (*integral*) is proportional to the number of protons in each group. In addition to providing the information on the identity of the analyzed substance (ethanol), the spectrum in Figure 1.9 indicates the presence of an impurity at ca. δ 4.6. The chemical shift value of the latter readily allows the assignment of the latter to H_2O.

1.6.1 CHEMICAL SHIFT

Since all protons have identical magnetic moments, one might expect all sorts of protons in a molecule to display resonance signals at the same field/frequency values. This is only true for isolated protons in a vacuum. In real systems, fortunately, the situation is not so trivial. When an atom is placed in a magnetic field, its electrons circulate about the direction of the applied magnetic field. This circulation induces an additional small local magnetic field at the resonating nucleus, which opposes the

Basic Principles of EPR and NMR Spectroscopy

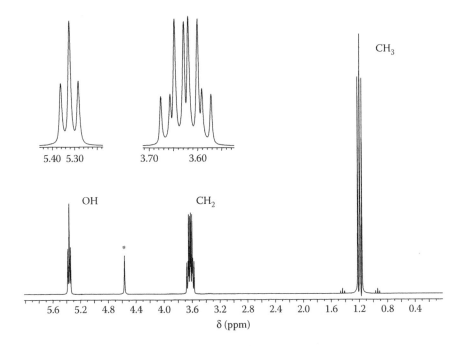

FIGURE 1.9 ^{1}H NMR spectrum of "neat" ethanol. Asterisk denotes the impurity (water).

external applied field. The resulting magnetic field at the nucleus is therefore generally smaller than the applied field by a factor σ, known as the *shielding constant*:

$$B = B_0(1-\sigma) \tag{1.19}$$

In some cases, the direction of the induced magnetic field coincides with the external applied field B_0. For example, the circulations of electrons in the aromatic π-orbitals create a magnetic field at the hydrogen nuclei, which enhances the B_0 field (deshielding). Overall, each nucleus in a molecule feels a unique magnetic field, which depends on the chemical environment of the nucleus. The parameter characterizing the relative change in the resonance frequency is called *chemical shift*, δ. It is accepted to define chemical shift as follows:

$$\delta = \frac{v - v_{ref}}{v_{ref}} \times 10^{-6} \tag{1.20}$$

where ν and v_{ref} are frequencies of the given resonance and the standard (reference molecule added in the sample), respectively. The chemical shift is expressed in *parts per million* (ppm) and is considered positive when the resonance is shifted *downfield* with respect to the reference. Typically, the ranges of known δ values for various nuclei vary from 10 ppm (for protons) to thousands of ppm for heavy nuclei. Various standards (usually their chemical shift is taken as zero, δ = 0) are used for various

14 Applications of EPR and NMR Spectroscopy in Homogeneous Catalysis

nuclei, for example, tetramethylsilane for ^{1}H, ^{13}C, and ^{29}Si, water for ^{17}O, 85% H_3PO_4 for ^{31}P, and CH_3NO_2 for ^{15}N.

1.6.2 SPIN–SPIN COUPLING

The resonances of OH, CH_2, and CH_3 protons (Figure 1.9) appear as multiplets due to the so-called *spin–spin coupling* (known also as *J*-coupling and *scalar coupling*). This coupling splits the NMR signals due to the interactions between the magnetic moments of different magnetic nuclei in the same molecule, transmitted through electrons involved in chemical bonding, and thus provides information on the connectivity of molecules. Scalar spin–spin coupling has no analog in classical electrodynamics. The energy of scalar spin–spin coupling between nuclei having spins I_1 and I_2 (having the corresponding magnetic moments $\vec{\mu}_1$ and $\vec{\mu}_2$) can be presented as follows:

$$E = J \cdot \vec{\mu}_1 \cdot \vec{\mu}_2 \tag{1.21}$$

where J is the coupling constant. Instead of using energy units, the magnitude of spin–spin couplings is commonly measured in frequency units (Hz). The number of splittings indicates the number of chemically bonded nuclei in the vicinity of the observed nucleus. Scalar coupling brings the most useful information for structural determination in one-dimensional (1D) NMR spectra and provides detailed insight into the connectivity of atoms in a particular molecule. Coupling to n equivalent ($I = 1/2$) nuclei splits the signal into an $(n + 1)$ multiplet with intensity ratios following Pascal's triangle. For example, the ^{1}H resonances of OH and CH_3 groups of ethanol are 1:2:1 triplets, while the CH_2 resonance is doublet of quartets due to splitting from one OH proton and three CH_3 protons (Figure 1.5). Note that coupling between nuclei that are magnetically equivalent (i.e., have the same chemical shift) does not manifest itself in the NMR spectra. Couplings between nuclei that are distant (more than three σ bonds) are typically too small (<1 Hz) to be observed in routine high-resolution NMR spectra. On the contrary, in cyclic and aromatic compounds, long-range couplings through more than three bonds can often be observed.

1.6.3 NMR LINE WIDTH

For the $I = 1/2$ magnetic nuclei (such as ^{1}H, ^{13}C, and ^{31}P), the widths of NMR resonances are usually determined by the so-called direct *dipole–dipole interaction*. This interaction is analogous to the classical through-space interaction of two magnetic moments $\vec{\mu}_1$ and $\vec{\mu}_2$, originating from the interaction of one of the magnetic moments with the magnetic field induced by the other magnetic moment. For example, the dipole $\vec{\mu}_1$ feels the magnetic field induced by the dipole $\vec{\mu}_2$, having the following magnitude:

$$\vec{B} = \frac{3(\vec{\mu}_2 \cdot \vec{r})}{r^5} \vec{r} - \frac{\vec{\mu}_2}{r^3} \tag{1.22}$$

Basic Principles of EPR and NMR Spectroscopy

where $\vec{r}$ is the radius vector connecting the nuclei with magnetic moments $\vec{\mu}_1$ and $\vec{\mu}_2$. According to the general expression for the energy of interaction of magnetic moment with external magnetic field ($E = -\vec{\mu} \cdot \vec{B}$), the energy of interaction of $\vec{\mu}_1$ with $\vec{\mu}_2$ will be

$$E = \frac{\vec{\mu}_1 \cdot \vec{\mu}_2}{r^3} - \frac{3(\vec{\mu}_1 \cdot \vec{r})(\vec{\mu}_2 \cdot \vec{r})}{r^5} \tag{1.23}$$

The magnitudes of direct dipole–dipole couplings are relatively large, of the order of kHz (cf. typically 1–250 Hz for scalar spin–spin couplings). In effect, dipole–dipole interactions appear to be the main reasons of dramatic broadening of NMR resonances in solid samples. In liquids, fortunately, dipole–dipole coupling is strongly suppressed; in most cases, it is averaged to zero as a result of fast rotational diffusion of molecules, due to the angular dependence of the value of E in Equation 1.23.

However, the molecular motions in solution result in random fluctuations of the dipole–dipole coupling, owing to fluctuations of the local magnetic field at each nucleus. Those fluctuations are one of the most important reasons of *nuclear spin relaxation*, which determines the line widths in the NMR spectra. ^{1}H isotope is the most abundant nuclei in aqueous and organic solutions and organic molecules, and its magnetic moment is the highest one of common (naturally occurring) nuclei. Therefore, protons usually make the major contribution into the dipole–dipole relaxation. This relaxation mechanism mostly determines the NMR line widths in ^{1}H and ^{13}C NMR spectra of compounds containing those isotopes. The typical widths of NMR resonances in this case range from <1 to several Hz. With slowing down, the molecular motions (which may be caused by using viscous liquids as NMR solvents or by cooling the solution), the relaxation due to *dipole–dipole* interactions gets faster, resulting in broadening of NMR resonances.

With increasing field strength, B_0, another phenomenon, called the *chemical shift anisotropy* (CSA), becomes more important relative to other relaxation processes. Generally, owing to the effect of shielding (see Section 1.6.1), the local magnetic field at a nucleus in a molecule placed in external magnetic field depends on how this molecule is oriented relative to the magnetic field. In effect, the chemical shift of a nucleus is a function of the orientation of the molecule in the magnetic field, and the shielding σ and chemical shift δ are tensors. Molecules in liquids rapidly experience all possible orientations relative to the direction of a magnetic field, so the local magnetic field at the nucleus is varying and the observed δ is an averaged value. However, those motions causing oscillations of the local magnetic field lead to spin relaxation. For some $I = 1/2$ nuclei (like ^{31}P, ^{19}F), the contribution of CSA to the relaxation may be more significant than the dipole–dipole interaction even in the commonly used magnetic fields (7–12 T).

Another important relaxation mechanism appears for nuclei with $I > 1/2$ because these have nonzero electric quadrupole moments Q, which is able to interact with *electric* field gradients. The electric field gradient is provided by an asymmetric distribution of electron density around the nucleus. In liquids, the effect of quadrupolar interactions on the position of NMR resonances is averaged due to diffusional motions, but these interactions effectively induce spin relaxation. For all quadrupolar

16   Applications of EPR and NMR Spectroscopy in Homogeneous Catalysis

nuclei, quadrupolar relaxation dominates over all other relaxation mechanisms. Therefore, the NMR resonances of quadrupolar nuclei (that are ubiquitous in inorganic and organometallic chemistry) can be several kHz wide, which complicates their detection and masks the observation of scalar couplings that the observed nuclei are involved in, as well as small chemical shift differences.

Fortunately, this rule has some important exceptions. If the quadrupole moment is small (as for ^{2}D, ^{6,7}Li, ^{9}Be, ^{14}N, and ^{17}O) or if the electric field gradient is small or zero (which is the case for sufficiently symmetric—usually tetrahedral or octahedral—molecules), the line broadening caused by quadrupolar relaxation is not too severe, and sharp (several Hz wide) resonances can be readily observed.

The application of NMR spectroscopy to homogeneous catalysis is restricted to measurements in liquid solutions since the NMR spectra of frozen solutions are extremely broad due to chemical shifts anisotropy and direct dipole–dipole and quadrupolar interactions. NMR measurements in liquid solutions allow the observation of averaged values for chemical shift and spin–spin coupling constant $\delta_0 = (\delta_{xx} + \delta_{yy} + \delta_{zz})/3$, $J_0 = (J_{xx} + J_{yy} + J_{zz})/3$, where $\delta_{xx}, \delta_{yy}, \delta_{zz}$ are the principal values of the chemical shift tensor, and J_{xx}, J_{yy}, J_{zz} are the principal values of the spin–spin coupling tensor.

For EPR spectroscopy, the situation is different. Although EPR resonances are much broader than NMR peaks (1–100 MHz vs. 1 Hz–10 kHz), the EPR spectra of frozen solutions can be well resolved due to a larger range of available magnetic fields (for X-band EPR, it is 0–20 GHz in frequency units). The analysis of the EPR spectra of frozen solutions (see also Section 1.4) often provides valuable information on the structure of transition metal complexes. Lowering the temperature improves the signal-to-noise ratio due to increased population difference (according to Equation 1.6). Moreover, for transition metal complexes, the so-called *spin–lattice relaxation time* T_1 (see Section 1.7) strongly depends on the temperature, often resulting in a situation when the spectra can only be detected at cryogenic temperatures (e.g., 77 or even 4 K), since at room temperature, their spin–lattice relaxation is too fast, resulting in unacceptably broad EPR resonances.

In the next section, the classical mechanics treatment of the physical principles of NMR and EPR spectroscopy, as developed by Felix Bloch, is presented. This approach provides a good description of the observed line shapes in CW and pulsed NMR and EPR spectra.

1.7 BLOCH EQUATIONS

The famous Bloch equations describe the time evolution of the total spin magnetization vector $\vec{M}$ of the sample in the presence of static and oscillating magnetic fields $\vec{B}_0$ and $\vec{B}_1$. We will consider below these equations in the context of observation of NMR; a similar approach is applicable to EPR.

Let us assume that the macroscopic sample contains an ensemble of identical molecules with nuclei of one type (with nuclear magnetic moment $\vec{\mu}_i$). The macroscopic magnetization of the sample will be

$$\vec{M} = \sum_i \vec{\mu}_i \tag{1.24}$$

Basic Principles of EPR and NMR Spectroscopy

Note that the macroscopic magnetization $\vec{M}$ is also related to the total spin angular momentum of the sample as $\vec{M} = \gamma \vec{P}$. The applied external magnetic field $\vec{B}_0$ will exert a torque $\vec{\tau}$ on the magnetic moment, resulting in the Larmor precession of the macroscopic magnetization about the direction of $\vec{B}_0$.

The torque is given by the following equation:

$$\vec{\tau} = \frac{d\vec{P}}{dt} = \vec{M} \times \vec{B}_0 \tag{1.25}$$

which, taking into account the ratio $\vec{M} = \gamma \vec{P}$, gives the equation for the time evolution of $\vec{M}$:

$$\frac{d\vec{M}}{dt} = \gamma \vec{M} \times \vec{B}_0 \tag{1.26}$$

or for individual components:

$$\frac{dM_x}{dt} = \gamma B_0 M_y$$

$$\frac{dM_y}{dt} = -\gamma B_0 M_x \tag{1.27}$$

$$\frac{dM_z}{dt} = 0$$

Let us introduce $\omega_0 = -\gamma B_0$ (*Larmor frequency*). Taking $\vec{B}_0$ parallel to the z axis ($\vec{B}_0 = \{0, 0, B_0\}$), and $M_x(0) = M_\perp^0, M_y(0) = 0$, and $M_z(0) = M_z^0$, the following solution of Equation 1.27 can be obtained:

$$M_x = M_\perp^0 \cos \omega_0 t$$

$$M_y = -M_\perp^0 \sin \omega_0 t \tag{1.28}$$

$$M_z = M_z^0$$

which describes the precession of $\vec{M}$ about the $\vec{B}_0$, the $M_\perp$ (the projection of $\vec{M}$ to the xy plane) rotating in the xy plane with the *Larmor frequency* ω_0, and M_z (the projection of $\vec{M}$ to the z axis) remaining constant. This is an idealized picture that does not take into account the phenomena (Section 1.6.3.) leading to *relaxation*; the

18   Applications of EPR and NMR Spectroscopy in Homogeneous Catalysis

latter consists of gradual approaching the equilibrium condition (when the transverse components of $\vec{M}$, M_x and M_y, are zero, and the parallel component M_z is equal to its equilibrium value M_0 defined by Equation 1.9).

Bloch had predicted [9] that the projections M_x, M_y, and M_z should approach their equilibrium values $M_x^0 = M_y^0 = 0$, $M_z^0 = M_0$ exponentially, with characteristic times T_2 (*transverse* or *spin–spin* relaxation time) and T_1 (*longitudinal* or *spin–lattice* relaxation time), respectively, such that Equation 1.27 transforms to the following equations:

$$\frac{dM_x}{dt} = \omega_0 M_y - M_x / T_2$$

$$\frac{dM_y}{dt} = -\omega_0 M_x - M_y / T_2 \tag{1.29}$$

$$\frac{dM_z}{dt} = -(M_z - M_0)/T_1$$

In CW NMR experiments, besides the static magnetic field $\vec{B}_0$ directed along the z axis, the sample is subjected to weak oscillating magnetic field $2B_1\cos\omega t$ directed along the x axis. This oscillating field can be formally presented as a sum of two magnetic fields with magnitude B_1 ($\vec{B}_1^{+\omega}$ and $\vec{B}_1^{-\omega}$), rotating in the xy plane in opposite directions with angular frequencies $+\omega$ and $-\omega$, respectively. In the coordinate system, rotating with angular frequency $+\omega$, the magnetic field $\vec{B}_1^{+\omega}$ is static against $\vec{M}$ and cause rotation of $\vec{M}$ around the direction of $\vec{B}_1^{+\omega}$, with angular frequency $-\gamma B_1$. The other component of magnetic field $\vec{B}_1^{-\omega}$ rotates around the direction of $\vec{M}$ with high frequency (2ω) and virtually does not interact with $\vec{M}$; this component may be excluded from further consideration. In effect, the overall magnetic field affecting the magnetization can be expressed as

$$\vec{B} = \vec{i} \cdot B_1 \cos \omega t - \vec{j} \cdot B_1 \sin \omega t + \vec{k} \cdot B_0 \tag{1.30}$$

where $\vec{i}$, $\vec{j}$, and $\vec{k}$ are unit vectors in the directions x, y, and z, respectively. In effect, in the laboratory frame, Equation 1.29 will transform to the following equations:

$$\frac{dM_x}{dt} = \gamma[B_0 M_y + B_1 \cdot M_z \sin \omega t] - M_x / T_2$$

$$\frac{dM_y}{dt} = -\gamma[B_0 M_x - B_1 \cdot M_z \cos \omega t] - M_y / T_2 \tag{1.31}$$

$$\frac{dM_z}{dt} = -\gamma[B_1 \cdot M_y \cos \omega t + B_1 \cdot M_x \sin \omega t] - (M_z - M_0)/T_1$$

Basic Principles of EPR and NMR Spectroscopy

This form is not that straightforward due to the oscillating components. Fortunately, the Bloch equations have more simple form in a frame of reference (x', y', z'), rotating with angular frequency ω around the z axis. This speed of rotation is the same as that of B_1, which is directed along x'. The projection of $\vec{M}$ along x' will be denoted as $\tilde{M}_x$, and those along y' and z' are $\tilde{M}_y$ and $\tilde{M}_z$, respectively (note that $\tilde{M}_z = M_z$). To transform vector $\vec{M}$ in the laboratory frame to vector $\tilde{M}$ in the frame, rotating about the z axis, one could apply the *rotation matrix* formalism known from linear algebra [10]:

$$\begin{bmatrix} \tilde{M}_x \\ \tilde{M}_y \\ \tilde{M}_z \end{bmatrix} = \begin{bmatrix} \cos\omega t & -\sin\omega t & 0 \\ \sin\omega t & \cos\omega t & 0 \\ 0 & 0 & 1 \end{bmatrix} \begin{bmatrix} M_x \\ M_y \\ M_z \end{bmatrix} \tag{1.32}$$

After applying Equation 1.32, the Bloch equations in the rotating frame will be expressed as

$$\frac{d\tilde{M}_x}{dt} = (\omega_0 - \omega)\tilde{M}_y - \tilde{M}_x/T_2$$

$$\frac{d\tilde{M}_y}{dt} = -(\omega_0 - \omega)\tilde{M}_x + \gamma B_1 \tilde{M}_z - \tilde{M}_y/T_2 \tag{1.33}$$

$$\frac{dM_z}{dt} = -\gamma B_1 \tilde{M}_y - (M_z - M_0)/T_1$$

where $\omega_0 = \gamma B_0$. The use of rotating frame notation is convenient for theoretical consideration. In practice (i.e., in pulsed NMR spectrometers), the so-called reference frequency ω_{ref} (ω_{ref} is close to the Larmor frequency ω_0) is subtracted from the NMR signal prior to FT. Mathematically, this procedure is equivalent to switch to the rotating frame. Both CW and pulsed NMR spectrometers detect $\tilde{M}_y$ (the projection of $\tilde{M}$ on the y' axis of the rotating frame). When both $\tilde{M}_x$ and $\tilde{M}_y$ are being detected simultaneously, it is called *quadrature detection*.

In CW spectroscopy, the ω frequency slowly passes through the resonance value ω_0, and the NMR signal (resonance absorption) is permanently detected. The steady-state solution is

$$\tilde{M}_x = \frac{(\omega_0 - \omega)\gamma B_1 T_2^2}{1 + (\omega_0 - \omega)^2 T_2^2 + T_1 T_2 \gamma^2 B_1^2} M_0$$

$$\tilde{M}_y = \frac{\gamma B_1 T_2}{1 + (\omega_0 - \omega)^2 T_2^2 + T_1 T_2 \gamma^2 B_1^2} M_0 \tag{1.34}$$

$$M_z = \frac{1+(\omega_0-\omega)^2 T_2^2}{1+(\omega_0-\omega)^2 T_2^2 + T_1 T_2 \gamma^2 B_1^2} M_0$$

When $\gamma B_1 \ll (T_1 T_2)^{-1/2}$, the oscillating magnetic field B_1 does not essentially change the Boltzmann population difference $N_1 - N_2$, and the last term in each denominator of Equation 1.34 may be neglected. In this case, $M_z \approx M_0$, and the transverse components $\tilde{M}_x$ and $\tilde{M}_y \ll M_0$. The shape of the NMR resonance $g(\omega_0 - \omega)$ corresponds to the value of detected $\tilde{M}_y$ and thus can be expressed as

$$g(\omega) = \frac{1}{\pi} \frac{T_2}{1+(\omega_0-\omega)^2 T_2^2} \tag{1.35}$$

where coefficient $1/\pi$ is introduced to normalize the function to a unit area under the curve. Function $g(\omega)$ describes the so-called *Lorentzian line shape*. Its important characteristic is *line width*, defined as width at half height $\Delta\omega_{1/2}$ (Figure 1.10), which is equal to $2/T_2$.

The physical meanings of the transverse (spin–spin) relaxation time T_2 and the longitudinal (spin–lattice) relaxation time T_1 lie outside the realm of the Bloch equations. In brief, T_2 is mostly a measure of the interactions of the individual spins, while T_1 determines the characteristic time of thermodynamic equilibration of the magnetization directed along the static magnetic field with its surroundings, achieved via interactions between the nuclear magnetic moments and other degrees

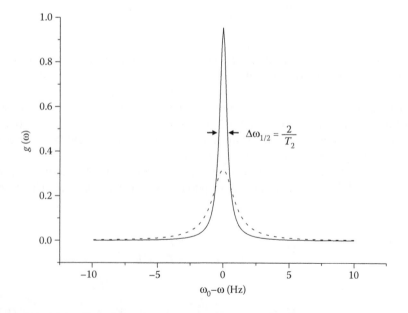

FIGURE 1.10 The Lorentzian line shapes. $T_2 = 3$ s (solid line) and $T_2 = 1$ s (dashed line).

Basic Principles of EPR and NMR Spectroscopy 21

of freedom of surrounding molecules (considered as "lattice"). In solid samples, direct dipole–dipole interactions between nuclei cause relatively fast relaxation of M_x and M_y without transfer of energy to the lattice, that is, $T_2 < T_1$. In nonviscous solvents, typically used for NMR measurements, T_2 and T_1 for ^1H have comparable values, typically lying in the range of 1–5 s. For ^{13}C nuclei, T_1 is typically several times as long as T_2.

In CW NMR experiments, the steady-state conditions are fulfilled when a spectral range of 1 Hz is recorded during ca. 1 s. In effect, it took several minutes to measure a typical ^1H NMR spectrum (for a 100 MHz spectrometer, the ^1H spectral range of 10 ppm is equal to 1000 Hz). Nowadays, CW NMR machines have been entirely replaced by pulsed FT instruments, providing a full-range spectrum or even accumulating many FIDs ("scans") within a limited time frame; the time required to measure one "scan" is usually as long as the longest of T_2 and T_1, typically a few seconds.

1.8 PULSED FT NMR SPECTROSCOPY

Let us place a sample incorporating one type of equivalent nuclei into the static magnetic field $\vec{B}_0$. At a particular moment of time, this sample is subjected to a short pulse of RF irradiation, generated by rapid opening of the RF transmitter with operating frequency ω, followed by its closing. Under resonance conditions ($\omega = \omega_0$), the rotating frame and the magnetic component $\vec{B}_1$ of the RF pulse rotate with Larmor frequency. In effect, in the rotating frame, the effect of B_0 vanishes, and the magnetization $\tilde{M}$ interacts only with $\vec{B}_1$ directed along the x' axis (the following equations are obtained from Equation 1.33 by applying $\omega = \omega_0$ and neglecting the relaxation terms):

$$\frac{d\tilde{M}_x}{dt} \approx 0$$

$$\frac{d\tilde{M}_y}{dt} \approx \gamma B_1 \tilde{M}_z \qquad (1.36)$$

$$\frac{dM_z}{dt} = -\gamma B_1 \tilde{M}_y$$

with the solution (obtained for starting conditions $M_x(0) = 0$, $M_y(0) = 0$, and $M_z(0) = M_0$)

$$\tilde{M}_x = 0$$

$$\tilde{M}_y = M_0 \sin \omega_1 t \qquad (1.37)$$

$$M_z = M_0 \cos \omega_1 t$$

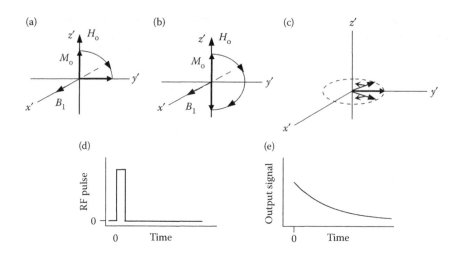

FIGURE 1.11 The effect of the RF pulse on magnetization M_0 (initially aligned along the z' axis) $90_x°$ pulse (a), $180_x°$ pulse (b), decay of $\tilde{M}_y$ after the $90_x°$ pulse (c), RF pulse corresponding to the case "a" (d), and output signal (FID) measured along the y' axis corresponding to the case "c" (e).

Just after switching on the transmitter, $\vec{M}$ starts to rotate in the $z'y'$ plane around the x' axis with angular frequency γB_1. During the pulse duration (τ), the magnetic moment $\vec{M}$ will turn by angle θ:

$$\theta = \gamma_N B_1 \tau \tag{1.38}$$

This is the main equation of pulsed NMR spectroscopy. The pulse τ, which turns the macroscopic magnetization by $\theta = \pi/2$, is called 90° pulse; the pulse of duration 2τ corresponding to $\theta = \pi$ is called 180° pulse (Figure 1.11a and b). Just after a 90° pulse, the magnetization vector $\vec{M}$ will be directed along the y' axis (Figure 1.11a). In NMR spectrometers, the intensity of the electric signal generated in the receiver coil is proportional to the projection of $\vec{M}$ into the xy plane. Therefore, the 90° pulse will ensure the maximum initial magnitude of the output signal.

$\vec{M}$ will remain static and directed along y', only if all static and fluctuating magnetic fields that the magnetic moments $\vec{\mu}_i$ are exposed to are the same. In practice, this is never the case, and the projection of $\vec{M}$ on the y' axis, $\tilde{M}_y$, will decay exponentially, via dephasing of the individual spin vectors (Figure 1.11c and e). In the ideally homogeneous magnetic field, the characteristic time of this exponential decay should be equal to the spin–spin relaxation time T_2. In reality, however, the inhomogeneity of the static magnetic field $\vec{B}_0$ may be the major factor determining the spin dephasing; the corresponding characteristic time T_2^* then will be shorter than T_2. If the sample contains equivalent nuclei, and the frequency of the RF pulse ω (equal to the rotation frequency of the rotating frame) coincides with the Larmor frequency of these nuclei ω_0, the output signal (FID) generated in the receiver

Basic Principles of EPR and NMR Spectroscopy 23

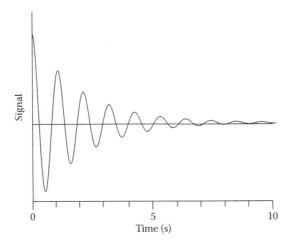

FIGURE 1.12 Model FID corresponding to an exponential function modulated with harmonic oscillations.

coil represents the exponential function decaying with the characteristic time T_2^* (Figure 1.11e). If ω and ω_0 are not equal, $\vec{M}$ will be directed along the y' axis only just after the 90° pulse, and then it will rotate about the z axis with the angular frequency $(\omega - \omega_0)$, and decay with characteristic time T_2^*. The resulting FID will be an exponential function modulated with harmonic oscillations with $v = (\omega - \omega_0)/2\pi$ (Figure 1.12). If the sample contains several groups of nuclei with various chemical shifts, the resulting FID will be a superposition of several decaying exponential functions modulated by harmonic oscillations. The frequency of those oscillations will correspond to the differences between the ω_0 and ω_i for each nuclei and hence to the chemical shift δ_i.

FID is a time-domain signal; to retrieve the frequency-domain spectrum from the FID, the operation of FT should be applied to the latter:

$$g(\omega) = \int_{-\infty}^{\infty} f(t) e^{i\omega t} dt \qquad (1.39)$$

where $f(t)$ is the time-dependent FID. FT returns a sum of individual peaks at particular frequencies; the latter correspond to the frequencies of harmonic oscillations modulating the decaying FID. Graphically, the effect of FT on the FIDs and the corresponding model spectra are presented in Figure 1.13.

The effect of RF pulse with the duration τ on the sample is approximately equivalent to the effect of polychromatic radiation with frequencies distributed in the range of $\Delta\omega \approx 1/\tau$, with the center of spectrum ω_0 corresponding to the operating frequency of the pulse generator. In effect, short pulses (several microseconds) result in excitation in a broad range of chemical shifts. For example, a 10 μs pulse effectively

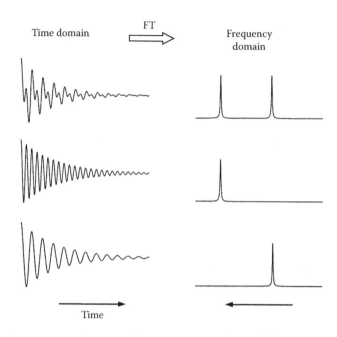

FIGURE 1.13 Effect of Fourier transform (FT) on the FID.

excites a spectral range of 10^5 Hz, which for 500 MHz spectrometer corresponds to a span of ^{1}H chemical shifts of 200 ppm. The time required for recording one FID is determined by T_2^*, and in the case of ^{1}H and ^{13}C nuclei, it typically takes 2–10 s. For accurate integration of the resonances, the macroscopic magnetization should return to its equilibrium value before the next pulse, which requires ca. $3–5T_1$, typically 10–60 s, which is much shorter than for CW NMR experiment (Section 1.7). This is one of the major advantages of pulsed FT NMR spectroscopy.

We have presented here a minimum basic introduction into NMR and EPR spectroscopies. More information on the principles and technical approaches of NMR spectroscopy can be found in Refs [5–15]. Additionally, some experimental techniques will be described in the next chapter.

REFERENCES

1. http://physics.nist.gov/cgi-bin/cuu/Category?view=html&Atomic+and+nuclear.x=79&Atomic+and+nuclear.y=13.
2. http://www-usr.rider.edu/~grushow/nmr/NMR_tutor/periodic_table/nmr_pt_frameset.html.
3. The presence of magnetic nucleus in a molecule does not guarantee the emergence of informative NMR spectra; the issues of *natural abundance* of the magnetic isotope and its gyromagnetic ratio (*NMR sensitivity* in a broader sense), *relaxation times*, *quadrupolar line broadening* (for nuclei with $I > 1/2$), etc. should also be taken into account.

Basic Principles of EPR and NMR Spectroscopy

4. The higher electron magnetic moment, compared to a nuclear magnetic moment, entails (1) higher resonance frequency (i.e., higher value of absorbed RF quanta) and (2) higher difference between the populations (1.8) of the electron spin states ($m_S = \pm 1/2$) as compared to nuclear spin states ($m_I = \pm 1/2$) ($\Delta E^e \gg \Delta E^N$).

5. Abragam, A., Bleaney, B. 1970. *Electron Paramagnetic Resonance of Transition Ions*, Oxford University Press, Oxford, UK.

6. Carrington, A., McLachlan, A. D. 1967. *Introduction to Magnetic Resonance*, Harper & Row, New York.

7. Weil, J. A., Bolton, J. R. 2007. *Electron Paramagnetic Resonance: Elementary Theory and Practical Applications*, Wiley, New York.

8. Ovchinnikov, I. V., Konstantinov, V. N. 1978. Extra absorption peaks in EPR spectra of systems with anisotropic g-tensor and hyperfine structure in powders and glasses. *J. Magn. Res.* 32: 179–190.

9. Bloch, F. 1946. Nuclear induction. *Phys. Rev.* 70: 470–474.

10. https://en.wikipedia.org/wiki/Rotation_matrix.

11. Harris, S. K. 1986. *Nuclear Magnetic Resonance Spectroscopy*, Longman, Harlow, UK.

12. Pople, J. A., Schneider, W. G., Bemstein, H. J. 1959. *High-Resolution Nuclear Magnetic Resonance*, McGraw-Hill, New York.

13. Slichter, C. P. 1989. *Principles of Magnetic Resonance*, 3rd ed., Springer, New York.

14. Deroume, A. E. 1990. *Modern NMR Techniques for Chemistry Research*, Pergamon Press, Oxford–New York–Beijing–Frankfurt–Sâo Paulo–Sydney–Tokyo–Toronto.

15. Claridge, T. D. W. 1999. *High-Resolution NMR Techniques in Organic Chemistry*, Pergamon Press, Amsterdam–Lausanne–New York–Oxford–Shannon–Singapore–Tokyo.

2 Some NMR Spectroscopic Techniques Used in Homogeneous Catalysis

Nowadays, NMR spectroscopic analyses are occurring everywhere in homogeneous catalysis. Homogeneous catalysts are usually soluble metal complexes or low-molecular-weight organic compounds; high-resolution NMR spectroscopy can provide valuable information on their composition and structure within a reasonably tight time frame. Organic compounds (typically, ligands for metal–complex catalysts) usually contain hydrogen and carbon atoms, which entails wide application of ^{1}H and ^{13}C NMR spectroscopy for routine analyses, as well as *in situ* detection and identification of chemical objects of interest for catalytic chemists.

The advantage of ^{1}H nucleus is its high sensitivity (the highest among naturally occurring isotopes). We note that the relatively small range of chemical shifts available for different protons in a molecule (for diamagnetic species, the chemical shifts in most cases fit within the range of $\delta + 12...0$, Figure 2.1 [1]) and the frequently observed overlapping of individual ^{1}H peaks can complicate the interpretation of 1D NMR spectra. At the same time, two-dimensional (2D) NMR techniques may "resolve" overlapping signals and provide useful additional information, such as visualize scalar (J) and dipolar couplings between different protons, identify protons belonging to the same molecules, etc.

^{13}C NMR spectra are usually much more informative than ^{1}H NMR spectra due to the wider range of typical chemical shifts ($\delta + 220...0$, Figure 2.1) [1]. Furthermore, there is much additional information that can be of use for structural analysis, such as the type of carbon (primary, secondary, tertiary, quaternary), $^1J_{CH}$ and long-range J_{CH} couplings and $^1J_{CC}$ couplings. Two-dimensional spectroscopy may be extremely informative for extracting qualitative information on the scalar couplings ("correlations") between ^{13}C and ^{1}H nuclei, as well as ^{13}C and ^{13}C nuclei. The major drawback of ^{13}C NMR is low natural abundance of the ^{13}C isotope (1.1%) and lower gyromagnetic ratio (3.98 times smaller than for ^{1}H), which results in much lower overall sensitivity of the ^{13}C nucleus; in effect, ^{13}C NMR spectra require relatively long acquisition times (or high concentrations). In some cases, ^{13}C enrichment (see Chapter 4) is required to solve the problem.

The aim of the first part of this chapter is, without getting too much into technical detail, to demonstrate the utility of various ^{1}H and ^{13}C NMR spectroscopic techniques, and to get insight into the information it can provide. To this end, we

27

¹H Chemical Shifts in Organic Compounds

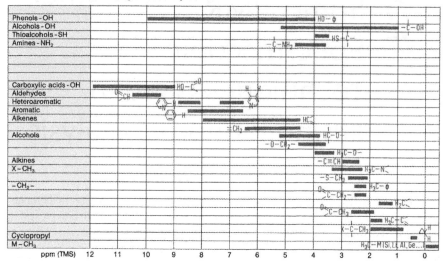

¹³C Chemical Shifts in Organic Compounds

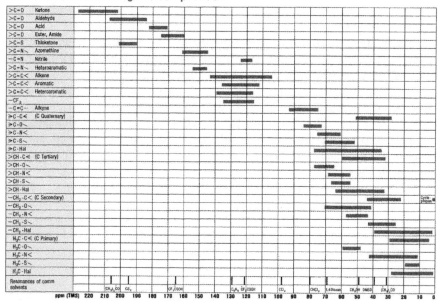

FIGURE 2.1 ¹H and ¹³C NMR chemical shifts diagram.

will briefly overview several NMR experiments (or techniques) that seem to be the most widely used in routine NMR characterizations, as well as in NMR analyses of metal complexes, ubiquitous in homogeneous catalysis. For more in-depth theoretical and technical descriptions, one can see References [2,3] and other literature cited in Chapter 1.

On the other hand, many transition metal complexes have open electron shells and are therefore paramagnetic, often with $S > 1/2$. Contrary to the existing opinion, such paramagnetic complexes may appear suitable to NMR spectroscopic characterization. Moreover, in some cases (e.g., for some non-Kramers ions with integer S), it is NMR spectroscopy that can provide valuable information on the complex structure, whereas EPR studies may be complicated or futile. That is why in the second part of this chapter, we will consider the application of NMR spectroscopy to the investigation of paramagnetic metal complexes. The idea is to give some practical examples and demonstrate how structural information can be obtained from NMR spectra of paramagnetic compounds.

2.1 MEASURING ^{1}H AND ^{13}C NMR SPECTRA OF A SAMPLE COMPOUND

2.1.1 ONE- AND TWO-DIMENSIONAL ^{1}H AND ^{13}C NMR SPECTRA

In this section, an NMR characterization of an organic compound will be described, starting from 1D ^{1}H NMR spectrum, following the order typically used by chemists in practice.

The vast majority of NMR machines working today in the laboratories are pulsed spectrometers, making use of pulsed Fourier transform NMR spectroscopy (Chapter 1). In effect, a simple ^{1}H NMR experiment may be considered as consisting of a short *pulse* of RF field, followed by a *relaxation delay*, during which *acquisition* is being performed (Figure 2.2). For achieving maximum sensitivity, the 90° ($\pi/2$) pulses may be used; typically, $\pi/2$ pulses have a duration of several microseconds. In turn, the duration of the acquisition + relaxation period should be sufficient for spin–lattice relaxation to occur (usually 3–5 T_1).

A sample ^{1}H NMR spectrum is given in Figure 2.3. For simplicity, only part of the whole spectrum (corresponding to the aliphatic part of the titanium(IV) complex [4]) is presented. Owing to the presence of many types of nonequivalent protons

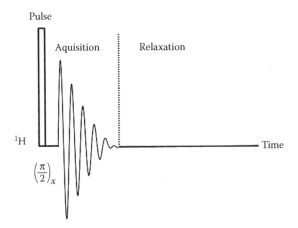

FIGURE 2.2 Schematic representation of a one-dimensional (1D) pulsed NMR experiment.

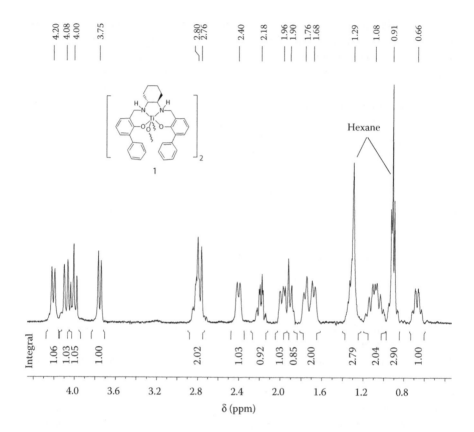

FIGURE 2.3 ¹H NMR spectrum (400.13 MHz, chloroform-d, 20°C) of a sample compound—dimeric titanium(IV) salan complex **1** (see inset). Only aliphatic part is shown.

(interacting with each other via scalar spin–spin coupling), the spectrum looks rather complicated. In addition, some of the peaks overlap, which introduces additional complication. Each resonance (or *peak*) has its characteristic chemical shift and integral intensity (or *integral*); the latter is proportional to the number of protons of this particular type in the molecule. In addition to the resonances of protons of the salan ligand, one can see the peaks attributable to hexane at δ 0.91 and 1.29. Hexane is used as a solvent in the course of the synthesis of the Ti–salan complexes, and it was shown to cocrystallize with them [4]. However, comparison of their integral values (ca. 1:1 instead of 4:3) suggests that there is most likely another peak (with integral intensity of ca. 1.0), overlapping with the hexane's triplet at δ 0.91. We note that all peaks have integral intensities of ca. 1.0 or 2.0; one could assume that in the latter case, there are most likely two overlapping peaks.

The next logical step would be to establish whether all considered peaks belong to the same molecule. Most straightforwardly, this can be achieved by using 2D homonuclear *COrrelation SpectroscopY* (¹H,¹H COSY). The scheme of the experiment is presented in Figure 2.4. The key feature of this sequence is the evolution time t_1: to obtain a 2D NMR spectrum, a series of experiments are performed, with different

Some NMR Spectroscopic Techniques Used in Homogeneous Catalysis 31

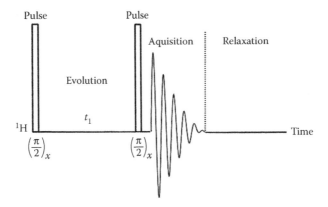

FIGURE 2.4 Schematic representation of a 2D ^{1}H,^{1}H COSY experiment.

t_1 for each experiment (incremented from one experiment to another), followed by the 2D Fourier transform. The resulting picture will be a three-dimensional (3D) surface, with two frequency axes (usually expressed in ppm) and the intensity axis. In practice, a COSY spectrum is usually represented as a cross section of this surface (Figure 2.5). For convenience, the 1D ^{1}H NMR spectrum is provided, playing the role of a "projection" of the main diagonal along the frequency axes.

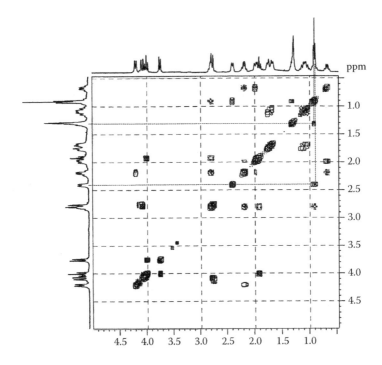

FIGURE 2.5 ^{1}H,^{1}H COSY spectrum (400.13 MHz, chloroform-d, 20°C) of a sample compound—dimeric titanium(IV) salan complex **1**. Only aliphatic part is shown.

32   Applications of EPR and NMR Spectroscopy in Homogeneous Catalysis

The most important information provided by the COSY spectrum is the *cross-peaks* placed off the main diagonal. Owing to the phenomenon called *magnetization transfer* [2], two symmetrical cross-peaks appear between two nuclei coupled to each other via scalar spin–spin coupling. In effect, the resulting picture should have twice as many cross-peaks as the number of scalar couplings between the protons of the sample. In practice, however, some cross-peaks may not appear if the coupling constant J is too small; typically, only peaks corresponding to $^2J_{HH}$ constants are well visualized, while those for longer-range couplings are not displayed or appear as trace peaks [5].

So, the homonuclear COSY spectrum shows the scalar couplings between the nuclei and thus can provide valuable structural information (how those nuclei are connected in the molecule). As an example of this kind, in Figure 2.5, the coupling between the hexane peaks is shown with dotted line. At the same time, it looks like that the δ 0.91 peak of hexane is coupled to the doublet at δ 2.40 (dotted line), which definitely belongs to the complex. The explanation is rather straightforward: as it was suspected before (on the basis of integral intensities), the hexane's triplet at δ 0.91 overlaps with some resonance from the titanium complex.

So far, we have considered ^1H NMR spectra, which could be acquired within a short time frame: several minutes for a 1D spectrum and 10–60 min for a 2D spectrum. At the next step, ^{13}C NMR spectra can be measured. Owing to the much lower sensitivity of the ^{13}C nucleus (see above), as well as longer delays between the pulses [6], the acquisition of a 1D ^{13}C NMR spectrum may require rather long times, up to many hours [7]. In most cases, researchers use ^1H-decoupled spectra [2,3] to get rid of splitting of ^{13}C resonances into multiplets owing to scalar coupling to protons [8]. ^1H decoupling is typically achieved by saturation of the sample at the ^1H frequency during the acquisition at the ^{13}C NMR frequency (Figure 2.6). In effect, the $^{13}C\{^1H\}$ NMR spectra appear as a set of singlets (Figure 2.7); the range of chemical shifts of ^{13}C NMR spectra is much broader than for ^1H NMR, which makes ^{13}C NMR data more informative and characteristic.

In the spectrum shown in Figure 2.7, one can readily distinguish the resonances of hexane (δ 14.3, 22.8, 31.7; the ^1H and ^{13}C chemical shifts of some common solvents and compounds can be found in References [1,9,10]). However, ^1H decoupling leads to the loss of some important information—on the multiplet structure of the ^{13}C peaks, reflecting the interaction of each ^{13}C carbon with its neighboring protons. To get such information quickly on a qualitative level, J-modulated ^{13}C NMR spectroscopy is usually used, which is essentially one of the many "spin-echo" experiments [2,3]. The corresponding pulse sequence is presented in Figure 2.8. The feature of the sequence is the 180° pulse (the ^{13}C channel) between the first 90° pulse and the acquisition; the 180° ^{13}C pulse is accompanied by switching on the ^1H decoupler [11]. Choosing the delay $\Delta = 1/J$ (where J is the $^1J_{CH}$ coupling constant [12]) leads to "J-edited" spectra, where ^{13}C peaks of carbons with odd and even numbers of attached protons appear as having opposite amplitudes (Figure 2.9).

J-modulated carbon spectra (JMOD) are very effective for establishing the substitution pattern of the carbons in the molecule. The method is almost as sensitive as conventional ^{13}C NMR (benefits from NOE signal enhancement) and provides the information on all carbons in the same spectrum [13], and using JMOD is on a routine

Some NMR Spectroscopic Techniques Used in Homogeneous Catalysis 33

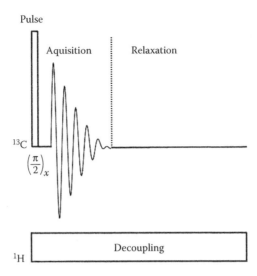

FIGURE 2.6 ^{13}C NMR experiment with ^{1}H decoupling.

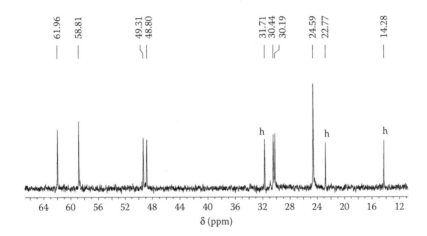

FIGURE 2.7 ^{13}C{^{1}H} NMR spectrum (400.13 MHz, chloroform-d, 20°C) of a sample compound—dimeric titanium(IV) salan complex **1**. Only aliphatic part is shown. Resonances of hexane are marked with "h."

basis technically more simple than using, for example, DEPT pulse sequence [2,3]. On the other hand, J-modulated ^{13}C spectra are proton-decoupled and therefore do not provide quantitative information on the values of the scalar coupling between the ^{13}C and ^{1}H nuclei ($^nJ_{CH}$). The latter constants may be collected by measuring gated-decoupled ^{13}C{^{1}H} NMR spectra, when ^{1}H-decoupling is switched off during acquisition (Figure 2.10a) in order to obtain the ^{13}C resonances as NOE-enhanced multiplets.

A general problem with ^{13}C NMR spectra is that the integral intensities of ^{13}C{^{1}H} NMR resonances, due to the NOE and long T_1, are *not* proportional to the number of

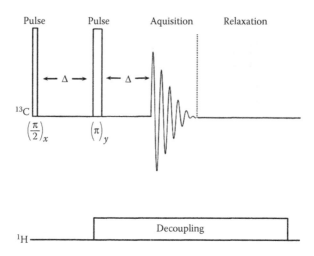

FIGURE 2.8 Pulse sequence for the *J*-modulated ^{13}C NMR experiment, with ^{1}H decoupling.

carbons. To avoid the NOE distortions, inverse-gated pulse sequence is used (Figure 2.10b). In the latter, the decoupler is switched on for a relatively short period of time during acquisition (which time is insufficient for NOE to establish) and then is switched off during the relaxation time. In effect, the ^{13}C NMR spectrum should appear as a set of singlet lines with correct integral intensities. To ensure correct integrals, the relaxation delay must be long enough (3–5 T_1) for the spin–lattice relaxation to occur completely after each pulse. In practice, the delay may be as long as several minutes.

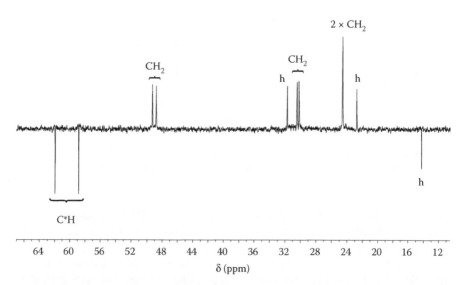

FIGURE 2.9 ^{13}C *J*-modulated NMR spectrum (400.13 MHz, chloroform-*d*, 20°C) of a sample compound—dimeric titanium(IV) salan complex **1** (aliphatic part). Multiplicities of carbons are shown. C* stands for asymmetric carbons. Resonances of hexane are marked with "h."

Some NMR Spectroscopic Techniques Used in Homogeneous Catalysis

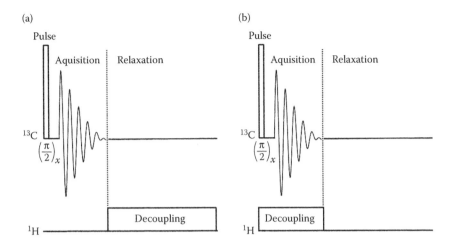

FIGURE 2.10 Gated- (a) and inverse-gated decoupled (b) ^{13}C NMR pulse sequences.

To establish the structure of the hydrocarbon framework of an organic compound, the researcher has to establish the correlation between its 1D ^{1}H and ^{13}C NMR spectra. This can be achieved in different ways. On the one hand, one can use *selective* ^{1}H decoupling [14], when only one ^{1}H resonance is irradiated: in the ^{13}C spectrum, only the corresponding carbon peak must appear as an NOE-enhanced singlet; all other ^{13}C resonances will be displayed as multiplets without NOE. However, this approach is poorly informative and too time-consuming since the number of selectively ^{1}H-decoupled ^{13}C NMR spectra should be equal to the number of resonances in the ^{1}H NMR spectrum.

Another, much more efficient, approach is the use of 2D heteronuclear correlation spectroscopy. The principles of heteronuclear H,X-correlation spectroscopy (often abbreviated as HXCO, or HETCOR, or H,X-COSY) are similar to those of homonuclear COSY (see above). The simplified pulse sequence for the H,X-COSY experiment (where X = ^{13}C) is presented in Figure 2.11. The parameter t_1 has the same meaning as in the homonuclear COSY experiment (incremented delay playing the role of variable evolution time). The delay Δ_1 (to let the evolution of heteronuclear coupling [2,3,16]) is set equal to $1/2J$ [12]. The other delay Δ_2 may in principle be varied for "editing" (manipulating with the intensities of CH, CH$_2$, and CH$_3$ resonances), but in most cases, it is set equal to $0.3/J$ for adequate reproduction of the intensities of peaks corresponding to differently substituted carbons.

The ^{1}H,^{13}C heteronuclear correlated spectrum (aliphatic part) of the sample compound **1** (Figure 2.3) is presented in Figure 2.12. With a ruler in hand, one can easily connect the correlating resonances of the 1D ^{1}H and ^{13}C NMR spectra (that are usually included as "projections" of the 2D plot (Figure 2.12). For example, the resonances of hexane are connected with solid lines; the HXCO spectrum readily confirms our initial suggestion that the broad ^{1}H resonance at δ 1.29 is in fact a mixture of two overlapping peaks, each correlating with two different ^{13}C NMR resonances.

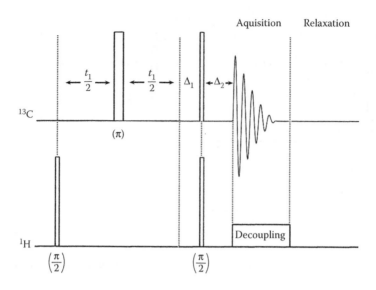

FIGURE 2.11 Pulse sequence for the H,X-COSY experiment (X = ^{13}C).

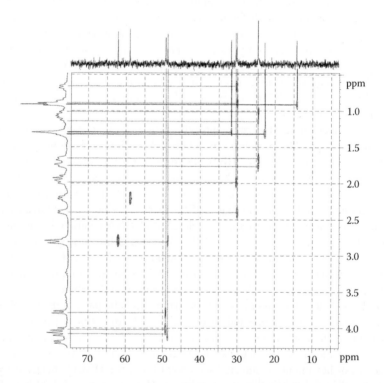

FIGURE 2.12 Two-dimensional ^{1}H,^{13}C HXCO spectrum (400.13 MHz, chloroform-d, 20°C) of a sample compound—dimeric titanium(IV) salan complex **1**. Only aliphatic part is shown. ^{1}H and ^{13}C resonances of hexane are connected with solid lines.

Some NMR Spectroscopic Techniques Used in Homogeneous Catalysis   37

The 2D ^{1}H,^{13}C-correlation spectra provide a useful tool for identifying pairs of diastereotopic protons—the protons that are bonded to the same carbon but are magnetically nonequivalent owing to the influence of the chiral perturbing center. In Figure 2.12, six pairs of diastereotopic protons are apparent: they exhibit themselves as correlations of one ^{13}C peak with two separate ^{1}H peaks (connected with dotted lines).

Two-dimensional ^{1}H,^{13}C-correlation NMR spectroscopy (HXCO) is a very powerful tool for the identification of the chemical structures, based on the determination of direct (^1J) C,H scalar couplings. Modern NMR spectrometers, taking advantage of pulsed magnetic field gradients, use other, gradient-dependent, techniques such as HSQC (*Heteronuclear Single Quantum Coherence spectroscopy*) or HMQC (*Heteronuclear Multiple Quantum Coherence spectroscopy*); the important feature of such programs is that detection is performed on the ^{1}H channel, which substantially reduces the acquisition time. Moreover, there are techniques optimized for detecting long-range correlations (^2J–^4J), such as COLOC (*COrrelation through LOng-range Coupling*), or HMBC (*Heteronuclear Multiple Bond Correlation spectroscopy*) for experiments with gradients. One could also mention the homonuclear, ^{13}C,^{13}C-correlation spectroscopy (INADEQUATE), which can also be very informative for structural assignments [15]. Further theoretical as well as practical aspects of various 2D NMR spectroscopic techniques can be found in the literature, see, for example, References [16,17].

2.1.2 ^{1}H COSY VERSUS ^{1}H TOCSY SPECTRA

In this section, we will briefly consider the difference between the two types of homonuclear correlation NMR experiments, COSY and TOCSY (*TOtal Correlation SpectroscopY*). The former experiment was discussed above; in general, it is used to reveal the scalar ^{1}H,^{1}H couplings. However, the intensities of the cross-peaks depend on the coupling constants: the smaller the J constants, the lower the intensity. For example, in Figure 2.5, mostly the peaks corresponding to ^2J constants are well represented, whereas those for longer-range couplings have lower intensities (and in some cases are not displayed at all). In practice, information on correlations between protons connected via more than three bonds is typically lost in ^{1}H,^{1}H-COSY spectra. In most cases, it is not a serious problem [18], and the combined information as provided by ^{1}H,^{1}H- and ^{1}H,^{13}C-COSY spectra is sufficient for the correct assignment. In some cases, however, it may be desirable to identify as many correlations between the protons of the molecule as possible [19]. In theory, TOCSY generates cross-peaks between all members of a coupled spin network due to a different mechanism of magnetization transfer [20]. In practice, TOCSY can efficiently identify long-range correlations (through 5–6 bonds). The following example illustrates a situation when TOCSY provides complementary information to COSY.

Figure 2.13a and b represents the ^{1}H-COSY and TOCSY spectra of the Ni-polymeryl species **2a**, which is the true catalytically active site (chain-carrying species) of ethylene polymerization by neutral salicylaldiminato nickel(II) complexes **2** (Scheme 2.1) [21]. Of particular interest is the polymeryl (Ni—CH$_2$—CH$_2$—CH$_2$—...) part of the intermediate **2a**. One can see that the COSY spectrum reflects only the

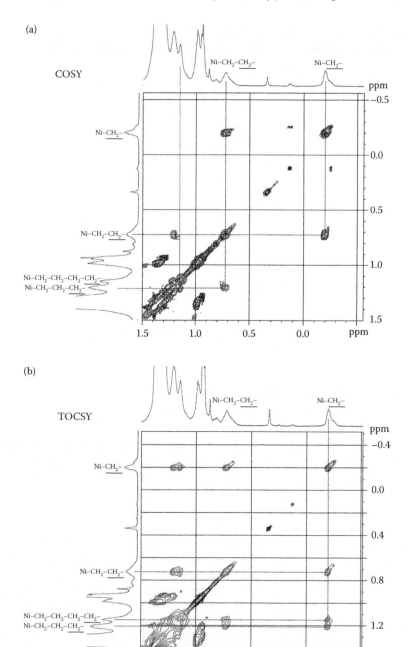

FIGURE 2.13 ^{1}H,^{1}H COSY (a) and TOCSY (b) NMR spectra (600 MHz, toluene-d_8, −20°C) of the Ni-polymeryl species **2a** obtained by the addition of 30 equivalents of C$_2$H$_4$ to 0.012 M solution of catalyst **2** and storing for 3 min at 60°C.

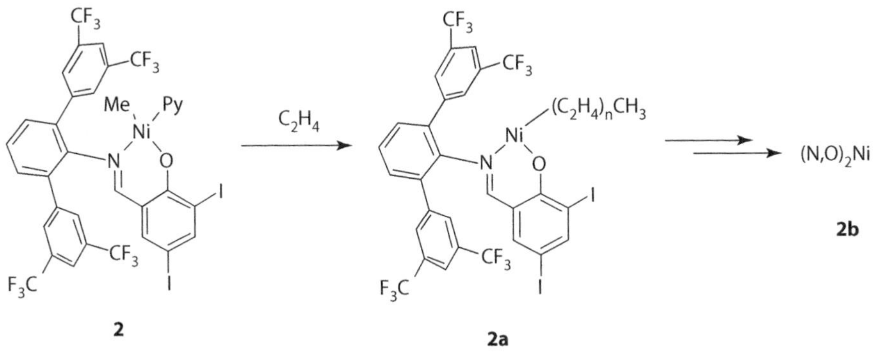

Some NMR Spectroscopic Techniques Used in Homogeneous Catalysis   39

SCHEME 2.1 Formation of the Ni-polymeryl species **2a** from the salicylaldiminato nickel(II) complexes **2** ($n > 2$).

correlations between protons of neighboring CH_2 groups. In principle, it is possible to establish the length of the polymeryl fragment "step by step"; in our case, however, only the correlations within the first three methylenes ($Ni-CH_2-CH_2-CH_2-$) are readily seen. On the contrary, the TOCSY spectrum clearly shows the correlations between the $Ni-CH_2$ protons and at least the next three methylene groups (Figure 2.13, bottom), that is, through five chemical bonds [22]. In Reference [21], this finding was used to distinguish the Ni-polymeryl intermediates from the previously known Ni-propyl and Ni-butyl species. In fact, the TOCSY and COSY spectra can provide complementary information: the former gives the general picture of scalar interactions existing within the same molecule, while the latter provides the "order" of these interactions and thus helps to establish the chemical structure. Additionally, the TOCSY spectra can be useful for separating signals from crowded spectral regions.

2.2 ^{1}H SPECTRA OF PARAMAGNETIC MOLECULES

Traditionally, NMR spectroscopy of paramagnetic molecules has been relatively underestimated by researchers. The authors of this book have an impression that the opinion that paramagnetism leads to unacceptable broadening of NMR resonances, and paramagnetic species are therefore unsuitable for NMR characterization, is still rather widespread (if not prevalent) within chemists. In fact, however, this opinion is only correct for $S = 1/2$ paramagnetic compounds [23,24], while for $S > 1/2$ paramagnetic species, the situation is not at all hopeless. Although the spectra of paramagnetic metal complexes are not so well resolved as those of diamagnetic compounds, and nucleus–nucleus couplings can rarely be observed in such NMR spectra, NMR studies of paramagnetic complexes can be a rich source of information, providing data on the structure of the complexes and the distribution of the unpaired electrons, as well as on exchange interactions the molecules are involved in, etc. [25].

On the other hand, an important drawback of paramagnetic NMR is the very narrow scope of suitable nuclei. In fact, only ^{1}H and ^{2}H are well available since only ^{1}H and ^{2}H NMR spectra exhibit measurable *paramagnetic shifts* (even for ^{1}H and ^{2}H, paramagnetic shifts may spread over several hundreds of ppm, cf. the typical

40   Applications of EPR and NMR Spectroscopy in Homogeneous Catalysis

range of ^{1}H chemical shifts of diamagnetic compounds is only ca. 10 ppm). Actually, the existence of the ^{2}H isotope provides a useful (and most convincing) tool for peak assignment of paramagnetic species, consisting in *selective deuteration* of the ligand. The replacement of hydrogen atom with the deuterium results in the disappearance of the corresponding peak in the ^{1}H NMR spectrum, accompanied by the emergence of the corresponding ^{2}H peak at the same (in the δ scale) position [27].

For other nuclei, the interactions between the electron spin and the nuclear spin most often lead to too high paramagnetic shifts. A fortunate exception is the case of ion pairs: as soon as the paramagnetic shift consists of the Fermi contact (through-bond) and dipolar (through-space) contributions, the solvent-separated counteranions may appear virtually unaffected by the presence of the paramagnetic center, providing sharp and informative resonances on heteronuclei (see below).

In this section, we will provide some illustrations how paramagnetic NMR can help researchers in getting the information on the nature of compounds that appear in catalytic studies, and on the interactions they are involved in, with a brief discussion.

2.2.1 Probing the Structure of an Unknown Ni(II) Complex by Multinuclear NMR

For the purposes of this chapter, a new complex of nickel(II) has been prepared by reacting the TPA ligand with nickel(II) acetate in acetonitrile, and ammonium hexafluorophosphate was added in the course of the synthesis (Scheme 2.2). While writing this sentence, the exact composition and the structure are unknown (the x-ray has not yet been acquired). Let us imagine ourselves real synthetic chemists: while awaiting the x-ray data, we have submitted our complex to the multinuclear NMR study.

The ^{1}H NMR spectrum of complex **3** is presented in Figure 2.14. The spectrum is characteristic of a paramagnetic species, which (1) corroborates that the Ni(II) metal center is indeed present in the structure and (2) indicates that the Ni(II) ion is in a nonplanar ligand environment [26]. The resonances fit within the spectral range of 143 to −3.3 ppm. The peaks are rather broad, and the integral intensities are not ideal; however, the latter well reflect the number of the corresponding protons with ±10% accuracy. This fact can be used for the line assignment (Figure 2.14 and Table 2.1).

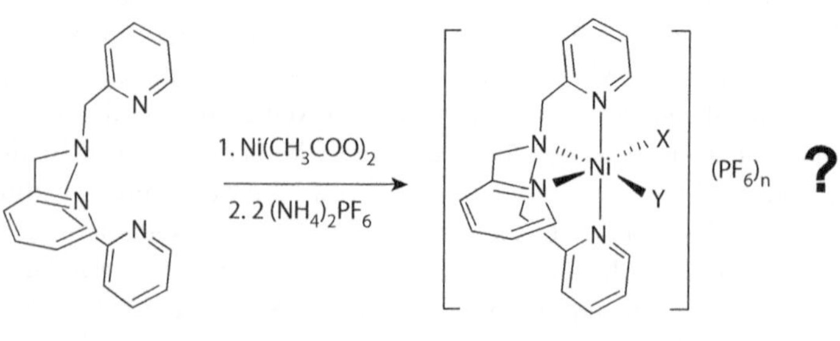

3

SCHEME 2.2   Synthesis of the Ni(TPA) complex.

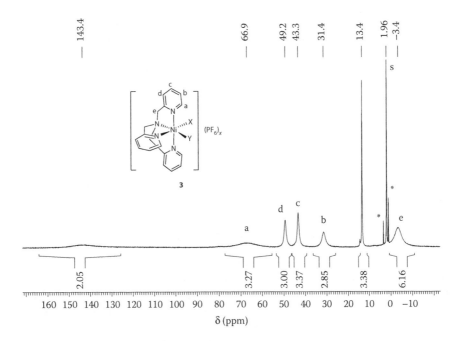

FIGURE 2.14 ^{1}H NMR spectrum (250 MHz, acetonitrile-d_3, 20°C) of the Ni complex **3** (inset), with tentative line assignment. Peak of residual acetonitrile-d_2 is marked with "s." Resonances of traces of diethyl ether (cosolvent used for crystallization) are marked with asterisks.

A complementary basis for the line assignment is the widths ($\Delta v_{1/2}$) of the paramagnetic resonances. In practice, the latter in most cases is determined by the pseudocontact interactions between the electron spin and the nuclear spin of the particular proton, which leads to the r^{-6} dependence of the $\Delta v_{1/2}$ on the distance between the nucleus and the paramagnetic center [24,27]. In Table 2.1, the data on the ^{1}H resonances of **3** are summarized, and the tentative assignment is given. The situation with ligands X and Y is not clear. On the basis of the integral values and the line widths, we can hypothesize that they are water molecule and acetate anion.

Let us now check whether there is counteranion and get insight into its composition. According to the synthetic procedure, we would expect that there should be hexafluorophosphate anion in the outer sphere; it is logical to measure the ^{31}P and ^{19}F

TABLE 2.1
^{1}H Spectral Data of Complex 3 (250 MHz, Acetonitrile-d_3, 20°C)

δ (ppm)	−3.4	13.4	31.4	43.3	49.2	66.9	143.4
Integral	6H	3H	3H	3H	3H	3H	2H
$\Delta v_{1/2}$ (Hz)	960	45	470	230	270	2300	3600
Assignment	e	OAc	b	c	d	a	H$_2$O

NMR spectra; both nuclei are rather sensitive and have 100% natural abundance [1]. The corresponding NMR spectra (Figure 2.15) give evidence that the phosphorus and fluorine are indeed present in the composition of compound 3; the ^{19}F NMR spectrum appears as a doublet, and the ^{31}P spectrum is septet, with the $^{1}J_{PF}$ coupling constant of 706 Hz. The chemical shifts and the coupling constants are typical for those of PF_6^- anions [28–30], indicating that paramagnetism of the cationic part of compound 3 does not essentially affect the counteranion (most likely efficiently separated from the cation by the acetonitrile molecules).

One can see that NMR data provide valuable information (yet insufficient for structural analysis) on the nature of our model paramagnetic compound. After establishing the x-ray structure (Figure 2.16), we can certify its consistency with our tentative assignment.

2.2.2 Temperature Dependence of the Paramagnetic Shift

Apart from the paramagnetic shift and paramagnetic line width, valuable data can be extracted from the temperature dependence of these values. First of all, the strong temperature dependence of paramagnetic shifts can be used to distinguish between the NMR resonances of paramagnetic species from those of diamagnetic compounds [31]; this is important if one has a mixture of diamagnetic and paramagnetic species in the sample. On the other hand, the character of the temperature dependence may provide information on the structure (monomeric/dimeric) of the paramagnetic species and on the chemical exchange processes it is involved in. In this section, several typical cases of this kind will be considered.

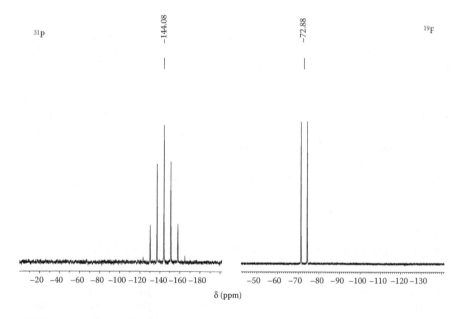

FIGURE 2.15 ^{31}P and ^{19}F NMR spectra (101.25 and 235.36 MHz, respectively, acetonitrile-d_3, 20°C) of the Ni complex 3.

Some NMR Spectroscopic Techniques Used in Homogeneous Catalysis   43

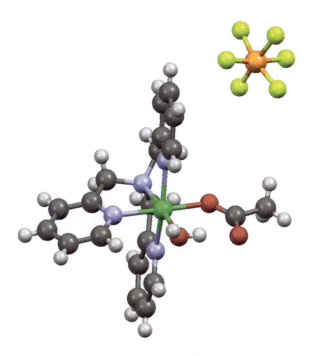

FIGURE 2.16 Molecular structure of compound **3**. Ni(II) green, O red, N violet, C gray, H light gray, F yellow, and P orange. CCDC 1481081 contains the supplementary crystallographic data for this figure. These data can be obtained free of charge from the Cambridge Crystallographic Data Centre via http://ww.ccdc.cam.ac.uk/data_request/cif and from the authors. The authors thank Dr D. G. Samsonenko for the x-ray measurements.

2.2.2.1 Temperature Dependence of the Paramagnetic Shift of Monomeric Compounds

Let us consider first the easiest and the most widespread situation, that is, the paramagnetic species is monomeric, and the paramagnetic shift is only determined by Fermi contact and pseudocontact (dipolar) interactions [27], Equation 2.1. Theoretical treatment predicts that the Fermi contact contribution to the paramagnetic shift has the $\sim T^{-1}$ temperature dependence [27,32]. In effect, the paramagnetic shift with prevailing Fermi contact contribution should demonstrate the simple Curie-type temperature dependence for the resulting paramagnetic shift (Equation 2.2; α may adopt both positive and negative values) [33]. In Figures 2.17 and 2.18, such a case is presented; one can see that in the range of 230–310 K, the temperature dependence is well described by the Curie law:

$$\delta = \delta_{dia} + \delta_{para} = \delta_{dia} + (\delta_{con} + \delta_{dip}) \qquad (2.1)$$

$$\delta_{para} \approx \delta_{con} = \frac{\alpha}{T} \qquad (2.2)$$

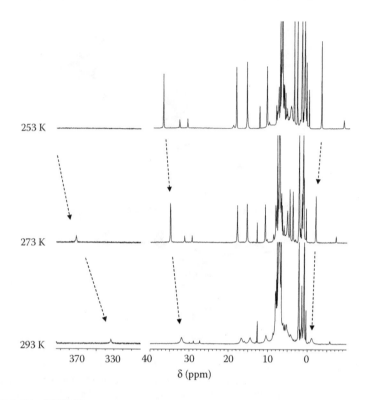

FIGURE 2.17 ¹H NMR spectra (600 MHz, toluene-d_8, 0°C) of the bis-ligated (Scheme 2.1) complex **2b**·(pyridine-d_5) (Py/Ni ca. 5:1) at different temperatures. Arrows show the evolution of positions of resonances with temperature.

Analyzing Equations 2.1 and 2.2, one can see that the paramagnetic contribution to the chemical shift vanishes at high temperature, so at $T \rightarrow \infty$, the overall shift δ must approach its "natural," diamagnetic value, which would be observed in the absence of Fermi contact and dipolar interactions. This fact can sometimes be exploited while making the peak assignments of ¹H NMR spectra of paramagnetic complexes: extrapolation of the experimental δ versus T^{-1} dependence to $T^{-1} \rightarrow 0$ should give the δ values close to those of the free ligand [34].

In practice, dipolar interactions can make a sizeable contribution to the paramagnetic shift. The dipolar contribution can have more complex temperature dependence, essentially a mixture of $\sim T^{-1}$ and $\sim T^{-2}$ terms, the former one originating from the g-tensor anisotropy and the latter stemming from the zero-field splitting D [32]. In effect, the experimental dependence of the overall paramagnetic shift can be represented [27,32] by the expression given by Equation 2.3. In References [27,35], examples of temperature dependence of the type (2.3) can be found. Experimental dependencies of the type (2.3) can be used for qualitative conclusions (i.e., that the contribution of dipolar interactions to the paramagnetic shift is significant), as well as for quantitative estimation of the zero-field splitting D [35].

Some NMR Spectroscopic Techniques Used in Homogeneous Catalysis

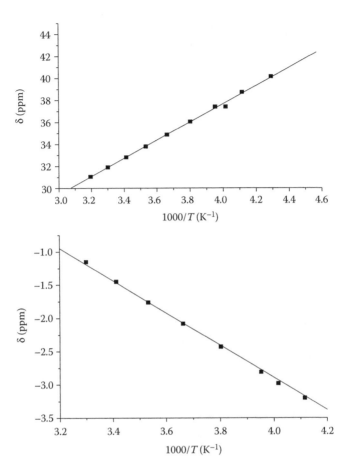

FIGURE 2.18 Linear (Curie) temperature dependence of paramagnetic shifts (δ vs. $1/T$) for the ^{1}H NMR resonances of **1c**′·(pyridine-d_5).

$$\delta_{para} = \frac{\alpha}{T} + \frac{\beta}{T^2} \qquad (2.3)$$

2.2.2.2 Temperature Dependence of Paramagnetic Shift of Antiferromagnetic Dimers

The temperature dependence of the paramagnetic shifts becomes more complicated when there are two paramagnetic centers (i.e., in binuclear complexes), their electron spins interacting with each other with the energy:

$$E = -2J \cdot \vec{S}_1 \cdot \vec{S}_2 \qquad (2.4)$$

Very often, this *exchange interaction* is *antiferromagnetic*: $J < 0$, which case is relevant to many important biological macromolecules [35]. In this case, the dependence of the paramagnetic shifts can deviate from the Curie law, the stronger the antiferromagnetic interaction (i.e., the larger the $|J|$ constant), the larger the deviation. In principle, analyzing the experimental dependence of δ_{para}, one can extract the J value. The Fermi contact shift for the coupled dimeric complex is given by the general equation [35]

$$\delta_{con} = \frac{P}{T} \frac{\sum A_i(2S_i+1)S_i(S_i+1)e^{-J[S(S+1)-S_1(S_1+1)-S_2(S_2+1)]/kT}}{\sum (2S_i+1)e^{-J[S(S+1)-S_1(S_1+1)-S_2(S_2+1)]/kT}} \quad (2.5)$$

where S is the spin of the dimer and S_1 and S_2 are the spins of the monomers constituting the dimer. The spin S adopts values $(S_1 + S_2)$, $(S_1 + S_2 - 1)$, …, $(S_1 - S_2)$. If we are fortunate and the hyperfine constants A_i are the same for all spin states [36], Equation 2.5 allows the determination of the exchange coupling constant J.

In Figure 2.19, the δ versus T plots for a dimeric, antiferromagnetically coupled ferric complex [37] are presented. For this case, $S_1 = S_2 = 5/2$, and S takes the values from 5 to 0; the $S = 0$ state corresponds to the lowest energy. It is archetypical for antiferromagnetically coupled dimers that the paramagnetic shift *increases* with increasing temperature (Figure 2.19), owing to the increasing population of excited, higher-spin states. Assuming that $A_i = A$, the temperature dependence of the observed shifts is expressed as

$$\delta_{con} = \delta_{dia} + \frac{\alpha}{T} \frac{\sum_{S=0}^{5}(2S+1)S(S+1)e^{-J[S(S+1)-17.5]/kT}}{\sum_{S=0}^{5}(2S+1)e^{-J[S(S+1)-17.5]/kT}} \quad (2.6)$$

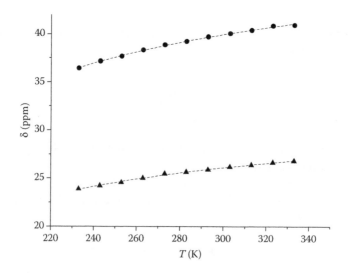

FIGURE 2.19 Experimental temperature dependence of chemical shifts of a diferric complex, fitted (dashed lines) to function given by Equation 2.6.

Some NMR Spectroscopic Techniques Used in Homogeneous Catalysis   47

where δ_{dia} is regarded to be temperature independent. Fitting the experimental curves to Equation 2.6 gave the estimate for J of -136 ± 6 cm^{-1} (Bryliakov, K. P. and Talsi, E. P. Unpublished data.), which is typical for μ-O-bridged diferric complexes [37]. In practice, however, the A_i constants may vary for different states (and even adopt different signs), so care must be taken when extracting the J values from experimental data. When J is large (comparable with kT), it may be useful to roughly evaluate the J by taking into account only the ground state and the first and second excited states (the contribution of the higher states is neglected):

$$\delta_{con} \approx \delta_{dia} + \frac{P}{T} \frac{6A_1e^{2J/kT} + 30A_2e^{6J/kT}}{1 + 3e^{2J/kT} + 5e^{6J/kT}} \tag{2.7}$$

In principle, Equation 2.7 can be used to evaluate the ratio A_1/A_2.

2.2.2.3 Measuring Magnetic Susceptibility (Evans Method) for Studying Spin Equilibrium

Measuring the magnetic susceptibility of transition metal complexes is a powerful tool that provides the information on the spin state of the complex, even if the complex is "EPR silent" or measuring the EPR is not possible. Usually, such measurements are conducted in the solid state [38]. However, the solution structure of metal complexes may differ substantially from their solid-state structure; therefore, the catalytic chemists dealing with transition metal complexes may sometimes be more interested in measuring the magnetic susceptibility of their complexes in solution.

To this end, they can use the Evans method [39]. The technique is based on the facts that (1) the chemical shift is based on the bulk magnetic susceptibility of the medium (solution) in which the molecule is situated and (2) the paramagnetic substance affects the bulk magnetic susceptibility of the solution in a known way. Typically, the paramagnetic substance is dissolved in an appropriate solvent, and the solution is placed in the NMR tube; at the same time, the pure solvent is placed in the sample in a thin narrow tube (Figure 2.20).

The solvent must contain an additive of a reference compound (e.g., tetramethylsilane), whose ^{1}H NMR shift is to be monitored. Owing to the presence of the paramagnetic complex, the reference sample in the inner tube and in the outer volume will resonate at different frequencies, resulting in different chemical shifts. The mass magnetic susceptibility χ_g, (cm^3 g^{-1}), can be calculated from the following equation:

$$\chi_g = -\frac{3}{4\pi m} \frac{\Delta f}{f} + \left[\chi_0 + \chi_0 \frac{d_0 - d_s}{m} \right] \tag{2.8}$$

where Δf is the frequency shift in Hz of the reference compound, f is the spectrometer frequency in Hz, m is the mass (in g) of the complex in 1 cm^3 of the solution, χ_0 is the mass susceptibility of the solvent in cm^3 g^{-1}, and d_0 and d_s are the densities of the solvent and solution, respectively [40]. The term in square brackets vanished because at the low concentrations used (10–15 mM), the solution density was

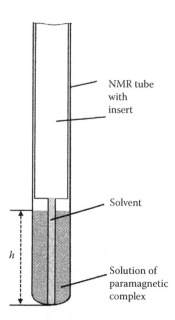

FIGURE 2.20 Preparation of the NMR sample for the Evans measurement.

reasonably approximated as $d_s = d_0 + m$. The factor $3/4\pi$ is applied when the axis of the cylindrical sample is parallel to the magnetic field [41,42]; for other sample configurations, different expressions should be used. To take into account the solution volume changes with temperature, correction was introduced in m values so that $m = m_0 h_0/h$, where m_0 is the mass of the complex in 1 cm³ of the solution at 298 K, h_0 and h are sample heights at 298 K and at given temperature, respectively (Figure 2.20).

The molar susceptibility χ_M (cm³ g⁻¹) can be obtained by multiplying the χ_g values by the molecular weight MW. The latter must be corrected for the diamagnetic contributions of the ligands, counteranions, and Fe^{II} core electrons using Pascal's constants to give the corrected molar susceptibility χ_M' [41]. The latter can be used to calculate the effective magnetic moment $\mu_{eff} = 2.828(\chi_M' T)^{1/2}$ (in units of μ_B), where T is the temperature in K [43].

In Reference [41], the paramagnetic susceptibility of the [Fe^{II}(BPMEN)(CH$_3$CN)$_2$](ClO$_4$)$_2$ complex (Scheme 2.3) in solution was measured. This study was inspired by the observation that the ¹H NMR spectrum of this complex in CD$_3$CN solution are typical for paramagnetic Fe(II) complexes at high temperature ($T > 320$ K) but convert to a diamagnetic spectrum at low temperatures ($T < 240$ K). The Evans measurements were conducted in this temperature range, and the effective magnetic moment μ_{eff} was found to depend on the temperature (Figure 2.21). It is seen that at low temperatures, the observed μ_{eff} is smaller than that expected for paramagnetic complexes (even for complexes with the lowest possible electron spin $S = 1/2$, the predicted μ_{eff} should be 1.73 μ_B [41]), while at high temperatures, the μ_{eff} is close to that corresponding to $S = 2$ (4.90 μ_B) [42]. The rational of such temperature dependence is the

Some NMR Spectroscopic Techniques Used in Homogeneous Catalysis

SCHEME 2.3 Structure of the [Fe^{II}(BPMEN)(CH$_3$CN)$_2$](ClO$_4$)$_2$ complex **4**.

coexistence (in equilibrium) of the d^6 iron(II) complex in the low-spin state ($S = 0$) and in the high-spin state ($S = 2$).

Interestingly, this dependence is not conventional *spin equilibrium* since the low-spin–high-spin transition is caused by a change in the inner coordination sphere of the iron(II) complex (where $n = 2$) [40,44]:

$$[\text{Fe}^{II}(\text{BPMEN})(\text{CH}_3\text{CN})_2]^{2+} \underset{\text{LS}}{\overset{K}{\rightleftharpoons}} [\text{Fe}^{II}(\text{BPMEN})(\text{CH}_3\text{CN})_{2-n}]^{2+}_{\text{HS}} + n\text{CH}_3\text{CN}$$

(2.9)

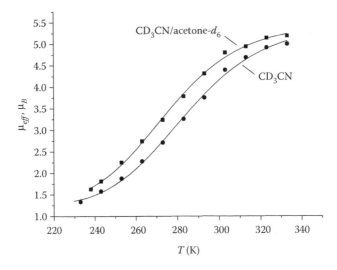

FIGURE 2.21 Plot of μ_{eff} versus 1000/T for complex **4**, in acetonitrile-d_3 (■), or in acetonitrile-d_3/acetone-d_6 (●), and their fits to Equation 2.10.

Since the dissociation of acetonitrile is a fast (on the NMR timescale) process, the measured μ_{eff} is an averaged value, whose temperature dependence can be expressed as follows [42]:

$$\mu_{eff}(T) = \frac{\{\mu_{LS}^2 + \mu_{HS}^2[e^{\Delta S°/R}e^{-\Delta H°/RT}]\}^{1/2}}{\{1 + e^{\Delta S°/R}e^{-\Delta H°/RT}\}^{1/2}} \quad (2.10)$$

where μ_{LS} and μ_{HS} are the magnetic moments of the low-spin and the high-spin states, respectively, and $\exp(\Delta S°/R)\exp(-\Delta H°/RT)$ is the observed equilibrium constant $K_{obs} = [\text{HS-Fe}^{II}]/[\text{LS-Fe}^{II}]$ [40]. Fitting the experimental points to Equation 2.10, one can obtain the μ_{LS} and μ_{HS}, as well as the thermodynamic parameters $\Delta S°$ and $\Delta H°$ for this LS–HS spin transition, associated with the dissociation of the bound acetonitrile [45,46]. In effect, for the above spin transition, $\Delta S°$ was found to be 135 J · mol^{-1} · K^{-1}, and $\Delta H°$ 39.5 kJ · mol^{-1}. Both values are substantially higher than those reported in the literature for solution spin crossover systems [42,47,48], thus supporting the hypothesis that the spin change in this system overlaps with the chemical processes of acetonitrile dissociation [40].

In pure acetone-d_6, no spin-equilibrium-type phenomena were observed, the paramagnetic shifts of **4** demonstrated the conventional Curie temperature dependence (see Reference [40]). This is not surprising since the absence of acetonitrile in solution must shift forward the equilibrium in Equation 2.9 (Le Châtelier's principle), leading to the pure high-spin-state Fe(II) complex.

The situation described above is a typical (yet rare) example of overlap of solvent effects with the true *spin crossover* (so that substitution or dissociation of coordinated ligands leads to the change of spin state). Phenomena like these should not be considered as true spin crossover interconversions [49]. Interestingly, the $\Delta S°$ and

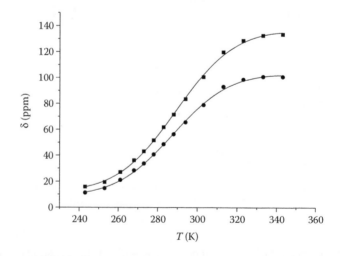

FIGURE 2.22 Plot of δ versus T variation for complex **4** (0.01 M solution in acetonitrile-d_3), fitted to Equation 2.11.

Some NMR Spectroscopic Techniques Used in Homogeneous Catalysis 51

$\Delta H°$ values can be obtained from the analysis of observed temperature dependence of the paramagnetic shifts of **4** in acetonitrile-d$_3$, taking into account the Fermi contact paramagnetic shift, as well as the LS–HS equilibrium [40]:

$$\delta_{obs}(T) = \delta_{LS} + \left(\delta_{LS} + \frac{C}{T}\right)\frac{e^{\Delta S°/R}e^{-\Delta H°/RT}}{1+e^{\Delta S°/R}e^{-\Delta H°/RT}} \tag{2.11}$$

When the experimental shifts (Figure 2.22) were fitted to Equation 2.11, the obtained parameters ($\Delta S° = 137\ \text{J} \cdot \text{mol}^{-1} \cdot \text{K}^{-1}$, $\Delta H° = 40\ \text{kJ} \cdot \text{mol}^{-1}$) were nearly the same as those estimated from the variable-temperature magnetic susceptibility measurements. This demonstrates the utility of the combined solution magnetic susceptibility/^{1}H NMR spectroscopic investigation for the detailed physicochemical investigations of the dynamic spin transition phenomena of this type.

REFERENCES

1. Bruker Almanac. 2014. Available online at http://3debooks.annikassociates.co.uk/?userpath=00000016/00020256/00082433.
2. Derome, A. E. 1990. *Modern NMR Techniques for Chemistry Research*, Pergamon Press, Oxford–New York–Beijing–Frankfurt–Sâo Paulo–Sydney–Tokyo–Toronto.
3. Claridge, T. D. W. 1999. *High-Resolution NMR Techniques in Organic Chemistry*, Pergamon Press, Amsterdam–Lausanne–New York–Oxford–Shannon–Singapore–Tokyo.
4. Talsi, E. P., Samsonenko, D. G., Bryliakov, K. P. 2014. Titanium salan catalysts for the asymmetric epoxidation of alkenes: Steric and electronic factors governing the activity and enantioselectivity. *Chem. Eur. J.* 20: 14329–14335.
5. On the other hand, various relaxation processes (i.e., due to chemical exchange or paramagnetism) may have a negative effect (destroy) on the magnetization transfer so that even $^2J_{HH}$ couplings may not be seen.
6. Typical delays are several seconds for ^{1}H NMR and 10–60 s for ^{13}C NMR spectra.
7. Fortunately, modern cryomagnets have strong fields, which increases the operating frequency and thus the sensitivity of the NMR spectrometers.
8. ^{1}H decoupling leads to coalescence of the ^{13}C multiplets into singlets; furthermore, upon irradiation at the ^{1}H frequency, the intensity of the resulting ^{13}C singlet increases (up to four times) due to the *nuclear Overhauser effect* (NOE). It is obvious that ^{1}H decoupling is only effective for carbons bonded to at least one proton.
9. Gottlieb, H. E., Kotlyar, V., Nudelman, A. 1997. NMR chemical shifts of common laboratory solvents as trace impurities. *J. Org. Chem.* 62: 7512–7515.
10. Fulmer, G. R., Miller, A. J. M., Sherden, N. H., Gottlieb, H. E., Nudelman, A., Stolz, B. M., Bercaw, J. E., Goldberg, K. I. 2010. NMR chemical shifts of trace impurities: Common laboratory solvents, organics, and gases in deuterated solvents relevant to the organometallic chemist. *Organometallics* 29: 2176–2179.
11. Which "freezes" the components of the multiplets by suppression of scalar coupling.
12. In practice, *J* is most often taken as 130–150 Hz, typically 145 Hz.
13. Setting $\Delta = 1/2J$ corresponds to a null for all protonated carbons [3], which may be additionally exploited for identifying quaternary carbons.
14. The so-called CW selective irradiation.
15. The sensitivity of this experiment is very poor, so it is most effectively performed on neat compounds.

52 Applications of EPR and NMR Spectroscopy in Homogeneous Catalysis

16. Croasmus, W. R., Carlson, R. M. K., Eds. 1994. *Two-Dimensional NMR Spectroscopy: Applications for Chemists and Biochemists* (2nd edition), John Wiley & Sons, New York–Chichester–Weinheim–Brisbane–Singapore–Toronto.
17. Heise, H., Metthews, S., Eds. 2013. *Modern NMR Methodology*. Springer, Heidelberg–New York–Dordrecht–London.
18. Sometimes, it is an advantage since the absence of peaks for long-range correlations facilitates the interpretation of the 2D NMR spectrum.
19. Such information is very important if one has to prove (or disprove) that some protons (even remote from each other) belong to the same molecule. TOCSY can also be useful when measuring $^1H,^{13}C$ heteronuclear correlation spectra is not possible, for example, because of insufficiently low concentration (in, e.g., the determination of structures of proteins).
20. TOCSY uses a "spin-lock" for coherence transfer. During the "spin-lock," all protons of a coupled system become "strongly coupled," leading to cross-peaks between all resonances of a coupled system.
21. Soshnikov, I. E., Semikolenova, N. V., Zakharov, V. A., Möller, H. M., Ölscher, F., Osichow, A., Göttker-Schnettmann, I., Mecking, S., Talsi, E. P., Bryliakov, K. P. 2013. Formation and evolution of chain-propagating species upon ethylene polymerization with neutral salicylaldiminato nickel(II) catalysts. *Chem. Eur. J.* 19: 11409–11417.
22. Further correlations are not clearly identified due to peak overlap.
23. Owing to relatively long electron spin relaxation time. For well-resolved NMR spectra to be obtained, the electron spin relaxation time T_{1e} should be of the order of 10^{-11} s or shorter [24].
24. Swift, T. J. 1973. In *NMR of Paramagnetic Molecules: Principles and Applications*, G. N. La Mar, W. DeW. Horrocks, Jr., R. H. Holm, Eds. Academic Press, New York and London.
25. Schwarzhans, K. E. 1970. NMR spectroscopy of paramagnetic complexes. *Angew. Chem. Int. Ed.* 9: 946–953.
26. Holm, R. H., Hawkins, C. J. 1973. For square planar or square pyramidal Ni(II) complexes, diamagnetic ($S = 0$) state is typical, Chapter 7. In *NMR of Paramagnetic Molecules: Principles and Applications*, G. N. La Mar, W. DeW. Horrocks, Jr., R. H. Holm, Eds. Academic Press, New York and London.
27. Paramagnetic shift is usually predominantly determined by the Fermi contact interactions (that depend on the electron spin density at the particular proton), which prevents establishing direct correlations between the electron–nuclear distance and the paramagnetic shift. Jesson, J. P. 1973. Chapter 1. In *NMR of Paramagnetic Molecules: Principles and Applications*, G. N. La Mar, W. DeW. Horrocks, Jr., R. H. Holm, Eds. Academic Press, New York and London.
28. Chang, H. W., Fang, C. S., Sarkar, B., Wang, J. C., Liu, C. W. 2010. Anion-templated syntheses of octanuclear silver clusters from a silver dithiophosphate chain. *Chem. Commun.* 46: 4571–4573.
29. Lehr, J., Lang, T., Blackburn, O. A., Barendt, T. A., Faulkner, S., Davis, J. J., Beer, P. D. 2013. Anion sensing by solution- and surface-assembled osmium(II) bipyridyl rotaxanes. *Chem. Eur. J.* 19: 15898–15906.
30. Makhoukhi, B., Villemin, D., Didi, M. A. 2015. Synthesis of bisimidazolium–ionic liquids: Characterization, thermal stability and application to bentonite intercalation. *J. Taibah Univ. Sci.* 10: 168–180.
31. Chemical shifts of diamagnetic species are weakly dependent on the temperature (unless the diamagnetic species is involved in exchange processes).
32. Kurland, R. J., McGarvey, B. R. 1970. Isotropic NMR shifts in transition metal complexes: The calculation of the Fermi contact and pseudocontact terms. *J. Magn. Res.* 2: 286–301.

Some NMR Spectroscopic Techniques Used in Homogeneous Catalysis   53

33. Equation 2.2 is a simplified (phenomenological) expression; for details, see References [27,32].
34. In most cases, such extrapolation is not very accurate and may be used only for qualitative treatment of the chemical shifts.
35. La Mar, G. N., Eaton, G. R., Holm, R. H., Walker, F. A. 1973. Proton magnetic resonance investigation of antiferromagnetic oxo-bridged ferric dimers and related high-spin monomeric ferric complexes. *J. Am. Chem. Soc.* 95: 63–75.
36. This is not at all *a priori* guaranteed.
37. Kurtz Jr., D. M. 1990. Oxo- and hydroxo-bridged diiron complexes: A chemical perspective on a biological unit. *Chem. Rev.* 90: 585–606.
38. O'Connor, C. J. 1996. Magnetic-susceptibility measurement techniques, Chapter 4. In *Molecule-Based Magnetic Materials.* ACS Symposium Series, Vol. 644, pp. 44–66. Washington, DC.
39. Evans, D. F. 1959. The determination of the paramagnetic susceptibility of substances in solution by nuclear magnetic resonance. *J. Chem. Soc.* 2003–2005.
40. Bryliakov, K. P., Duban, E. A., Talsi, E. P. 2005. The nature of the spin state variation of $[Fe^{II}(BPMEN)(CH_3CN)_2](ClO_4)_2$ in solution. *Eur. J. Inorg. Chem.* 72–76.
41. Drago, R. S. 1992. *Physical Methods for Chemists*, 2nd ed. Surfside Scientific Publishers, Gainesville, FL.
42. Turner, J. W., Schultz, F. A. 2001. Solution characterization of the iron(II) bis(1,4,7-triazacyclononane) spin-equilibrium reaction. *Inorg. Chem.* 40: 5296–5298.
43. Naklicki, M. L., White, C. A., Plante, L. L., Evans, C. E. B., Crutchley, R. J. 1998. Metal–ligand coupling elements and antiferromagnetic superexchange in ruthenium dimers. *Inorg. Chem.* 37: 1880–1885.
44. Fundamentally, this LS–HS transition occurs due to *weakening* of the ligand field upon dissociation of the coordinated acetonitrile, which draws closer the energy levels of the transition metal ion.
45. Equation 2.10 assumes a four-parameter nonlinear fit, which may be of insufficient accuracy. In Reference [41], Equation 2.10 was used essentially to find the μ_{LS} and μ_{HS} values (1.40 μ_B and 5.45 μ_B, respectively, which is higher than spin-only values for $S = 0$ and $S = 2$ states, indicating substantial orbital contribution). The precise measurements of the ΔS° and ΔH° parameters was accomplished by the Crawfors and Swanson approach [47], which allows obtaining the linear fit for the dependence of the observed equilibrium constant versus $1/T$ [41].
46. Crawford, T. H., Swanson, J. 1971. Temperature dependent magnetic measurements and structural equilibria in solution. *J. Chem. Educ.* 48: 382–386.
47. König, E. 1991. Nature and dynamics of the spin-state interconversion in metal complexes. *Struct. Bonding* 76: 51–152.
48. Lemercier, G., Bousseksou, A., Verlst, M., Varret, F., Tuchagues, J. P. 1995. Dynamic spin-crossover in $[Fe^{II}(TRIM)_2]Cl_2$ investigated by Mössbauer spectroscopy and magnetic measurements. *J. Magn. Magn. Mater.* 150: 227–230.
49. Toftlund, H. 2001. Spin equilibrium in solutions. *Monatsh. Chem.* 132: 1269–1277.

3 NMR and EPR Spectroscopy as a Tool for the Studies of Intermediates of Transition Metal–Catalyzed Oxidations

The direct reaction of triplet oxygen $^3O_2$ with singlet organic molecules to give singlet products is a spin-forbidden process. In terms of chemical kinetics, this means that uncatalyzed reaction of $^3O_2$ with organic substrates will proceed very slowly. At the same time, such reactions may be facilitated by transition metal complexes; the role of the metal center is (1) to "activate" the dioxygen molecule and convert it to reactive metal–oxygen species, which will then (2) transfer the oxygen atom to the organic substrate. In this chapter, we will particularly focus on the second process; typically, it is conducted by metal–oxygen intermediates (such as superoxo, peroxo, hydroperoxo, alkylperoxo, and oxo complexes of transition metals). The unfading interest in experimental and theoretical investigations of metal–oxygen reactive species and oxidation reactions they conduct is to a large degree accounted for by the participation of related metal–oxygen intermediates in reactions occurring in living organisms in the presence of metalloenzymes. The corresponding branch of catalytic science, focused on the investigation of synthetic metal complexes, modeling the catalytic functions of natural metalloenzymes, is called *biomimetic catalysis*. On the other hand, the attention to metal-catalyzed oxidations is caused by the growing demand in novel, more efficient, and "green" catalyst systems for chemo-, regio-, and stereoselective oxidation of organic substrates, which is mostly dictated by toughening economic and environmental constraints. Nowadays, it is generally accepted that the design of novel catalyst systems should be based on the understanding of the detailed mechanism of their catalytic action, which in turn requires deep insight into the nature of the reactive species.

This chapter will present the NMR and EPR spectroscopic detections and characterizations of various metal–oxygen intermediates. The only exception is the last section, dedicated to oxygen-free Co(III) species, capable of directly abstracting an electron from alkanes and alkylbenzenes.

56   Applications of EPR and NMR Spectroscopy in Homogeneous Catalysis

3.1   SUPEROXO COMPLEXES

The early studies of the superoxide anion $O_2^{\cdot-}$ and its complexes with transition metals of the type $LM(O_2^{\cdot-})$ (where M is a metal, and L is an organic ligand) reported that $O_2^{\cdot-}$ and $LM(O_2^{\cdot-})$ were not highly reactive toward organic substrates [1,2]. The only fairly well-understood preparative catalytic reaction incorporating superoxo complex as an active intermediate was the oxidation of 2,6-disubstituted phenols to the corresponding quinones, promoted by cobalt(II) complexes L_1Co^{II} (where L_1 = Schiff base ligand) [3–8].

3.1.1   SUPEROXO COMPLEXES OF Co(III)

Nishinaga and coworkers demonstrated that the monomeric superoxo complex $Co^{III}(Salpr)(O_2^{\cdot-})$ (1) (Salpr = bis(3-(salicylidenamino)propyl)methylamine) (Figure 3.1) can abstract a hydrogen atom from 2,4,6-trisubstituted phenol: the authors detected a simultaneous disappearance of the EPR signal of 1 and the buildup of the signal of the phenoxy radical when the oxygen supply is limited (Scheme 3.1, reaction 1) [4]. It was assumed that the phenoxy radical reacted with 1 to afford the peroxy-p-quinolato cobalt(III) complex 2 (Figure 3.1, Scheme 3.1, reaction 2). In the case of 2,4,6-tri-*tert*-butyl phenol as a substrate ($R_1 = R_2 = t$Bu), 2 was isolated, and its x-ray structure was established [6]. When 2,6-dimethylphenol was used as a substrate, the decay of 2 resulted in the formation of benzoquinone (Scheme 3.1, reaction 3) [7].

$Co^{II}(L_1)(B)$ and $Co^{III}(L_1)(B)(O_2^{\cdot-})$ ($S = 1/2$) species (L_1 is tetradentate N,O-donor ligand, and B is Py, H_2O, DMF, etc.) display informative EPR spectra. As an example, the EPR spectrum of frozen solution of $Co^{II}(acacen)(py)$ (3) in toluene is presented

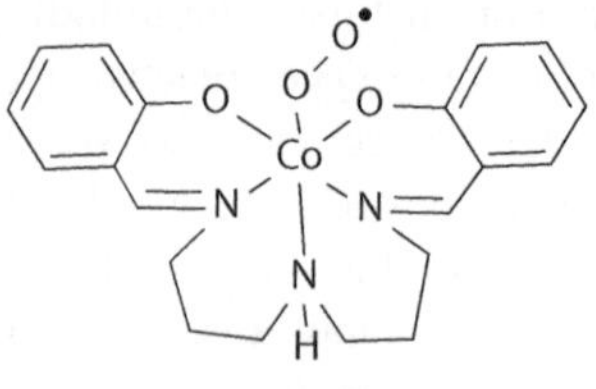

(Salpr)Co^{III}(O$_2$)

1

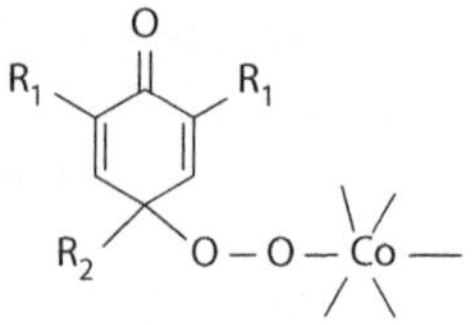

2

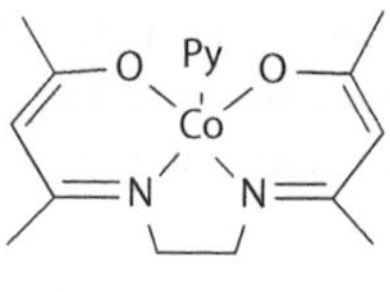

(acacen)Co^{II}(Py)

3

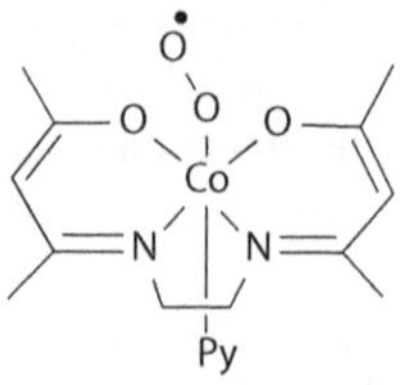

(acacen)Co^{III}(Py)(O$_2$)

4

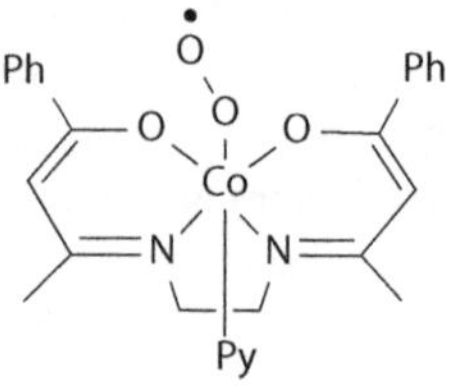

(benacen)Co^{III}(Py)(O$_2$)

5

FIGURE 3.1   Cobalt superoxo and related complexes.

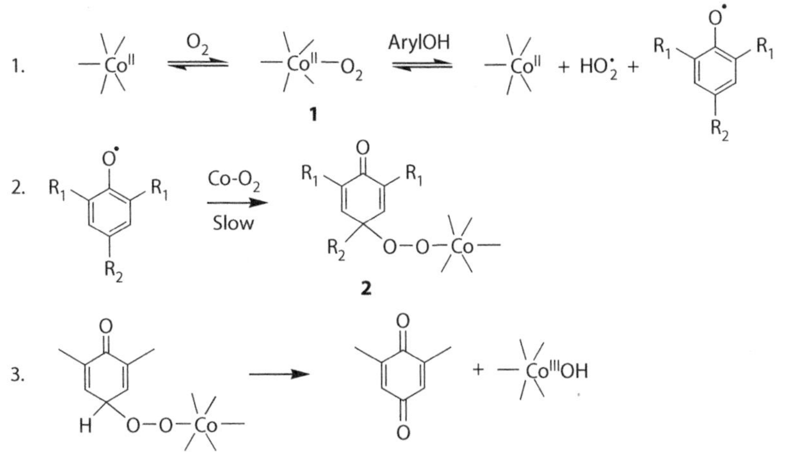

NMR and EPR Spectroscopy

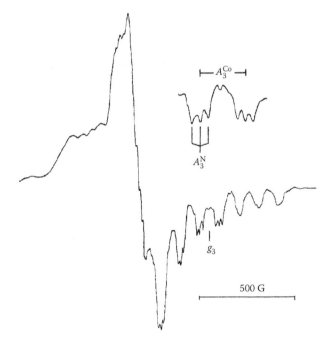

SCHEME 3.1 Proposed mechanism for the oxidation of substituted phenols with Co(Salpr).

(acacen = N,N'-ethylenebis(acetylacetoniminide)) (Figure 3.2, Table 3.1). The values of g_3 and A_3 can be easily derived from the experimental spectrum (see corresponding notations in Figure 3.2).

The values of g_1, g_2, A_1, A_2 can be determined with lower accuracy because of poor spectral resolution (Figure 3.2). Some of the hfs resonances corresponding to

FIGURE 3.2 EPR spectrum of Co(acacen)Py (**3**) in frozen toluene solution at −196°C. (Reproduced with permission from Hoffman, B. M. et al. 1970. *J. Am. Chem. Soc.* 92: 61–65.)

58 Applications of EPR and NMR Spectroscopy in Homogeneous Catalysis

TABLE 3.1

EPR Parameters for Some $Co^{III}(L)(O_2^{\cdot-})$, $Pd^{II}(L)(O_2^{\cdot-})$, and $Ni^{II}(L)(O_2^{\cdot-})$ Complexes in Frozen Solutions (Toluene for Co and Ni Species, and Toluene/Chloroform = 1:1 for Pd Species)

Compound	g_1	g_2	g_3	A_1, G	A_2, G	A_3, G	Reference
$Co^{II}(acacen)(Py)$ (3)	2.440	2.225	2.011	43.5	11.6	97.5[a]	[9]
$Co^{II}(benacen)(Py)$	2.440	2.23	2.014	48.7	13	98	[3]
$Co^{III}(acacen)(Py)(\eta^1\text{-}O_2^{\cdot-})$ (4)	2.080	1.998	1.998	19.3	10.4	10.4	[3]
$Co^{III}(benacen)(Py)(\eta^1\text{-}O_2^{\cdot-})$ (5)	2.077	1.998	1.998	18.3	10.4	10.4	[3]
$Co^{III}(Salpr)(\eta^1\text{-}O_2^{\cdot-})$ (1)	2.085	2.007	2.000	22.7	13.3	13.5	[3]
$Pd^{II}_3(OAc)_6(\eta^1\text{-}O_2^{\cdot-})$ (6)	2.107	2.013	2.001	6.7	3	4.5	[11]
$Pd^{II}(OAc)_2(S)(\eta^2\text{-}O_2^{\cdot-})$ (7)	2.077	2.027	2.006	–	–	–[b]	[11]
$[Ni^{II}(PhTt^{Ad})(\eta^2\text{-}O_2^{\cdot-})]$ (10)	2.24	2.19	2.01	–	–	–	[17]
$[Ni^{II}(14\text{-}TMC)(\eta^1\text{-}O_2^{\cdot-})]^+$ (11)	2.29	2.21	2.09	–	–	–	[18]
$[Ni^{II}(dkim)(\eta^2\text{-}O_2^{\cdot-})]$ (12)	2.138	2.116	2.067	–	–	–	[20]
$[Ni^{II}(13\text{-}TMC)(\eta^1\text{-}O_2^{\cdot-})]^+$ (13)	2.25	2.21	2.06	–	–	–	[22]
$[Ni^{III}(13\text{-}TMC)(\eta^2\text{-}O_2^{2-})]^+$ (14)	2.19	2.19	2.07	–	–	–	[22]

[a] Some of resonances in the g_3 region display shf splitting from one nitrogen nucleus $A_3^N = 15G$.
[b] Not observed.

A_3 exhibit superhyperfine (shf) splitting from one pyridine ^{14}N ($I = 1$) nucleus (Figure 3.2, inset). The absence of shf splitting from in-plane nitrogen atoms gives evidence that the unpaired electron is associated with the $3d_{z^2}$ orbital of Co [9].

The EPR spectrum of $Co^{III}(acacen)(py)(O_2^{\cdot-})$ (4) (Figure 3.3a) displays much smaller hfs from cobalt than the parent complex 3 (Figure 3.2, Table 3.1) [9]. It is logical to conclude that the unpaired electron is mainly located at the $O_2^{\cdot-}$ ligand of 4. In the early studies, the EPR spectrum of 4 was interpreted in assumption of axial anisotropy of the g- and A-tensors [9]. However, for complex 1, rhombically anisotropic g- and A-tensors were reported [3] (Table 3.1). Computer simulation shows that the axial EPR spectrum with parameters presented in Table 3.1 markedly differed from the experimental one (Figure 3.3a). A closer agreement between the simulated and experimental spectra of 4 can be obtained upon the assumption of rhombic anisotropy of g- and A-tensors (Figure 3.3b). The parameters of the EPR spectrum of Figure 3.3b are close to those of 1 (Table 3.1).

The x-ray data available for superoxo complex $(benacen)Co^{III}py)(O_2^{\cdot-})$ (5) (benacen = N,N'-ethylenebis(benzoylacetoniminide)) show that the superoxide anion $O_2^{\cdot-}$ is coordinated to cobalt in the end-on (η^1) fashion (Figure 3.1) [10]. By analogy, end-on coordination of $O_2^{\cdot-}$ is proposed for superoxo complexes 1 and 4 (Figure 3.1) [3].

NMR and EPR Spectroscopy

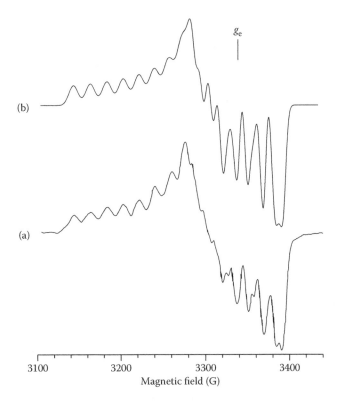

FIGURE 3.3 EPR spectrum of Co^{III}(acacen)(Py)(η^1-O$_2^{\cdot-}$) (**4**) in frozen toluene solution at −196°C (a). (Reproduced with permission from Hoffman, B. M. et al. 1970. *J. Am. Chem. Soc.* 92: 61–65.) Simulated spectrum $g_1 = 2.01305$, $g_2 = 1.99320$, $g_3 = 2.08100$, $A_1 = 4.25$, $A_2 = 9.70$, and $A_3 = 23.25$ G (b).

3.1.2 Superoxo Complexes of Pd(II)

The palladium superoxo complex formed in the reaction of hydrogen peroxide with palladium(II) acetate represents another example of LM(O$_2^{\cdot-}$) complexes displaying remarkable reactivity toward organic substrates. It was found that reacting Pd(OAc)$_2$ with 30% H$_2$O$_2$ in a toluene/chloroform = 1:1 mixture at room temperature affords a relatively stable superoxo complex of palladium with the proposed structure Pd$_3$(OAc)$_6$(η^1-O$_2^{\cdot-}$) (**6**) (Figure 3.4) [11,12]. In the frozen solution (−196°C), this complex displays rhombic EPR spectrum (Figure 3.5) with resolved hfs (marked with asterisks) from one [105]Pd nucleus ($I = 5/2$, natural abundance 22%) and with g-values ($g_1 = 2.107$, $g_2 = 2.013$, $g_3 = 2.001$) similar to those for cobalt superoxo complex **1** (Table 3.1). The EPR spectrum of **6** at room temperature displayed a single signal with shoulders stemming from unresolved hfs from [105]Pd [11,12]. The frozen-solution EPR spectrum, similar to that of **6**, can be obtained via the reaction of Pd$_3$(OAc)$_6$ with KO$_2$ ($g_1 = 2.083$, $g_2 = 2.012$, $g_3 = 2.001$), which supports the assignment of **6** to

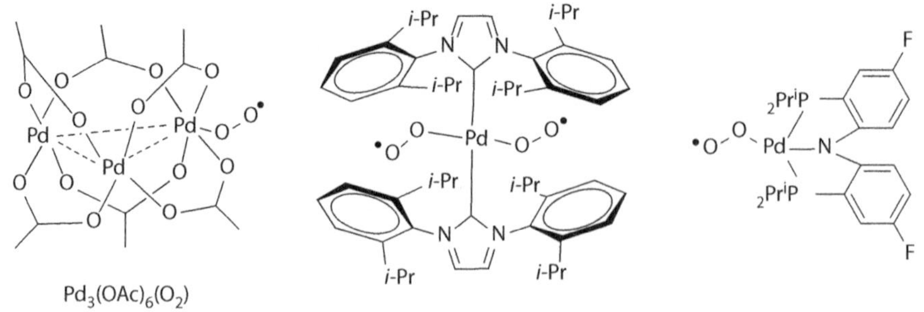

![Structures 6, 8, 9]

FIGURE 3.4 Proposed structure of palladium(II)-superoxo complex $Pd^{II}_3(OAc)_6(\eta^1-O_2^{\cdot-})$ (**6**) and molecular structures of complexes $[(IPr)_2Pd(\eta^1-O_2^{\cdot-})_2]$ (**8**) and $[(^FPNP)Pd(\eta^1-O_2^{\cdot-})]$ (**9**).

the superoxo complex of palladium [12]. The addition of ethylene ($[C_2H_4] = 0.15$ M) to a solution of **6** ($[6] = 0.005$ M) in $CDCl_3$ resulted in the drop of the concentration of **6** (evidenced by EPR) and formation of ethylene oxide (confirmed by ^{1}H NMR). The observed rate constants for the decomposition of the superoxo complex and ethylene oxide formation were estimated at 25°C, $k = (3.5 \pm 0.5) \times 10^{-3}$ s^{-1} M^{-1}. Ethylene oxide was the only observed product of this reaction [12].

In nonpolar solvents, palladium(II) acetate exists in the trimeric form $Pd_3(OAc)_6$, whereas in strongly coordinating solvents (S), it exists as a monomer $Pd(OAc)_2S_2$ [13]. It was found that the addition of CH_3CN or DMSO (dimethyl sulfoxide) to the solution of **6** in $CHCl_3$ leads to the conversion of **6** into a new superoxo complex **7** with more pronounced rhombicity of g-tensor (Table 3.1). In contrast to **6**,

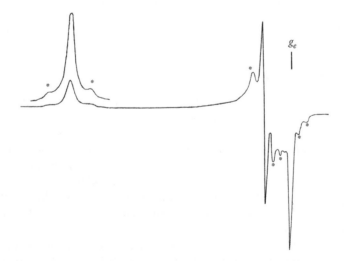

FIGURE 3.5 EPR spectrum ($CHCl_3$, −196°C) of Pd-superoxo complex **6**, observed in the course of reaction of palladium(II) acetate with H_2O_2 in $CHCl_3$. Asterisks mark the partially resolved hfs from ^{105}Pd nucleus ($I = 5/2$, natural abundance 22%). (Reproduced with permission from Talsi, E. P. et al. 1987. *Inorg. Chem.* 26: 3871–3878.)

NMR and EPR Spectroscopy

superoxo complex **7** was inert toward ethylene. It was assumed that **6** is trimer $Pd_3(OAc)_6(\eta^1\text{-}O_2^{\bullet-})$ with end-on coordination of the superoxide anion, while **7** is a monomer $Pd(OAc)_2S(\eta^2\text{-}O_2^{\bullet-})$ with side-on coordination of superoxide anion [12]. Valentine and Selke reported the strictly different reactivity of the ferric porphyrin peroxo complexes with η^1- and η^2-coordination of the peroxo moiety [14]. For a long time, palladium(II) superoxo complexes found in the systems $Pd(OAc)_2/H_2O_2$ and $Pd(OAc)_2/KO_2$ remained the only spectroscopically characterized examples of palladium superoxo complexes. In 2011, it was shown that Pd^0 complex with sterically hindered carbene ligands $[Pd(IPr)_2]$ (IPr = 1,3-bis(diisopropyl)phenylimidazol-2-ylidene) reacted with O_2 at low temperature ($-78°C$) to generate a bis-superoxo complex $[(IPr)_2Pd(\eta^1\text{-}O_2^{\bullet-})_2]$ (**8**) (Figure 3.4) [15]. According to x-ray crystallographic data, **8** was a bis-superoxo Pd(II) complex with η^1-superoxide anion binding. No data on the reactivity of **8** were provided [15]. At the same time, Ozerov with coworkers reported the x-ray structure of the superoxo complex $[(^FPNP)Pd(\eta^1\text{-}O_2^{\bullet-})]$ (**9**, ^FPNP = bis[2-iPr$_2$P(4-fluoro-phenyl)amido], Figure 3.4) [16]. This complex displayed a single line in the EPR spectrum with g-factor $g_0 = 2.06$ (toluene, 20°C), similar to that of **6** ($g_0 = 2.041$); no hfs from palladium was observed. The reactivity of **9** was relatively low. In the presence of 2% of **9**, 95% conversion of Ph_3P to Ph_3PO was observed after 3 days at ambient temperature in C_6D_6 [16].

The results presented above show that the superoxo complexes of cobalt and palladium can display noticeably high reactivity toward organic substrates. However, there are no industrial catalyst systems involving metal-superoxo complexes as the active oxidizing species. At the same time, there continues to be the interest in superoxo complexes of first-row transition metal ions (e.g., Cu, Fe, Ni, Cr, Co, Mn), which is mainly inspired by their participation as intermediates in the reaction cycles of various enzymes or their models.

3.1.3 Superoxo Complexes of Ni(II)

The first nickel superoxo complexes with η^2- and η^1-coordination of the superoxide anion were reported by Riordan et al.; see structures $[Ni^{II}(PhTt^{Ad})(\eta^2\text{-}O_2^{\bullet-})]$ (**10**) and $[Ni^{II}(14\text{-}TMC)(\eta^1\text{-}O_2^{\bullet-})]^+$ (**11**) in Figure 3.6 [17,18]. Complexes **10** and **11** were obtained by one-electron reduction of O_2 by the corresponding Ni^I precursors at low temperature and were characterized by UV–Vis (UV-visible spectroscopy), FT-IR (Fourier transform infrared spectroscopy), EPR, ^1H NMR, XAS (X-ray adsorption spectroscopy)/EXAFS (extended X-ray absorption fine structure), and DFT (density functional theory) calculations. The analysis of their EPR data indicates that in complexes **10** and **11**, a high-spin Ni^{II} (d^8, $S = 1$) paramagnetic center is antiferromagnetically coupled with an $O_2^{\bullet-}$ anion radical ($S = 1/2$), resulting in the d_z^2 doublet ground state (Table 3.1). Samples of **10** and **11** prepared from $^{17}O_2$ ($I = 5/2$) show no or minor EPR line broadening due to hyperfine interaction, indicating that very little unpaired spin density resides on the superoxide anion ligand [17,18]. Complexes **10** and **11** oxidize PPh_3 to $OPPh_3$. Samples of **11** prepared from $H_2^{18}O_2$ produce $^{18}OPPh_3$, confirming that hydrogen peroxide is the true oxygen source. The oxidation of PPh_3 to $OPPh_3$ by **11** proceeds in quantitative yield, the reaction rate exhibiting a first order in PPh_3 [18,19]. The reactivity of **10** and **11** toward other organic compounds was not reported.

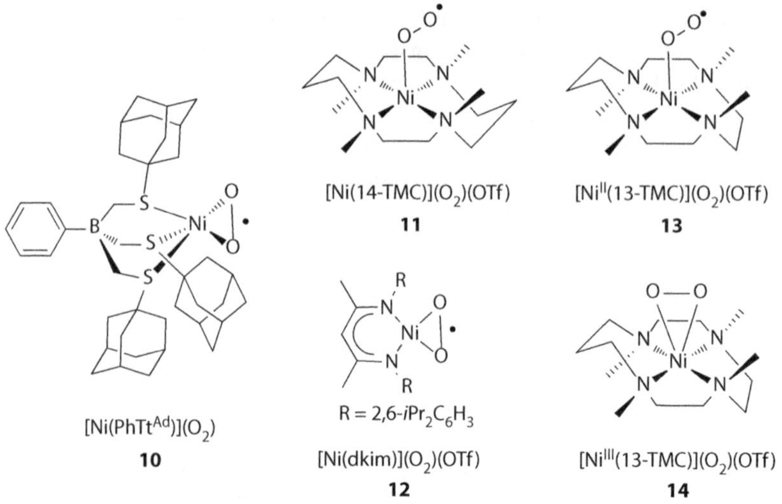

62 Applications of EPR and NMR Spectroscopy in Homogeneous Catalysis

FIGURE 3.6 Superoxo (**10–13**) and peroxo (**14**) complexes of nickel. Otf counteranion is not shown.

In 2008, Driess with coworkers reported the preparation, isolation, and reactivity study of the novel paramagnetic nickel(II) superoxo complex $[Ni^{II}(dkim)(\eta^2\text{-}O_2{}^{\bullet-})]$ (**12**) (Figure 3.6) [20]. It was fully characterized by several spectroscopic techniques (X-ray, FT-IR, EPR, ^{1}H NMR, EI-MS) [20]. In the absence of substrates, **12** was stable at room temperature; it oxidized PPh_3 to $OPPh_3$, but appeared inert toward olefins, sulfides, and aldehydes. Complex **12** was also unreactive toward alkanes bearing weak C—H bonds, such as 9,10-dihydroanthracene or triphenylmethane. However, reacting **12** with cyclohexanol led to the formation of cyclohexanone in 18% yield (with respect to the starting nickel complex). Similar reactions with 1-phenylethanol and 2,4-di-*tert*-butylphenol afforded acetophenone (in 22% yield) and 2,6-di-*tert*-butyl-1,4-benzoquinone (in 50% yield), respectively. The ability of **12** to activate N—H bonds was probed by the reaction with 1,2-diphenylhydrazine as substrate, and azobenzene formation was reported [21]. So, it seems clear that **12** is unable to perform oxygen atom transfer to alkenes, sulfides, or alkanes, but it can oxidize O–H and N—H groups.

It is assumed that in **12** the unpaired electron is centered on the superoxide moiety because the Ni^{II} site (d^8) is diamagnetic [20]. In this respect, **12** differs from **10** and **11**, where a high-spin Ni^{II} (d^8, $S = 1$) paramagnetic center is antiferromagnetically coupled to an $O_2{}^{\bullet-}$ anion radical ($S = 1/2$), and the unpaired electron is predominantly located in the d_{z^2} orbital of the Ni center [17–19].

EPR measurements of **10–12** in frozen toluene solutions reveal rhombic g-tensors, typical for paramagnetic $S = 1/2$ ground state (Table 3.1). The g-tensor anisotropy of **12** $(g_{max}-g_{min}) = 0.071$ is smaller than the corresponding values for **10** and **11** $(g_{max}-g_{min} = 0.20\text{–}0.23)$. This agrees with the proposed different electronic structures (different location of the unpaired electron) of **10**, **11**, and **12**.

NMR and EPR Spectroscopy

To understand the roles of Ni-superoxo versus Ni-peroxo species in oxidation chemistry, it is important to compare the structures and reactivities of known Ni-superoxo versus Ni-peroxo species. Recently, the synthesis of end-on Ni^{II} superoxo complex $[Ni^{II}(13\text{-}TMC)(\eta^1\text{-}O_2^{\cdot-})]^+$ (13) and side-on Ni^{III} peroxo complex $[Ni^{III}(13\text{-}TMC)(\eta^2\text{-}O_2^{2-})]^+$ (14), bearing common macrocyclic ligand, was reported [22] (Figure 3.6, Table 3.1). These intermediates were fully characterized by various spectroscopic techniques, as well as by x-ray crystallography and DFT calculations. The reactivities of 13 and 14 in electrophilic and nucleophilic reactions were compared. The end-on superoxo complex 13 was capable of conducting oxidative electrophilic reaction (oxidation of PEt_3 to $OPEt_3$ at $-40°C$ in $CH_3CN/CH_3OH = 1:1$, $k = 3.6 \times 10^{-2}$ M^{-1} s^{-1}). Contrariwise, the peroxo complex 14 was inert toward PEt_3 even at 25°C, but it conducted nucleophilic oxidation reactions. For example, it reacted with 2-phenylpropionaldehyde in CH_3CN at 25°C ($k = 4.2 \times 10^{-3}$ M^{-1} s^{-1}) to afford acetophenone [22]. The EPR parameters of 13 are close to those of the related complex 11 (Table 3.1).

3.1.4 Superoxo Complexes of Copper(II)

In the past decade, superoxo complexes of copper have attracted the most significant interest among related complexes of other metals because cupric superoxo species are assumed to be responsible for initiating the oxidation of various organic substrates by copper enzymes [23–36]. For example, they can hydroxylate substrates with weak C–H bonds (ca. 88 kcal mol^{-1}) in copper enzymes such as peptidylglycine-α-hydroxylating monooxygenase (PHM) and dopamine-β-monooxygenase (DβM) [25,26,30] or initiate N–H oxidation in copper-containing amine oxidases (CAO) [33].

An important step in understanding the mode of action of DβM and PHM was made in 2004 with the publication of the x-ray structure of the dioxygen-bound form of PHM, soaked with the substrate (peptide containing N-acetyl-diiodo tyrosyl-D-threonin) [37]. O_2 is bonded in an end-on (η^1) fashion to the Cu_B atom of PHM (Cu_B–O–O angle is 110°C and O–O distance is 1.23 Å). This geometry corresponds to the coordination of superoxide anion to Cu_B. Other coordination sites of the tetrahedrally coordinated Cu_B are occupied by N atoms of His^{242} and His^{244}, and sulfur atom of Met^{314}. The revealed geometry of the PHM-substrate complex is suitable for abstracting the substrate hydrogen atom by the $O_2^{\cdot-}$ ligand of PHM.

There have been two more structurally characterized cupric superoxo complexes. One of them is synthetic complex $[Cu^{II}(HB(3\text{-}tBu\text{-}5\text{-}iPrpz)_3)(\eta^2\text{-}O_2^{\cdot-})]$ (15) with a sterically hindered 3,5-disubstituted tris(pyrazolyl)borate ligand and a side-on $O_2^{\cdot-}$ moiety (Figure 3.7) [38]. This species has a singlet ground state ($S = 0$). The third example is complex $[(TMG_3tren)Cu^{II}(\eta^1\text{-}O_2^{\cdot-})]^+$ (16) with η^1-bound $O_2^{\cdot-}$ fragment {TMG_3tren = (1,1,1-tris[2-[N^2-(1,1,3,3-tetramethylquanidino)]ethyl]amine} (Figure 3.7) [39]. The electronic structure of 16 was elucidated by Solomon and coworkers [34,40]. It was shown to possess an $S = 1$ ground state, with ferromagnetically coupled spins at the Cu(II) ion and at the superoxo fragment. Further studies of the other spectroscopically characterized end-on cupric superoxo complexes $[Cu^{II}(L_1)(\eta^1\text{-}O_2^{\cdot-})]$ (17) [41], $[Cu^{II}(L_2)(\eta^1\text{-}O_2^{\cdot-})]$ (18) [30], $[Cu^{II}(L_3)(\eta^1\text{-}O_2^{\cdot-})]$ (19) [31],

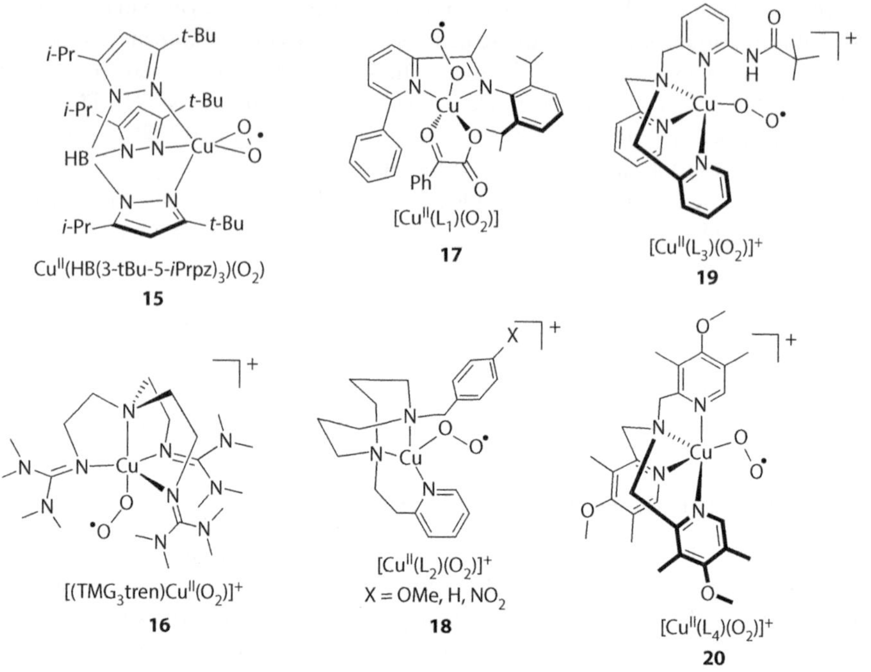

64   Applications of EPR and NMR Spectroscopy in Homogeneous Catalysis

FIGURE 3.7   Superoxo complexes of copper(II).

and $[Cu^{II}(L_4)(\eta^1\text{-}O_2^{\cdot-})]$ (**20**) [35] (Figure 3.7) revealed $S = 1$ ground state for all those complexes. Complexes **17–20** were EPR-silent in perpendicular microwave excitation mode. However, in agreement with the triplet ground state, they displayed paramagnetically shifted NMR resonances [36]. Furthermore, complex **18** exhibited an EPR signal at $g \approx 4$ in the parallel microwave excitation mode. This signal corresponds to "$\Delta m_S = 2$" transition, forbidden in the conventional perpendicular mode, thus corroborating the triplet ground state [30].

Very recently, a detailed mechanistic insight into the oxidation of a series of *para*-substituted 2,6-di-*tert*-butylphenols (*p*-X-DTBPs) by **20** was reported. Kinetic investigations revealed that **20** abstracted the phenolic hydrogen from *p*-X-DTBPs. This is the first and the rate-determining step for the oxidation of *p*-X-DTBPs. The mechanisms of *p*-X-DTBPs oxidations with copper and cobalt superoxo complexes (Scheme 3.1) are similar [35].

In biomimetic studies, modeling natural metalloenzymes, synthetic Cu(II)-superoxo complexes have appeared reactive in the oxidation of ligand C–H groups and O–H groups of substrates [27–31,35]. However, these data cannot be directly transferred to DβM and PHM enzymes; for those, very limited evidence (obtained by theoretical and indirect kinetic studies) in favor of the involvement of the cupric superoxo species in the hydrogen atom abstraction process is available. Mononuclear species such as cupric hydroperoxide $LCu^{II}OOH$ or copper oxyl $LCu^{II}\text{-}O^{\cdot}$ have been considered as alternative reactive intermediates, capable of participating in copper enzyme-mediated biotransformations [42].

NMR and EPR Spectroscopy

3.1.5 Superoxo Complexes of Iron(III)

The formation of an iron(III) superoxo species upon O_2 binding to an iron(II) center is the first step proposed for the activation of O_2 by cytochrome P450 enzymes, the true oxidizing species being high-valent metal-oxo intermediates [43–45]. However, for some non-heme iron enzymes, ferric-superoxo complexes can by itself initiate catalytic transformations via hydrogen atom abstraction from substrates with relatively week C–H bonds. The known example is *myo*-inositol oxygenase (MIOX), carrying out the oxidation of *myo*-inositol [cyclohexane-(1,2,3,4,5,6-hexa)ol] (MI) to D-glucuronate (DG). MIOX activates dioxygen at a mixed-valent diiron(II/III) cluster, to afford an $S = 1/2$ diiron(III/III)-superoxo species **21** [45,46] (Scheme 3.2). This intermediate was trapped and characterized by EPR ($g_1 = 2.05$, $g_2 = 1.98$, $g_3 = 1.90$). The rhombic g-tensor and observed hfs from one ^{57}Fe nucleus are rationalized in terms of the proposed diiron(III/III)-superoxo structure with coordination of the superoxide to a single iron. The observed primary kinetic isotope effect $k_H/k_D \geq 5$ shows that the reaction indeed proceeds via the cleavage of the C–H(D) bond of *myo*-inositol by **21** [46].

Another mechanism of interaction of iron superoxo complex with the substrate was documented for homoprotocatechuate 2,3-dioxygenase (2,3-HPCD), an enzyme that cleaves the aromatic ring of the substrate (3,4-dihydroxyphenylacetate) [47,48]. It was proposed that simultaneous binding of the substrate and O_2 to the iron allows electron transfer from the substrate to oxygen, with the formation of ferrous superoxo complex and substrate cation radical. Recombination of those two radicals would trigger ring cleavage and oxygen insertion, to yield a muconic semialdehyde adduct as a product (Scheme 3.3). The EPR spectroscopic data show that ferric-superoxo

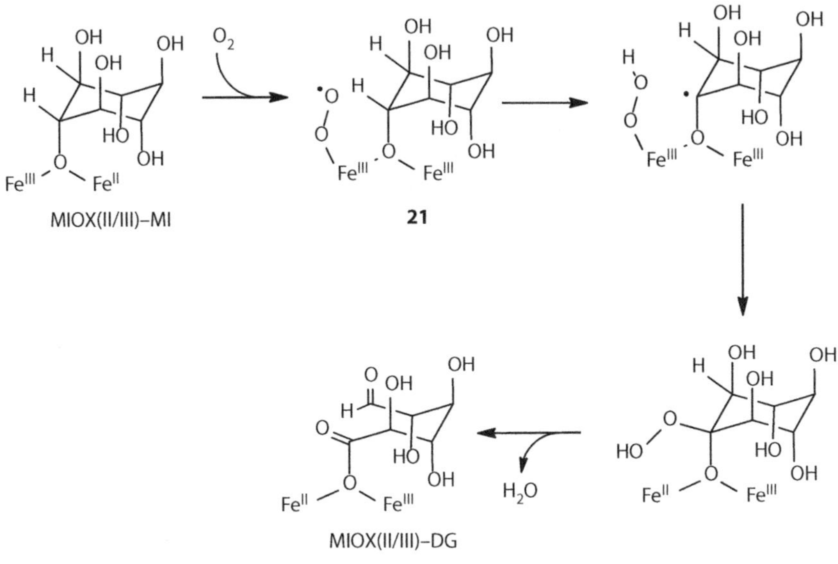

SCHEME 3.2 Proposed mechanism for the conversion of MI to DG initiated by (superoxo) diiron(III/III) complex.

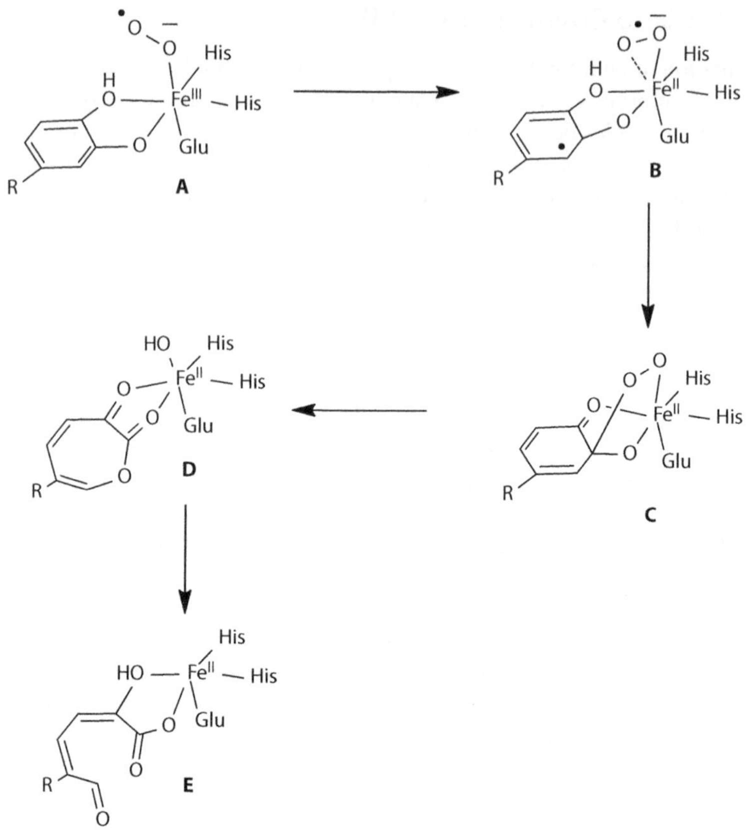

SCHEME 3.3 The proposed mechanism of action of homoprotocatechuate 2,3-dioxygenase.

complex **A** is formed at the initial stage of the catalytic transformations presented in Scheme 3.3. **A** has $S = 2$ ground state, resulting from antiferromagnetic coupling of high-spin Fe(III) ($S = 5/2$) with the superoxide anion ($S = 1/2$), and exhibits an EPR signal near $g \approx 8.2$ in parallel mode [48]. The reaction scheme depicted in Scheme 3.3 is strongly supported by x-ray characterization of intermediates formed during the reaction of the 4-nitrocatechol (4NC) with O_2, catalyzed by 2,3-HPCD. The following intermediates were trapped and characterized (see Scheme 3.3): ternary complex of enzyme with 4CN and side-on bond dioxygen species (**B**), alkylperoxo intermediate (**C**), and complex with cleaved-ring product (**E**) [47].

In spite of extensive studies of non-heme iron enzymes, there have been no unambiguous data on the structures and reactivities of ferric-superoxo complexes participating in their catalytic cycles. For synthetic non-heme iron complexes, the information on the structure and reactivity of iron-superoxo complexes is also very restricted. Hitherto, there have been two reports on synthetic non-heme dinuclear iron-superoxo complexes with the proposed diiron(μ-hydroxo) structures $Fe^{III}Fe^{III}$–$O_2^{\cdot-}$ (**22**, Figure 3.8) [49] and $Fe^{II}Fe^{III}$–$O_2^{\cdot-}$ [50], and one report on mononuclear superoxo complex with the proposed structure Fe^{III}–$O_2^{\cdot-}$ (**23**, Figure 3.8) [51].

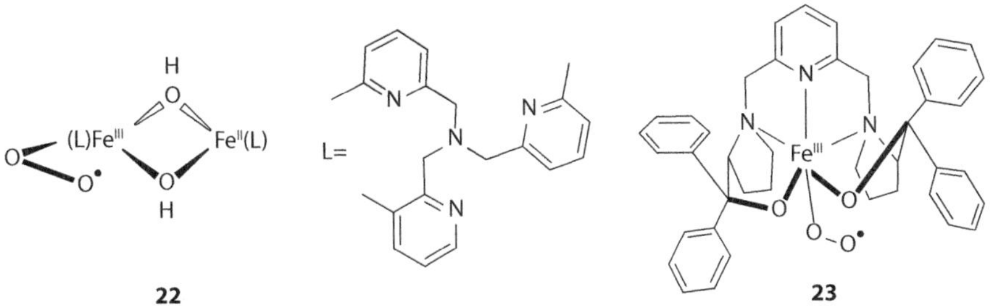

NMR and EPR Spectroscopy

FIGURE 3.8 Proposed structures of synthetic superoxo complexes of iron.

Resonance Raman and Mössbauer spectroscopy were employed to unambiguously assign **23** as a species containing a mononuclear iron(III)-superoxide core. Complex **23** displays a resonance-enhanced vibration at 1125 cm^{-1} corresponding to the O—O bond of the superoxo moiety, and the Mössbauer spectrum corresponds to a high-spin Fe^{III} center, exchange-coupled to the superoxo ligand. Complex **23** oxidizes dihydroanthracene to anthracene, supporting the assumption that Fe^{III}–O$_2$$^{\cdot-}$ species can carry out H-atom abstraction [51]. The main drawback hampering more detailed studies of iron-superoxo complexes in non-heme iron enzymes and model systems is their low stability.

The presented data show that the major current interest in the studies of metal-superoxo complexes is focused on the enzymic systems and their models (superoxo complexes of copper and iron). It is firmly established that for some non-heme iron and copper enzymes, superoxo complexes can initiate substrate transformation by direct abstraction of hydrogen atom from substrates with relatively week C—H bonds [27,28,32,45,46]; moreover, for some non-heme iron enzymes, the substrate transformation can be initiated by recombination of substrate cation radical and superoxo complex [47,48]. As for synthetic catalyst systems, the only example of preparative catalytic reaction incorporating superoxo complex as an active interme-diate remains the oxidation of 2,6-disubstituted phenols to quinones catalyzed by Co(II) complexes [3–8].

3.2 ALKYLPEROXO COMPLEXES

As was noted above, metal-superoxo complexes participate in some oxidative transfor-mations of organic substrates, catalyzed by copper and iron enzymes. However, there are no examples of their key role in industrial oxidations. Contrariwise, it is generally accepted that molybdenum- and titanium-alkylperoxo complexes are the key inter-mediates of such important homogeneous catalytic processes as the Halcon epoxida-tion of propylene [52–57] and the Katsuki–Sharpless epoxidation of allylic alcohols [58–61]. The titanium-hydroperoxo complexes are proposed to be the active species of heterogeneous epoxidation of propylene on titanium silicalites (TS-1) [62,63]. The existing examples of practical applications as well as the growing demand in indus-trial epoxides stimulate the extensive search for novel catalyst systems based on vari-ous complexes of molybdenum and titanium, and inspire their mechanistic studies.

3.2.1 Alkylperoxo Complexes of Molybdenum

Molybdenum(VI) complexes have been among the best catalysts for the epoxidation of olefinic substrates with alkylhydroperoxides as terminal oxidants. The most frequently used and tested monomeric complexes are $Mo(CO)_6$ and $MoO_2(acac)_2$. It was observed that the rate constants for olefin epoxidation observed with these two catalysts were the same [64]. The absence of dependence of the reaction rate on the structure of the molybdenum precatalyst was ascribed to the formation of Mo(VI)-1,2-diol complexes in the course of molybdenum-catalyzed epoxidations [64,65]. It was found that Mo(VI)-1,2-diol complexes can exist in the reaction solution in the form of mononuclear and dinuclear species. In the case of cyclohexene as substrate, these species can be schematically presented by the structures **24** and **25**, respectively (Figure 3.9) [64]. In agreements with these results, the ^{95}Mo NMR spectra of the catalyst systems $MoO_2(acac)_2/tBuOOH/cyclohexene$ and $Mo(CO)_6/tBuOOH/cyclohexene$ in benzene display resonances from two complexes (Figure 3.10a). The relatively sharp resonance at δ 277 ($\Delta v_{1/2} = 100$ Hz) can be assigned to the mononuclear complex **24**, and the broader resonance at δ 56 ($\Delta v_{1/2} = 250$ Hz) most likely belongs to the dinuclear complex **25** [54]. To support this assignment, the ^{95}Mo NMR spectrum of the system $MoO_2(acac)_2/1,2$-*trans*-cyclohexane diol in benzene was recorded (Figure 3.11). This spectrum displays intense peak from complex **25** and weak resonance from complex **24** (Figure 3.11) [55]. The ^{17}O NMR spectrum of the same sample exhibits ^{17}O resonance at δ 1034 from the Mo=O groups of $MoO_2(acac)_2$, and four resonances with the ratios 2:2:1:1 at δ 956, 917, 892, and 878 from complex **25** (Figure 3.12a). The Mo=O and Mo–O–Mo groups of $MoO_2(acac)_2$ and **25** were enriched with ^{17}O isotope by the addition of $H_2^{17}O$ (3% enrichment in ^{17}O) to the sample. The dinuclear

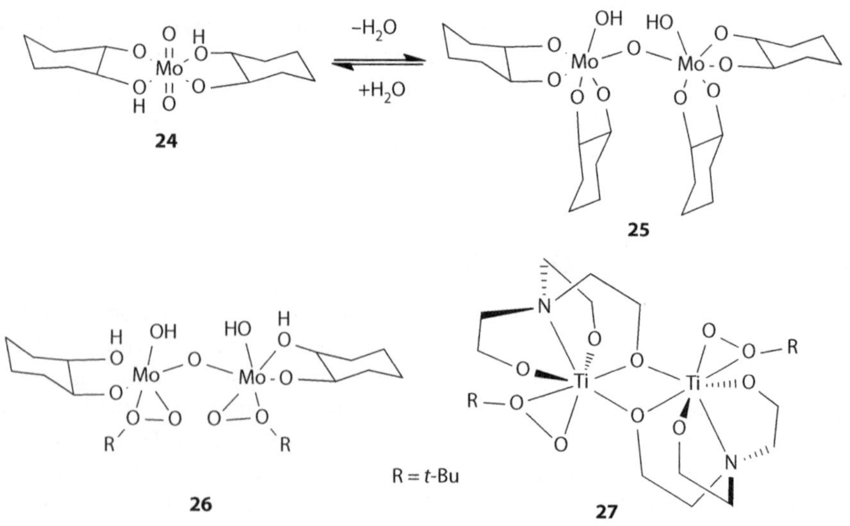

FIGURE 3.9 Proposed structures of Mo(VI) complexes with *trans*-cyclohexane-1,2-diol and *t*BuOOH (**24–26**), and the structure of x-ray-characterized 2-(*tert*-butylperoxo)titanatrane dimer of Ti(IV) (**27**).

NMR and EPR Spectroscopy

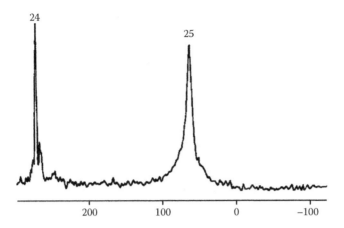

FIGURE 3.10 ^{95}Mo NMR spectrum (benzene, 20°C) of the sample MoO$_2$(acac)$_2$/tBuOOH/ C$_6$H$_{10}$ = 1:30:30 in benzene recorded 15 min after heating at 80°C ([MO$_2$(acac)$_2$] = 0.05 M. (Reproduced with permission from Talsi, E. P. et al. 1993. *J. Mol. Catal.* 83: 329–346.)

complex **25** presented in Figure 3.9 should display two ^{17}O NMR peaks 2:1 from the O=Mo—O—Mo=O moiety. Most likely, the observed 2:2:1:1 pattern corresponds to two structural isomers of **25** (the ratio of isomers is 1:1). The detailed structures of isomers are unknown. However, it is well established that in the catalyst system MoO$_2$(acac)$_2$/tBuOOH/cyclohexene, the initial molybdenum(VI) complex converts in the course of epoxidation into molybdenum(VI) complexes with 1,2-alkanediols, the by-products of epoxidation.

The ^{13}C, ^{95}Mo, and ^{17}O NMR spectroscopic studies of the catalyst systems MoO$_2$(acac)$_2$/tBuOOH/cyclohexene and MoO$_2$(acac)$_2$/tBuOOH/*trans*-1,2-cyclohexanediol in benzene have shown that *t*BuOOH replaces the 1,2-diolate ligands in **25** to form the alkylperoxo complex **26** (Figure 3.9). After the addition of

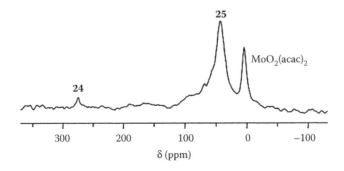

FIGURE 3.11 ^{95}Mo NMR spectrum (CHCl$_3$, 25°C) of the MoO$_2$(acac)$_2$/1,2-*trans*-cyclohexane-diol = 1:2; [MoO$_2$(acac)$_2$] = 0.1 M. (Reproduced with permission from Dorosheva, T. S. et al. 2003. *Mendeleev Commun.* 13: 8–9.)

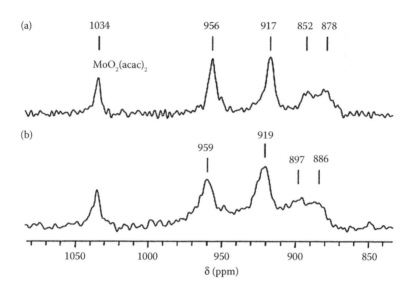

FIGURE 3.12 ^{17}O NMR spectra (CHCl$_3$, 20°C, [MoO$_2$(acac)$_2$] = 0.1 M) of MoO$_2$(acac)$_2$/1.2-*trans*-cyclohexanediol = 1:2 (a) and sample (a) 30 min after the addition of *t*BuOOH ([*t*BuOOH] = 2 M. This sample contained 0.003 mL of H$_2$^{17}O (3% enrichment in ^{17}O) in 1.5 mL of solution. (Reproduced with permission from Dorosheva, T. S. et al. 2003. *Mendeleev Commun*. 13: 8–9.)

20 equivalents of *t*BuOOH to the solution containing MoO$_2$(acac)$_2$ and **25** in CHCl$_3$ at room temperature, the ^{13}C NMR peaks of the coordinated OO*t*Bu moiety of the alkylperoxo complex **26** appeared. The resonance at δ 26.0 and two resonances at 83.3 and 82.9 were assigned to the primary and quaternary carbon atoms of the coordinated alkylperoxo moiety of **26**, respectively. Free *t*BuOOH exhibited the corresponding ^{13}C peaks at δ 25.4 and 80.2 [55]. The chemical shifts of peaks at δ 83.3 and 82.9 are close to the corresponding peak for η^2-(*tert*-butylperoxo)titanatrane dimer **27** (δ 83.93 in CD$_2$Cl$_2$) (Figure 3.9) [66]. The peaks of **26** disappear when *t*BuOOH decomposes to *t*BuOH upon storing the sample at room temperature. The two peaks at δ 83.3 and 82.9 for **26**, probably, reflect the existence of two isomers of the starting complex **25**. The concentration of **26** reached 70% of the starting molybdenum complex. A comparison of the intensities of the corresponding ^{13}C NMR peaks shows that **26** incorporates one –OO*t*Bu and one 1,2-cyclohexanediolate ligand per one molybdenum. The alkylperoxo ligand in **26** is most likely η^2-coordinated, as in η^2-(*tert*-butylperoxo)titanatrane dimer **27**.

The ^{17}O NMR spectra of the system MoO$_2$(acac)$_2$/1,2-cyclohexane diol/*t*BuOOH bear evidence that the O=Mo—O—Mo=O fragment of **25** remains intact upon the formation of **26**. The addition of *t*BuOOH to the sample of Figure 3.12a gave rise to a small (about 2 ppm) increase of the chemical shifts of M=O oxygens, and a more sizeable (by 5–8 ppm) increase of the chemical shifts of the Mo—O—Mo oxygens, whereas the ratio between the peak intensities of M=O and Mo—O—Mo moieties remained unchanged (Figure 3.12b). It is natural to conclude that the formation of **26** occurs via the replacement of 1,2-cyclohexanediolato ligands in **25** by *t*BuOO$^-$ [55].

NMR and EPR Spectroscopy

Hence, the key epoxidizing agent of the catalyst system MoO$_2$(acac)$_2$/olefin/tBuOOH is likely a dinuclear μ-oxo-bridged alkylperoxo complex of molybdenum(VI) (**26**) (Figure 3.9).

The independence of the epoxidation rate on the structure of the starting molybdenum complex was only demonstrated for molybdenum complexes with labile or oxidation-unstable ligands (e.g., MoO$_2$(acac)$_2$, Mo(CO)$_6$), which could be replaced with the 1,2-alkanediolate ligands, formed in the course of olefin epoxidation. Complexes of the type [MoO$_2$Cl$_2$L], where L is polypyridyl (**28–38** [56,67]), 1,4-diazabutadiene (**39, 40** [68]), or salen-type ligand (**41–42** [69]) (Figure 3.13), do not lose the starting ligand in the course of epoxidation and thus exhibit strong dependence of the catalytic activity on the nature of the ligand L. Furthermore, the catalytic activities of the systems **28–42**/tBuOOH/alkene are much lower than that of the system MoO$_2$(acac)$_2$/tBuOOH/alkene, so one can expect that more stable catalysts **28–42** are more suitable for mechanistic studies.

Kühn and coworkers suggested that the active species of epoxidation of the catalyst systems **28–34**/tBuOOH/alkene is a η^1-alkylperoxo complex of molybdenum(VI) (**43**, Figure 3.14). This assumption was based on the detection of a new IR band in the reaction of [MoO$_2$Cl$_2$L] with tBuOOH (L = 4,4′-bis(n-hexyl)-2,2′-bipyridine) [56].

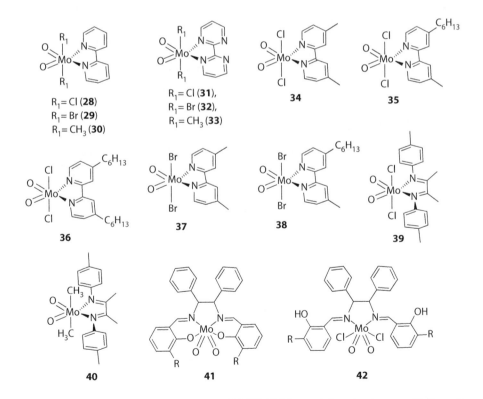

FIGURE 3.13 Molybdenum complexes [Mo^{VI}O$_2$X$_2$L] tested in selective epoxidation of alkenes with tBuOOH.

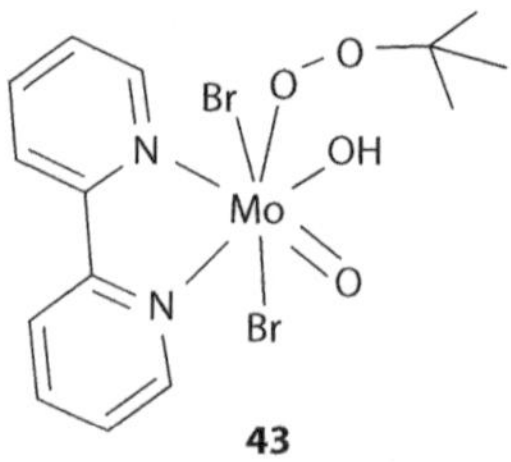

72   Applications of EPR and NMR Spectroscopy in Homogeneous Catalysis

43

FIGURE 3.14 Proposed structure of alkylperoxo complex of molybdenum(VI) formed in the catalyst system $MoO_2Br_2(2,2'$-bipy)/tBuOOH.

This band at 885 cm^{-1} was similar to the ν (O–O) stretching band at 890 cm^{-1} for the X-ray crystallographically characterized vanadium(V) alkylperoxo complex, (dipic) $VO(\eta^2$-OOtBu) (dipic = 2,6-pyridinedicarboxylate) [70]. The authors proposed the η^1-coordination mode for OOtBu group in **43**; however, the η^2-coordination may be more probable since known X-ray-characterized alkylperoxo complexes of vanadium [70] and titanium [66] display η^2-coordination of the alkylperoxo moiety.

Until now, there have been no reported examples of isolated and X-ray-characterized alkylperoxo complexes of molybdenum or reactivity data for alkylperoxo complexes of molybdenum toward stoichiometric epoxidation of alkenes. Nevertheless, it is widely accepted that the active species of the catalyst systems based on molybdenum complexes and alkylhydroperoxides are the corresponding alkylperoxo complexes of molybdenum. This assumption is based on the spectroscopic detection of alkylperoxo complexes of molybdenum [55–57] and analogy with the alkylperoxo complexes of titanium and vanadium (see below).

3.2.2 ALKYLPEROXO COMPLEXES OF TITANIUM

Alkylperoxo complexes of titanium were proposed to be the active species of the practically important Katsuki–Sharpless asymmetric epoxidation system [71–73] and Kagan–Modena asymmetric sulfoxidation system [74,75]. The Katsuki–Sharpless epoxidation system consists of Ti(OiPr)$_4$, diisopropyl tartrate (DIPT), tBuOOH, and allylic alcohol (R$_1$OH) as a substrate. Owing to the fluxional exchange of the tartrate, substrate, and oxidant molecules, the ^{1}H and ^{13}C NMR spectra of the catalyst system Ti(OiPr)$_4$/DIPT/tBuOOH/R$_1$OH are broadened and therefore poorly informative. Therefore, the actual structure of the active species operating in this catalyst system remains unclear. On the basis of indirect experimental data, dinuclear η^2-alkylperoxo complex **44** (Figure 3.15) is postulated as the active species of asymmetric epoxidation [73].

So far, the only X-ray-characterized Ti(IV) alkylperoxo complex has been the achiral η^2-$tert$-butylperoxotitanatrane dimer **27** [66]. Although this complex was shown to oxidize amines and sulfides to the corresponding oxides, it was not active in olefin epoxidation. Bonchio et al. have generated the asymmetric sulfoxidation catalysts from titanium(IV) isopropoxide and enantiopure trialkanolamine ligands. Electrospray ionization mass spectrometry (ESI-MS), in combination with

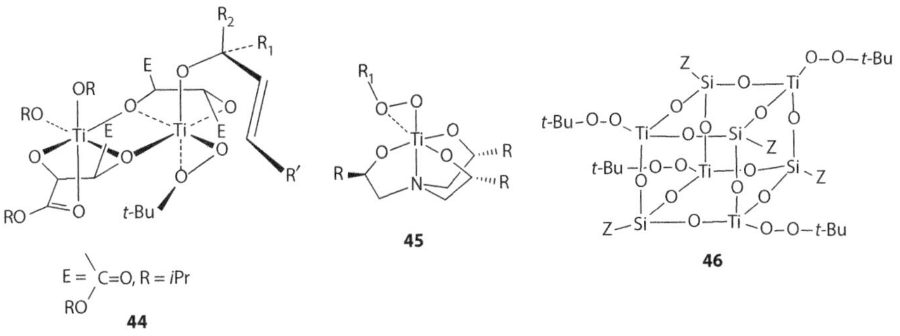

NMR and EPR Spectroscopy

73

FIGURE 3.15 Proposed structures of alkylperoxo complexes of Ti(IV).

low-temperature ^{1}H NMR technique, has shown that in the presence of excess tBuOOH, the mononuclear alkylperoxo complex **45** predominates in the reaction solution (Figure 3.15). This complex serves as the active species of enantioselective sulfoxidation [76].

Fujiwara et al. have shown that cubic silicon–titanium μ-oxo complex reacted with 4 equivalents of tBuOOH in hexane at −20°C to form a novel complex **46** as a white solid. This complex reacted with cyclohexene to afford cyclohexene oxide. The chemical shift of the peak of methyl carbons of the coordinated OOtBu group in complex **46** ($\delta = 26.5$) was very close to that of the X-ray-characterized complex **27** ($\delta = 26.6$). On the basis of this similarity and reactivity data, **46** was assigned to the alkylperoxo complex (Figure 3.15) [77].

Using ^{1}H and ^{13}C NMR spectroscopy, Babushkin and Talsi have monitored the reaction of Ti(OiPr)$_4$ with tBuOOH in CH$_2$Cl$_2$ (CDCl$_3$) at −70°C to −30°C [78,79]. It has been unambiguously shown that alkoxo ligands of Ti(OiPr)$_4$ are successively displaced by alkylperoxo ones with the increase in the tBuOOH:Ti(OiPr)$_4$ ratio from 1:1 to 10:1 to form an equilibrium mixture of alkylperoxo complexes Ti(η^2-OOtBu)$_n$(OiPr)$_{4-n}$, where $n = 1$–4. The equilibrium (3.1) between all these complexes and Ti(OiPr)$_4$ is rapidly established even at −70°C:

$$\mathrm{Ti(O\textit{i}\,Pr)_4} + n(t\mathrm{BuOOH}) \leftrightarrow \mathrm{Ti(OO\textit{t}Bu)}_n(\mathrm{O\textit{i}\,Pr})_{4-n} + n(i\,\mathrm{Pr\,OH}), \quad n = 1-4 \quad (3.1)$$

The molecular composition of the alkylperoxo complexes formed was determined by comparison of the relative intensities (NOE corrected) of the ^{13}C NMR signals for the quaternary carbons of the OOtBu ligands and the ternary carbon atoms of the OiPr ligands, which were 1:3, 2:2, and 3:1 for Ti(OOtBu)(OiPr)$_3$ (**47**), Ti(OOtBu)$_2$(OiPr)$_2$ (**48**), and Ti(OOtBu)$_3$(OiPr) (**49**), respectively. Only the signals of the OOtBu ligands correspond to complex Ti(OOtBu)$_4$ (**50**) in the ^{1}H and ^{13}C NMR spectra (Figure 3.16). As an example, Figure 3.17 shows the ^{13}C NMR spectrum of the system Ti(OiPr)$_4$ + 2(tBuOOH) in CH$_2$Cl$_2$ at −30°C. The signals of the quaternary carbons of the OOtBu ligands (δ 87.4...86.7) and the tertiary carbons of the OiPr ligands (δ 79.0...77.3) of complexes **47–49** are clearly seen.

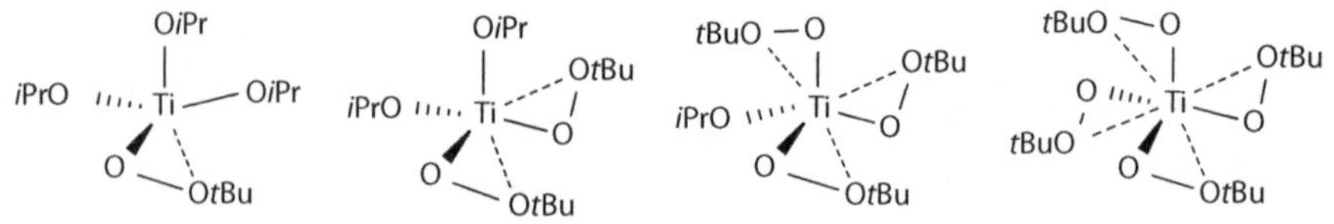

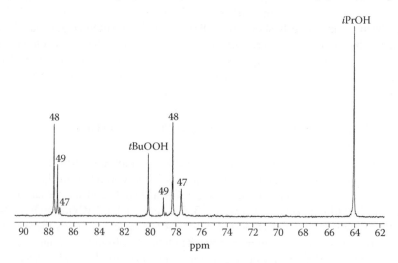

FIGURE 3.16 Proposed structures of titanium(IV) alkylperoxo complexes **47–50**.

The difference in the chemical shifts for the quaternary carbon atom of the *t*Bu group in Ti(OO*t*Bu)$_n$(O*i*Pr)$_{4-n}$ (*n* = 1–4) and in *t*BuOOH is 6.9–7.5 ppm. The corresponding difference for the primary carbon atoms of the *t*Bu group is 0.25–0.75 ppm (Table 3.2). To define the mode of the OO*t*Bu ligand coordination to titanium atom in complexes **47–50**, the above differences were compared with those for the separately prepared alkoxo complexes Ti(O*t*Bu)$_n$(O*i*Pr)$_{4-n}$, where *n* = 1–4 [78]. The corresponding differences for the α-carbon of the O*t*Bu groups in Ti(O*t*Bu)$_n$(O*i*Pr)$_{4-n}$, and *t*BuOH were 11.2–12.2 ppm (Table 3.3). Thus, in the case of η^2-coordination of the OO*t*Bu ligand, one would expect a difference in the chemical shifts for the quaternary carbon atom of the *t*Bu group in **47–50** and in *t*BuOOH up to 11–12 ppm. The experimentally observed difference of 6.9–7.5 ppm (Table 3.3) supports the η^2-coordination of OO*t*Bu moieties in the alkylperoxo complexes **47–50** (Figure 3.16). It is noteworthy that only one set of ^{13}C and ^{1}H NMR signals was observed for ligands of any particular complex Ti(OO*t*Bu)$_n$(O*i*Pr)$_{4-n}$ with *n* = 1–4. This fact provides evidence in favor of fast (even at −70°C) intramolecular rearrangements in the coordination sphere, leading to apparent magnetic equivalence of the ligands and to fast interconversion of all possible structural isomers of each alkylperoxo complex.

FIGURE 3.17 ^{13}C NMR spectrum (CH$_2$Cl$_2$ at −30°C) of the system Ti(O*i*Pr)$_4$ + 2 *t*BuOOH ([Ti(O*i*Pr)$_4$] = 0.3 M, [*t*BuOOH] = 0.6 M). The signals of complexes **47–49** are marked. *i*PrOH is the product of the ligand substitution. (Reproduced with permission from Babushkin, D. E., Talsi, E. P. 2003. *J. Mol. Catal. A Chem.* 200: 165–175.)

NMR and EPR Spectroscopy

TABLE 3.2
^{13}C and ^{1}H NMR Chemical Shifts[a] for Titanium(IV) Alkylperoxo Complexes Ti(OOtBu)$_n$(OiPr)$_{4-n}$

Compound	$\underline{C}$(CH$_3$)$_3$	$\underline{C}$H(CH$_3$)$_2$	C(C$\underline{H}_3$)$_3$	C$\underline{H}$(CH$_3$)$_2$	CH(C$\underline{H}_3$)$_2$	C(C$\underline{H}_3$)$_3$	CH(C$\underline{H}_3$)$_2$
Ti(OiPr)$_4$	–	76.59	–	4.491	1.270	–	26.73
47	86.93	77.51	1.444	4.564	1.276	26.26	26.52
48	87.42	78.18	1.450	4.656	1.269	26.41	26.01
49	87.21	78.95	1.421	4.724	1.286	26.63	25.89
50	87.04	–	1.426	–	–	26.80	–

[a] ^{13}C NMR spectra were measured at $-30°C$ in CH$_2$Cl$_2$, and ^{1}H NMR spectra were measured at $-40°C$ in CDCl$_3$.

The reaction of **47–50** with benzyl phenyl sulfide was studied by ^{1}H NMR spectroscopy in chloroform-d at $-30°C$ [79]. At this temperature, alkylperoxo complexes **47–50** are stable, and their concentrations do not change noticeably during several hours. The ^{1}H and ^{13}C NMR data show that at the ratio of the reactants [tBuOOH]/[Ti(OiPr)$_4$] ≤ 0.5, alkylperoxo complex **47** is the predominant titanium species in the reaction solution. The reactivity of **47** toward the sulfide can be readily determined by the ^{1}H NMR monitoring of the concentrations of **47**, sulfide and sulfoxide. The reaction was performed directly in an NMR tube. Thus, the obtained rate constant k_{47} for the reaction of complex **47** with benzyl phenyl sulfide was $(8 \pm 2) \times 10^{-4}$ M^{-1} s^{-1} (at $-30°C$). At the initial concentration ratio [tBuOOH]/[Ti(OiPr)$_4$]≈2, complex **48** is the predominant species in the system Ti(OiPr)$_4$ + tBuOOH in CH$_2$Cl$_2$. The reactivity of complex **48** toward benzyl phenyl sulfide was noticeably higher than that of **47**, $k_{48} = (3.2 \pm 0.6)$ 10^{-3} M^{-1} s^{-1} (at $-30°C$). At the reactants ratio [tBuOOH]/[Ti(OiPr)$_4$] > 10, complexes **49** and **50** predominated in solution. They

TABLE 3.3
^{13}C and ^{1}H NMR Chemical Shifts[a] for Ti(IV) Alkoxo Complexes Ti(OtBu)$_n$(OiPr)$_{4-n}$

n	$\underline{C}$(CH$_3$)$_3$	$\underline{C}$H(CH$_3$)$_2$	C(C$\underline{H}_3$)$_3$	C$\underline{H}$(CH$_3$)$_2$	CH(C$\underline{H}_3$)$_2$	C(C$\underline{H}_3$)$_3$	CH(C$\underline{H}_3$)$_2$
0	–	76.59	–	4.491	1.253	–	26.73
1	81.28	76.36	1.316	4.488[b]	1.248[c]	32.35[c]	26.71[c]
2	80.98	76.14	1.313	4.484[b]	1.248[c]	32.35[c]	26.71[c]
3	80.68	75.85	1.308	4.476	1.243	32.35[c]	26.71[c]
4	80.33	–	1.298	–	–	32.35[c]	–

[a] ^{13}C NMR spectra were recorded at $0°C$ in CH$_2$Cl$_2$, and ^{1}H NMR spectra were recorded at $20°C$ in CDCl$_3$.

[b] Partially resolved.

[c] Unresolved.

76 Applications of EPR and NMR Spectroscopy in Homogeneous Catalysis

were much more reactive toward benzyl phenyl sulfide; for their reaction rate constants, only the lower estimate was obtained: k_{49}, $k_{50} \geq 1.5 \times 10^{-2}$ M^{-1} s^{-1} (at $-30°$C) [79]. So, the reactivity of alkylperoxo complexes of titanium Ti(OOtBu)$_n$(OiPr)$_{4-n}$, $n = 1$–4 toward benzyl phenyl sulfide increases with the increase in the number of alkylperoxo ligands. Moreover, the reactivity per each alkylperoxo group also increases in this row.

Complex **47** is far less reactive toward olefins than toward sulfides (so that its interaction with olefins was studied at much higher temperature, 20°C). At this temperature, self-decomposition of **47** takes place with the rate constant $k_d = 4 \times 10^{-4}$ s^{-1}. In the course of this decomposition, one molecule of acetone is formed due to the oxidation of one of the OiPr ligands by the OOtBu moiety. For the formation of the epoxide to be detected, the rate of the reaction of **47** with the olefin must be competitive with the rate of this self-decay. For the sample with [$cyclo$-C$_6$H$_{10}$] = 1.4 M, [Ti(OiPr)$_4$] = 0.24 M, and [tBuOOH] = 0.14 M, the molar ratio of the oxidation products [acetone]/[epoxide] was about 3. The rate constant for the reaction of **47** with cyclohexene was estimated as 10^{-4} M^{-1} s^{-1} (at 20°C). Complex **47** was unreactive toward stilbene and styrene under those conditions. Allylic alcohols react very rapidly with **47** at 20°C, to form the corresponding epoxides. In this case, self-decomposition of **47** with acetone formation can be entirely suppressed.

3.2.3 ALKYLPEROXO COMPLEXES OF VANADIUM

Since the first publications on vanadium alkylperoxo complexes, the latter had been implicated as the active species of transition metal–catalyzed epoxidation of olefins with alkyl hydroperoxides [80].

Mimoun and coworkers isolated and x-ray-characterized the stable (dipicolinato) vanadium(V) η^2-alkylperoxo complex **51** (Figure 3.18) [70]. However, **51** was found to be inert toward olefins. Later, the same group synthesized alkylperoxo complexes **52a–e** with tridentate N-(2-oxidophenyl)-salicylidenaminato Schiff base ligands (R′-OPhsal-R″) (Figure 3.18) [81]. Complexes **52a–e** were characterized by elemental analysis and IR and ^{1}H NMR spectroscopy. Those complexes were the first well-defined examples of alkylperoxo complexes of d^0-transition metals, capable

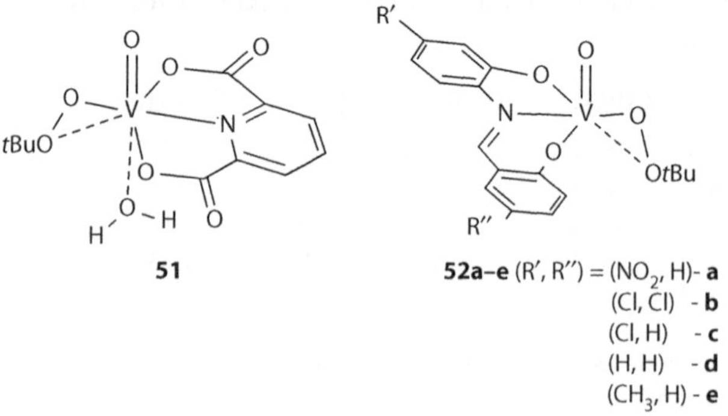

51

52a–e (R′, R″) = (NO$_2$, H)- **a**
(Cl, Cl) - **b**
(Cl, H) - **c**
(H, H) - **d**
(CH$_3$, H) - **e**

FIGURE 3.18 Alkylperoxo complexes of vanadium with dipicolinato (**51**) and salicylidenaminato (**52a–e**) ligands.

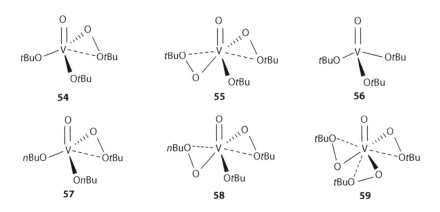

FIGURE 3.19 Proposed structures of alkylperoxo and alkoxo complexes of vanadium 54–59.

of epoxidizing olefins with high selectivity. The reaction was highly stereospecific: *cis*-olefins and *trans*-olefins were selectively transformed to *cis*-epoxides and *trans*-epoxides, respectively [81].

Vanadium(V) species formed in the catalyst system VO(acac)$_2$/*t*BuOOH/olefin in benzene-d_6 were studied by ^{1}H and ^{51}V NMR spectroscopy [82]. It was found that alkylperoxo complex VO(acac)$_2$(OO*t*Bu) (**53**) is formed at the initial stage of reaction. Then, it rapidly decays, and three new vanadium(V) complexes **54–56** appear in the catalyst system studied (Figure 3.19). The concentrations of **54** and **55** decline with time, while that of **56** increases. Independent synthesis showed that **56** is complex VO(O*t*Bu)$_3$. The structures of **54** and **55** were not clearly established [82]. A subsequent ^{51}V, ^{17}O, ^{13}C, and ^{1}H NMR spectroscopic study of the vanadium species formed in the catalyst system VO(O*n*Bu)$_3$/*t*BuOOH unambiguously revealed the composition of complexes **54** and **55** [83].

Multinuclear NMR spectroscopic monitoring of the interaction of VO(O*n*Bu)$_3$ with *t*BuOOH in CH$_2$Cl$_2$ at −40°C showed that the alkoxo ligands of VO(O*n*Bu)$_3$ were successively replaced by alkylperoxo ones with the increase in the *t*BuOOH:VO(O*n*Bu)$_3$ ratio from 1:1 to 10:1, to form an equilibrium mixture of alkylperoxo complexes VO(OO*t*Bu)$_k$(O*n*Bu)$_{3-k}$, where k = 1, 2, 3 [83]:

$$\text{VO(O}n\text{Bu)}_3 + k t\text{BuOOH} \leftrightarrow \text{VO(OO}t\text{Bu)}_k(\text{O}n\text{Bu})_{3-k} + k t\text{BuOH} \qquad (3.2)$$

Each alkylperoxo complex displays individual signals in the ^{51}V, ^{17}O, ^{13}C, and ^{1}H NMR spectra (Table 3.4). The composition of alkylperoxo complexes was determined from relative intensities of the ^{13}C resonances of the OO*t*Bu ligands and the primary carbon atoms of the O*n*Bu ligands of VO(OO*t*Bu)$_k$(O*n*Bu)$_{3-k}$ for k = 1 and 2. The relative intensities (NOE corrected) of these resonances for VO(OO*t*Bu)(O*n*Bu)$_2$ (**57**) and VO(OO*t*Bu)$_2$(O*n*Bu) (**58**) were 1:2 and 2:1, respectively. Only the signals of the OO*t*Bu ligands correspond to the VO(OO*t*Bu)$_3$ complex (**59**) in the ^{1}H and ^{13}C NMR spectra. The observed chemical shifts for the quaternary carbon atom

78   Applications of EPR and NMR Spectroscopy in Homogeneous Catalysis

TABLE 3.4
^{13}C, ^{1}H, ^{51}V, and ^{17}O NMR Chemical Shifts (CH$_2$Cl$_2$, −40°C) for Vanadium(V) Alkylperoxo Complexes VO(OOtBu)$_k$(OnBu)$_{3-k}$

Compound	$\underline{C}$(CH$_3$)$_3$	C($\underline{C}$H$_3$)$_3$	C(C$\underline{H}_3$)$_3$	O$\underline{C}$H$_2$	OC$\underline{H}_2$	^{51}V	V = ^{17}O
VO(OnBu)$_3$	–	–	–	84.3[a]	5.04[b]	−593[c,d]	1162[b,g]
57	87.0[c]	26.05	1.32	84.90[a]	5.07[b]	−554[e]	1158[b]
58	87.9[c]	26.22	1.36	85.85	5.04	−530 ± 3[c,f]	1190
59	89.9	26.60	1.45	–	–	−545 ± 2[c]	1236
tBuOOH	81.0	25.85	1.23	–	–	–	–
nBuOH	–	–	–	62.85	3.63	–	–

[a] Unresolved at −40°C due to fast ligand exchange.
[b] Unresolved.
[c] Chemical shifts depend on the tBuOOH concentration.
[d] Chemical shift in the presence of 2 M tBuOOH in CH$_2$Cl$_2$. In neat CH$_2$Cl$_2$, the chemical shift is −557 ppm, temperature dependence −1.4 ± 0.1 ppm/K.
[e] Temperature dependence of the chemical shift is less than 0.02 ppm/K.
[f] Temperature dependence of the chemical shift is −0.15 ± 0.01 ppm/K.
[g] ^{17}O NMR signal of nBuO ligands was at 350 ppm.

of the tBu groups of **57–59** (Table 3.4) is indicative of the η^2-coordination of the alkylperoxo moiety.

In the same solvent, ^{51}V chemical shifts of **55** and **58** almost coincided, confirming that **55** is alkylperoxo complex VO(OOtBu)$_2$(OtBu). ^{51}V chemical shifts of **54** and **57** were also very close, thus suggesting that **54** is alkylperoxo complex VO(OOtBu)(OtBu)$_2$[82,83]. One can conclude that three types of vanadium(V) complexes exist and operate in the catalyst system V^{IV}O(acac)$_2$/tBuOOH/olefin: VO(OOtBu)(OtBu)$_2$ (**54**), VO(OOtBu)$_2$(OtBu) (**55**), and VO(OtBu)$_3$ (**56**) (Figure 3.19). In the case of allylic alcohol R$_1$OH as a substrate, one of the OtBu groups of **54** and **55** can be replaced by R$_1$O group. There are still no direct data on the reactivity of **54** and **55**. Apparently, by analogy with titanium congeners, alkylperoxo complexes **54** and/or **55** are the active epoxidizing species of the catalyst system V^{IV}O(acac)$_2$/tBuOOH.

In 1973, Sharpless and Michaelson reported on the remarkable reactivity and selectivity of vanadium catalysts toward the epoxidation of allylic alcohols with alkylhydroperoxides [84]. Later, the same group found that the combination of VO(acac)$_2$ and chiral hydroxamic acid **60** (Figure 3.20) was capable of epoxidizing allylic alcohols by tBuOOH with up to 50% *ee* [85]. The enantioselectivity and efficiency of the vanadium-based catalyst systems for the asymmetric epoxidation of allylic alcohols were improved using specially designed hydroxamic acids **61, 62** (up to 96% *ee* [86,87]), and **63** (up to 71% *ee* [88,89] (Figure 3.20)).

The structure and reactivity of vanadium(V) complexes formed *in situ* in the catalyst system VO(OiPr)$_3$/hydroxamic acid **63**/tBuOOH were examined by means of ^{51}V, ^{13}C, and ^{17}O NMR spectroscopy. For the first time, reactive vanadium(V) alkylperoxo intermediates in vanadium/hydroxamic acid epoxidations (Scheme 3.4)

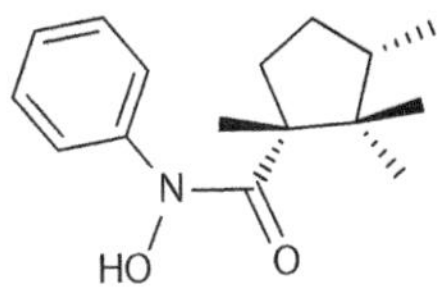

NMR and EPR Spectroscopy

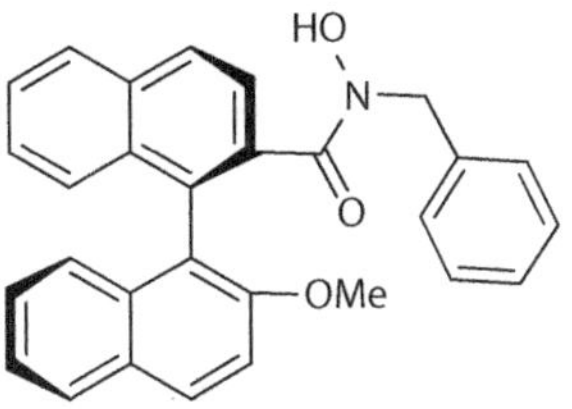

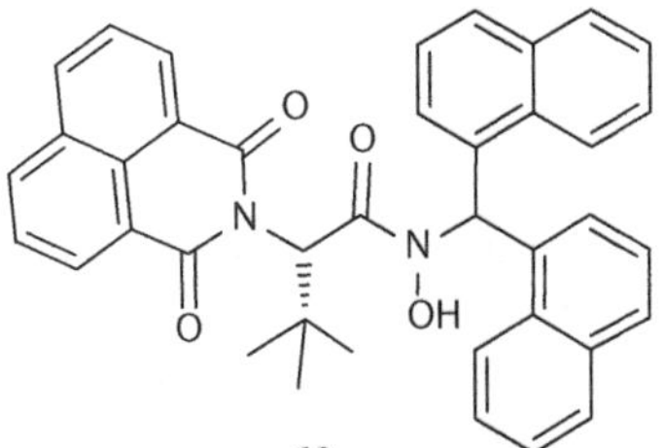

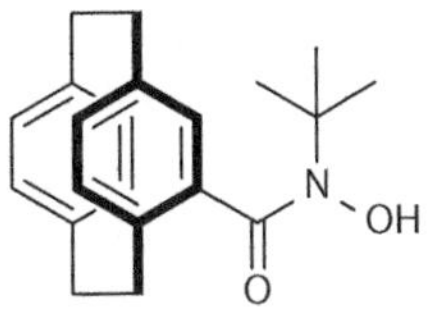

FIGURE 3.20 Chiral hydroxamic acids used for vanadium-catalyzed enantioselective epoxidation of allylic alcohols with alkyl hydroperoxides.

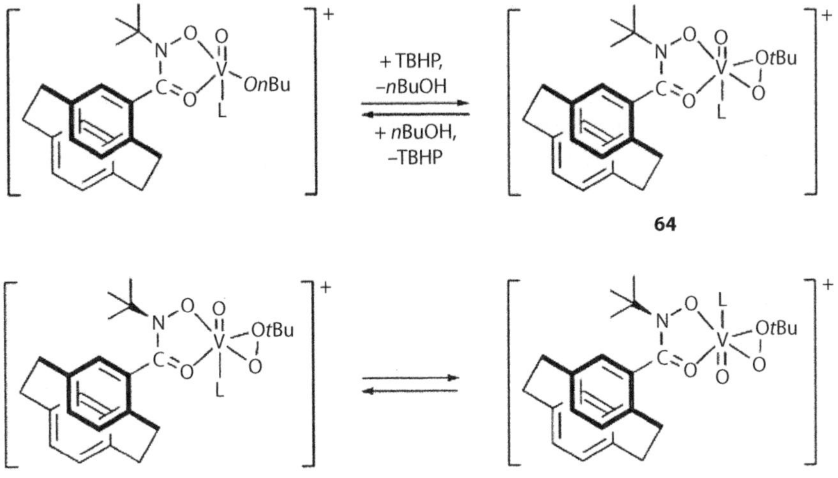

SCHEME 3.4 Formation of diastereomeric vanadium(V) *tert*-butyl peroxo complexes. (Reproduced from Bryliakov, K. P. et al. 2003. Multinuclear NMR study of the reactive intermediates in enantioselective epoxidation of allylic alcohols catalyzed by a vanadium complex derived from a planar-chiral hydroxamic acid. *New J. Chem.* 27: 609–614. With permission from the Centre National de la Recherche Scientifique (CNRS) and The Royal Society of Chemistry.)

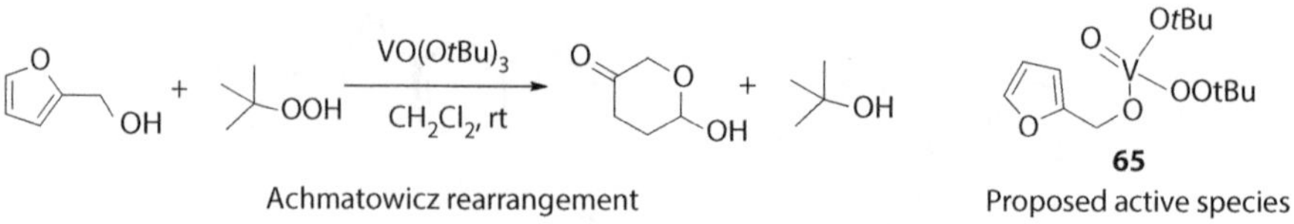

80 Applications of EPR and NMR Spectroscopy in Homogeneous Catalysis

SCHEME 3.5 Achmatowicz rearrangement of furfurol.

were observed and spectroscopically characterized [90]. With a planar-chiral [2.2] paracyclophane-derived hydroxamate **63** as the chelating ligand, two diastereomeric alkylperoxo vanadium(V) complexes **64a,b** were formed in a 3:1 ratio, differing in the relative positioning of the V=O group and the planar-chiral aromatic part (Scheme 3.4). Upon the addition of geraniol as substrate, complexes **64a,b** disappeared in a parallel manner, and geraniol epoxide was formed. Probably, the existence of those two diastereomeric complexes and their comparable reactivities account for the observed moderate enantioselectivity level (71% *ee*) and hence sets a fundamental limitation for the use of such ligands in this asymmetric catalysis [90].

Very recently, ^{51}V NMR spectroscopy was used for the characterization of vanadium species formed during the transformation of furfurol into the synthetically important lactol (6-hydroxy-2*H*-pyran-3(6*H*)-one, Achmatowicz rearrangement, Scheme 3.5). By monitoring the reaction mixture under the reaction conditions by ^{51}V NMR, the authors were able to identify various vanadium species, including the alkylperoxo intermediate VO(O*t*Bu)(OO*t*Bu)(furfurol) (**65**) (Scheme 3.5) [91].

3.3 PEROXO COMPLEXES

Hydrogen peroxide has been considered as one of the most attractive "green" oxidants, affording water as the only by-product in oxidation reactions, which stimulated the interest in catalyst systems relying on transition metals and H_2O_2, in both the industry and academia. Particularly, peroxo complexes of molybdenum, vanadium, and titanium have been intensively studied as possible active species of catalysts based on the above transition metals.

3.3.1 PEROXO COMPLEXES OF MOLYBDENUM

In 1970, Mimoun, Roch, and Sajus discovered that diperoxo molybdenum complex $Mo(O_2)_2$·HMPA (**66**, Figure 3.21) (HMPA = hexamethylphosphoric triamide) is capable of stoichiometrically oxidizing alkenes to epoxides in good yields at room temperature in aprotic solvents [92]. ^{18}O labeling studies provided evidence that the epoxide oxygen originated exclusively from the peroxo ligands [93]. Theoretical calculations suggested that the epoxidation occurred via direct nucleophilic attack of the olefin at the electrophilic peroxidic oxygen through a transition state of spiro structure (**67**, Figure 3.21) [94]. Using hybrid density functional approach, Rösch and coworkers proposed different peroxide intermediates formed in the systems H_3NMoO_3/H_2O_2, and H_3NOMoO_3/H_2O_2 [95]. They have studied mono- and bisperoxo intermediates as well as hydroperoxo derivatives as the potentially reactive species. The inspection

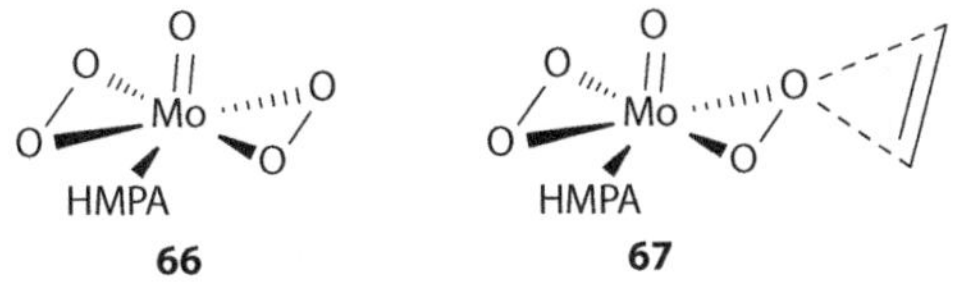

NMR and EPR Spectroscopy

FIGURE 3.21 Diperoxo complex of molybdenum(VI) and proposed transition state of its reaction with ethylene.

of the energy pattern of various intermediates showed that the hydroperoxo intermediates may be competitive (and even superior) to the peroxo intermediates as the active species of epoxidation [95].

^{1}H, ^{17}O, and ^{95}Mo NMR spectroscopic monitoring of the intermediates formed in the catalyst systems $MoO_2(acac)_2/H_2O_2$ and MoO_3/H_2O_2 in acetonitrile have shown that molybdenum complex **68** (Figure 3.22) is the major molybdenum species present in the reaction solution ($[H_2O_2]/[Mo] > 3$) [96]. The ^{17}O NMR spectra of the catalyst system $MoO_2(acac)_2/H_2O_2$ (Figure 3.23a) and complex $MoO(O_2)_2$·HMPA (Figure 3.23b) (natural abundance of the ^{17}O isotope, solvent CH_3CN) show that molybdenum complexes present in both samples display similar ^{17}O resonances from Mo=O (δ 857, $\Delta v_{1/2} = 200$ Hz) and $Mo(O_2)_2$ (δ 450, $\Delta v_{1/2} = 2500$ Hz) moieties, with relative intensities 1:4. It is logical to assign **68** to a diperoxo molybdenum complex $MoO(O_2)_2$·H_2O (Figure 3.22). With the increase in the concentration of water in acetonitrile from 4 to 25 M, the ^{95}Mo NMR peak of **68** shifted 45 ppm upfield, while the ^{17}O chemical shift of Mo=O shifted 16 ppm downfield. Similar changes in ^{95}Mo and ^{17}O chemical shifts were observed when $MoO(O_2)_2$·HMPA converted into $MoO(O_2)_2$·HMPA·H_2O. Apparently, with the increase in H_2O concentration, the $MoO(O_2)_2$·(H_2O) complex (**68**) transformed into molybdenum complex with the proposed structure $MoO(O_2)_2$·($H_2O)_2$ (**69**) (Figure 3.22). Complex **69** was inert toward alkenes. In contrast, **68** reacted with cyclohexene and allylic alcohols (in dioxane and acetonitrile) to afford the corresponding epoxides.

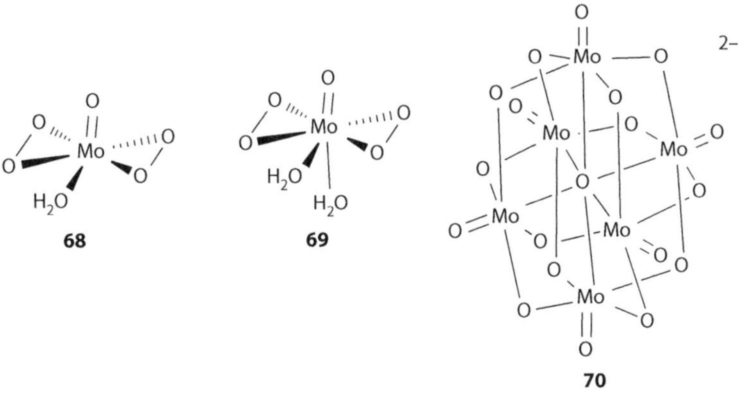

FIGURE 3.22 Molybdenum complexes formed in the catalyst system $MoO_2(acac)_2/H_2O_2/C_6H_{10}$ in acetonitrile.

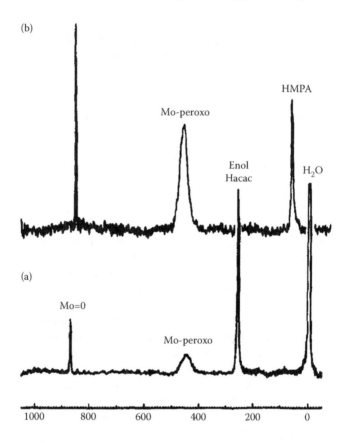

FIGURE 3.23 ^{17}O NMR spectra (acetonitrile, 20°C) of (a) Mo-peroxo complex **68** and (b) 0.5 M solution of MoO(O$_2$)$_2$·HMPA. Complex **68** was prepared by the addition of 70% H$_2$O$_2$ to a 0.3 M solution of MoO$_2$(acac)$_2$ in acetonitrile to obtain [H$_2$O$_2$] = 5 M. HMPA = hexamethylphosphoramide. (Reproduced with permission from Talsi, E. P. et al. 1993. *J. Mol. Catal. A Chem.* 83: 347–366.)

The ^{95}Mo NMR spectrum of the sample MoO$_2$(acac)$_2$/[H$_2$O$_2$] ([MoO$_2$(acac)$_2$] = 0.1 M, [H$_2$O$_2$] = 1 M), recorded 10 min after the addition of H$_2$O$_2$, displays predominantly the resonance of **68** (δ −122, $\Delta v_{1/2}$ = 900 Hz) (Figure 3.24a). After the addition of cyclohexene to the sample of Figure 3.24a and 2-min storing of the resulting solution at 80°C, the resonance of **68** disappeared, and those of the 1,2-diolo complex **25** (Figure 3.11) and Mo$_6$O$_{19}^{2-}$ (**70**) (Figure 3.22) were observed (Figure 3.24b). The addition of new portions of H$_2$O$_2$ restored the concentration of **68** (Figure 3.24c) [96]. When the new portion of cyclohexene was added to the sample of Figure 3.24c, the concentration of **70** was partially restored (Figure 3.24d). Most probably, the bis(peroxo) complex **68** is the active species of epoxidation. Complex **68** was shown to stoichiometrically epoxidize alkenes in good yields at room temperature in poorly coordinating solvents (while being inert in strong σ-donor solvents, e.g., DMF, DMSO, HMPA, and Py) [92]. This effect was explained by the decrease of

NMR and EPR Spectroscopy

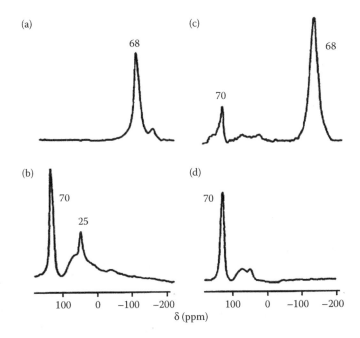

FIGURE 3.24 ^{95}Mo NMR spectra (acetonitrile, 20°C) of samples prepared by various treatments of a 0.1 M solution of MoO$_2$(acac)$_2$ in acetonitrile: (a) 10 min after the addition of 70% H$_2$O$_2$ ([H$_2$O$_2$] = 1 M) at 20°C, (b) 5 min after the addition of cyclohexene ([C$_6$H$_{10}$] = 1 M) to the sample (a) and storing at 80°C for 2 min, (c) 5 min after the addition of another portion of H$_2$O$_2$ ([H$_2$O$_2$] = 2 M) to the sample in (b) at 20°C, and (d) 5 min after the addition of further portion of H$_2$O$_2$ ([H$_2$O$_2$] = 2 M) to the sample in (c) and heating at 80°C for 2 min. (Reproduced with permission from Talsi, E. P. et al. 1993. *J. Mol. Catal. A Chem.* 83: 347–366.)

electrophilic properties of the coordinated peroxo groups upon the introduction of additional σ-donor ligand [94–96].

Recent data on the synthesis and catalytic activity of vanadium and molybdenum peroxides were reviewed by Conte and Floris [97]. A number of well-characterized molybdenum and vanadium peroxides have so far been reported, along with the elucidation of factors governing the reactivity and selectivity of the peroxo complexes of transition metals; however, examples of commercial applications of catalyst systems of this type have not been reported thus far.

3.3.2 Peroxo Complexes of Vanadium

To date, numerous vanadium mono- and bis(peroxo) complexes have been prepared, which oxidized olefins and sulfides with hydrogen peroxide [89,97–99]. The oxidation selectivity by vanadium compounds is usually lower than for their molybdenum congeners, due to the contribution of nonselective radical oxidation. For example, complex VO(O$_2$)(Pic)(H$_2$O)$_2$ (**71**) prepared by Mimoun and coworkers (Figure 3.25) is stable and inactive in strong σ-donor solvents (DMF, H$_2$O, MeOH) and decomposes

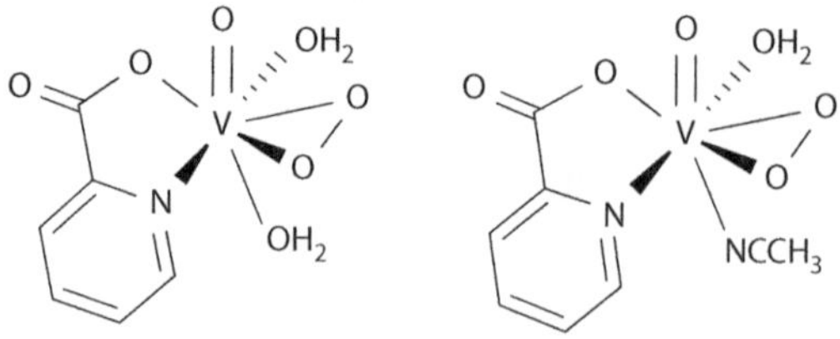

Applications of EPR and NMR Spectroscopy in Homogeneous Catalysis

FIGURE 3.25 Monoperoxo complexes of vanadium, active (**72**) and inert (**71**) toward organic substrates.

in less coordinating solvents (CH_3CN, CH_3NO_2). This decomposition can initiate radical oxidation of benzene and alkanes [100]. Various groups tried to elucidate the decomposition mechanism of complex **71**; however, this question is yet to be answered [100–103]. ^{51}V NMR studies of the solutions of **71** in CH_3CN, containing various amounts of water, showed the coexistence of complexes $VO(O_2)(Pic)(H_2O)_2$ (**71**) (δ −565) and $VO(O_2)(Pic)(H_2O)(CH_3CN)$ (**72**) (δ −522) in CH_3CN/H_2O mixtures (Figure 3.25). At high H_2O concentrations in CH_3CN, **72** converts to **71** [103]. Complex **72** is unstable and initiates radical oxidation of organic substrates, whereas **71** is stable and inert. The radical nature of species formed upon decomposition of **72** in CH_3CN was confirmed by trapping $HO^{\bullet}$ radicals in the reaction solution using 3,3,5,5-tetramethyl-pyrroline-N-oxide (TMPO) as a spin trap [103]. The observed deactivation of **72** upon substitution of CH_3CN ligand by stronger σ-donor ligands (e.g., H_2O, MeOH, DMF) may be caused by the suppression of electron transfer from the peroxo group to the metal to give a radical-like species V^{IV}–O–O$^{\bullet}$ [100,103].

Pombeiro, Shulpin, and coworkers demonstrated that vanadate anion, VO_3^-, efficiently catalyzes the oxidation of organic compounds with H_2O_2 in acetonitrile in the presence of pyrazine-2-carboxylic acid (PCA), added to the solution as a cocatalyst [104 and references therein]. Detailed ^{51}V NMR and kinetic studies of the catalyst systems vanadium complex/PCA/H_2O_2 confirmed the key role of $HO^{\bullet}$ and $HO_2^{\bullet}$ radicals in the oxidation of organic substrates; most likely, vanadium complexes containing the V(PCA)(OO)(OOH) moiety are responsible for the generation of these radicals [104].

From the practical perspective, the design of highly chemo- and stereoselective oxidizing catalyst systems requires that the contribution of radicals, initiating nonselective pathways, should be reduced to a minimum. Mizuno and coworkers demonstrated that divanadium-substituted γ-Keggin polyoxotungstates $TBA_3[\gamma\text{-}PW_{10}V_2O_{38}(OH)_2]$ (**73**, TBA = tetra-n-butylammonium) and $TBA_4[\gamma\text{-}SiW_{10}V_2O_{38}(OH)_2]$ (**74**) (Figure 3.26) are efficient catalysts for a range of selective liquid-phase oxidations with H_2O_2 [105–108]. Complex **74** selectively catalyzes olefin epoxidation [105], whereas **73** catalyzes the epoxidation of electron-deficient alkenes [106], hydroxylation of alkanes [108], and alkylbenzenes [107]. ^1H, ^{51}V, ^{183}W NMR and cold-spray ionization mass spectrometric studies of the systems **73**/H_2O_2 and **74**/H_2O_2 indicate that intermediates with $[OV\text{-}(\mu\text{-}OH)(\mu\text{-}OOH)\text{-}VO)]$ (**75**) and $[OV\text{-}(\mu\text{-}\eta^2{:}\eta^2\text{-}OO)\text{-}VO)]$ (**76**) core are formed in these systems, and they can be responsible for the selective oxidations [105–108] (Scheme 3.6).

NMR and EPR Spectroscopy

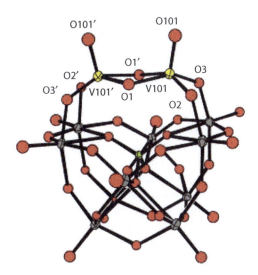

FIGURE 3.26 Atomic representation of $[\gamma\text{-PV}_2\text{W}_{10}\text{O}_{38}(\text{OH})_2]^{3-}$ (**74**). (Reproduced with permission from Kamata, K. et al. 2010. *Nat. Chem.* 2: 478–483.)

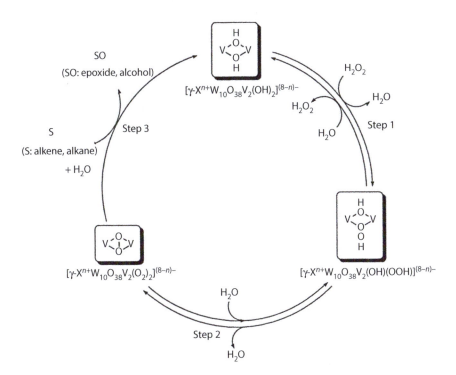

SCHEME 3.6 Proposed mechanism for the oxidation of hydrocarbons with H_2O_2 catalyzed by divanadium-substituted polyoxotungstates **74** ($X = Si^{4+}$) and **73** ($X = P^{5+}$). (Reproduced with permission from Kamata, K. et al. 2010. *Nat. Chem.* 2: 478–483.)

86 Applications of EPR and NMR Spectroscopy in Homogeneous Catalysis

Kholdeeva and coworkers reported that phosphotungstate **74** is an efficient catalyst for the selective oxidation of industrially important alkylphenols/naphthols with 35% H_2O_2 [109]. By applying the optimized reaction conditions, 2,3,5-trimethyl-p-benzoquinone (TMBQ, vitamin E key intermediate) was obtained in a nearly quantitative yield via the oxidation of 2,3,6-trimethylphenol (TMP). The efficiency of H_2O_2 utilization reached 90%. The active vanadium peroxo complex, responsible for the oxidation of TMP to TMPQ, was identified by ^{51}V and ^{31}P NMR spectroscopy. The ^{51}V spectrum of **73** in CH_3CN displayed a single ^{51}V resonance at −581 ppm and a ^{31}P resonance at −14.2 ppm. After treatment of **73** with 1–2 equivalents of 35% H_2O_2, new ^{51}V and ^{31}P resonances at δ −557 and −14.8 appeared. The ^{51}V resonance at δ −557 had previously been assigned to a vanadium μ-hydroxo–μ-hydroperoxo complex **75** [105]. After the subsequent addition of a 10-fold excess TMP, the resonances of **75** rapidly disappeared. The decomposition of the hydroperoxo complex in the absence of the phenolic substrate was significantly slower (2 h vs. 20 min in the presence of TMP at 25°C). The new portions of H_2O_2 partially restored the concentration of **75**. These data allowed the authors to propose that it was **75** that conducted the selective oxidation of TMP to TMBQ [109].

Peroxo complexes of vanadium, capable of enantioselective oxidation of organic substrates with high efficiency, have been very rare. One of the major achievements of this kind has been the catalyst system for the asymmetric oxidation of sulfides, reported by Bolm and Bienewald [110]. It employed a catalyst formed *in situ* at room temperature from [VO(acac)$_2$] and chiral Schiff base of type **77**, and 30% aqueous H_2O_2 as the oxidizing agent (Figure 3.27). The catalyst system afforded chiral sulfoxides (with enantioselectivities up to 85% *ee*), demonstrating exceptionally high efficiency, which allowed the reduction of the catalyst loading down to 0.01 mol.%. Subsequently, several other research groups investigated related vanadium systems (see Reference [89] and references therein). Even though the enantioselectivities of

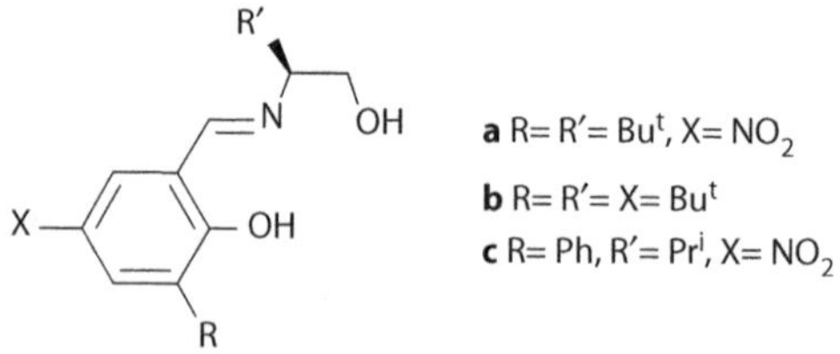

a R= R′= Bu^t, X= NO$_2$

b R= R′= X= Bu^t

c R= Ph, R′= Pr^i, X= NO$_2$

77

78 **79**

FIGURE 3.27 The Schiff base ligands **77** proposed by Bolm and Bienewald [110] and the proposed structures of the peroxovanadium(V) intermediates **78** and **79**.

catalyst systems based on vanadium complexes with Schiff base ligands are not perfect yet, such systems occurred to be useful for mechanistic studies because the vanadium(V) intermediates responsible for the enantioselective oxidation were easily detectable by ^{51}V NMR and displayed informative ^{13}C NMR spectra.

Using ^{51}V and ^{13}C{^{1}H} NMR spectroscopy, the structure and reactivity of vanadium peroxo complexes formed in the catalyst system [VO(OiPr)$_3$]/Schiff base ligand/H$_2$O$_2$ were studied [111]. Two types of monoperoxo vanadium(V) complexes **78** and **79** (Figure 3.27) bearing one Schiff base ligand per vanadium predominated in this system at −12°C. Those complexes were unstable at room temperature and decomposed with a half-life time of 20 min. The rate of this decomposition markedly increased in the presence of methyl phenyl sulfide. The new portions of H$_2$O$_2$ partially restored the concentration of monoperoxo complexes **78** and **79**. Analysis of the ^{51}V and ^{13}C{^{1}H} NMR data led Bryliakov et al. to the conclusion that the monoperoxo complexes **78** and **79** corresponded to complexes with bidentate and tridentate coordination of the Schiff base ligand (Figure 3.27) [111].

3.3.3 Peroxo Complexes of Titanium

In contrast to molybdenum and vanadium peroxo complexes, all isolated and well-characterized peroxo complexes of titanium appeared to be inactive toward the oxidation of organic substrates (except triphenylphosphine) in stoichiometric reactions [112–114]. Kholdeeva and coworkers have succeeded in preparing the first protonated peroxotitanium complex, [Bu$_4$N]$_4$[HPTi(O$_2$)W$_{11}$O$_{39}$] (**80**), via the reaction of the μ-oxo dimeric heteropolytungstate [Bu$_4$N]$_8$[(PTiW$_{11}$O$_{39}$)$_2$O] (**81**) with a 15-fold excess of 35% aqueous H$_2$O$_2$ in CH$_3$CN [115]. The atomic representation of the polyoxoanion [PTiW$_{11}$O$_{39}$]$^{3-}$ is shown in Figure 3.28. Complex **80** was isolated and well characterized by various techniques. Potentiometric titration with methanolic TBAOH confirmed the presence of one acidic proton in the molecule of **80**. The addition of 1 equivalent of OH$^-$ to **80** (^{31}P NMR: δ −12.4) resulted in the formation of the well-known complex [Bu$_4$N]$_5$[PTi(O$_2$)W$_{11}$O$_{39}$] (**82**) (^{31}P NMR: δ −13.0). Complexes **80** and **82** displayed sharply different

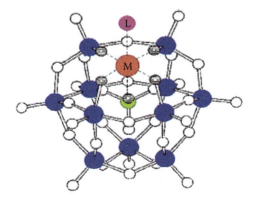

FIGURE 3.28 Atomic representation of polyoxoanion [PTiW$_{11}$O$_{39}$]$^{3-}$. (Reproduced with permission from Kholdeeva, O. A. et al. *Inorg. Chem.* 39: 3828–3837.)

88   Applications of EPR and NMR Spectroscopy in Homogeneous Catalysis

activities toward organic substrates. Complex **82** was inactive toward the oxidation of both olefins and thioethers, whereas **80** stoichiometrically oxidized methyl phenyl sulfide to the corresponding sulfoxide (the concentration of **80** was monitored by ^{31}P NMR, and that of sulfoxide was followed by gas chromatography) [115]. The latter experiment was the first unequivocal demonstration of a direct stoichiometric reaction between a peroxotitanium complex and an organic substrate. Based on the product and kinetic studies, an outer-sphere electron transfer mechanism involving the formation of a thioether radical cation has been proposed [115]. Importantly, the presence of proton in the molecule of **80** is crucial for its reactivity. DFT and Raman studies on the monoprotonated peroxo complex **80** had indicated that the proton is located at one of the Ti—O—W bridging oxygen atoms. One of the possible explanations of the higher reactivity of **80** as compared to **82** is the increase in the redox potential of the peroxo group upon protonation. Cyclic voltammetry study revealed that **80** has higher redox potential ($E_{1/2}$ = 1.25 V) compared to **82** ($E_{1/2}$ = 0.88 V) [116].

Another important result concerning the role of protons in the oxidation of organic substrates with peroxo complexes of titanium was obtained in 2005 [117]. It was shown that **80**, which is not able to mediate the heterolytic oxidation of alkenes to epoxides, became a good epoxidizing agent by the addition of one more proton to its cationic part to form diprotonated complex $[H_2PTi(O_2)W_{11}O_{39}]^{3-}$ (**83**). DFT calculations showed that the energy barrier for heterolytic oxygen transfer (i.e., epoxidation) decreases significantly upon protonation (ca. 6 kcal mol^{-1}). This finding can explain why the presence of the second proton switches the oxidation mechanism from a homolytic to a heterolytic one [118].

3.4   OXO COMPLEXES

In the past 20 years, the studies focused on biomimetic oxidations with environmentally friendly oxidants H_2O_2 and O_2 have expanded greatly, resulting in the emergence of numerous bioinspired catalysts, capable of mimicking the reactivity of natural metalloenzymes in selective oxidation reactions. Today, some of the bioinspired catalyst systems may be regarded as promising, more selective, efficient, and clean alternatives to traditional systems [119,120]. This direction is rapidly progressing now, bringing out non-heme iron and related manganese complexes that are regarded as the most prominent prototypes for the future "green" catalyst systems for chemo-, regio-, stereoselective, and stereospecific oxidations of alkanes and alkenes with H_2O_2. Oxocomplexes of iron and manganese are often proposed as active species of these catalyst systems.

3.4.1   OXOCOMPLEXES $[Cr^V = O(SALEN)]^+$

In 1985, Kochi and coworkers reported that $[Cr^{III}(salen)]^+$ complexes (**84**) were capable of catalyzing the epoxidation of unfunctionalized alkenes using PhIO as the stoichiometric oxidant (Figure 3.29) [121]. Good yields were achieved for many olefins; it was noted that the addition of oxygen donor ligands, such as pyridine N-oxide (PyO) and triphenylphosphine oxide (Ph$_3$PO), increased the reaction rates. It was shown that the reaction of cationic complexes $[Cr^{III}(salen)]^+$ (**84**) with PhIO

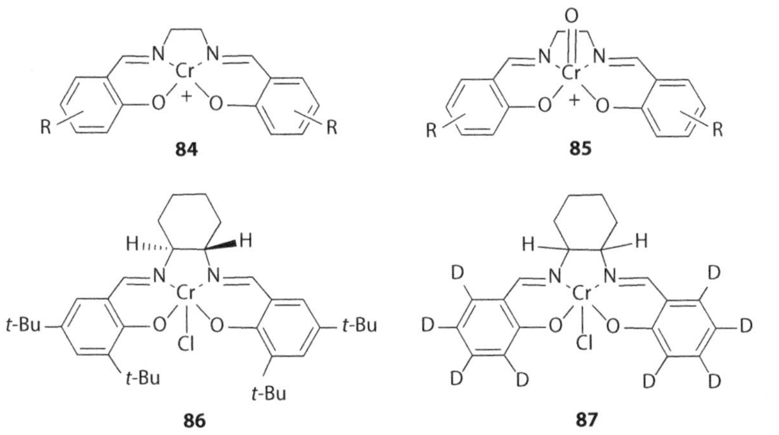

NMR and EPR Spectroscopy

FIGURE 3.29 Chromium(III)-salen catalysts (**84**, **86**, **87**) and oxochromium(V) complex **85**—the active species of selective epoxidation of olefins.

in CH_3CN leads to the formation of oxochromium(V) complexes $[Cr^V = O(salen)]^+$ of the type **85**—the active species of epoxidation (Figure 3.29). The EPR spectra of the d^1 complex **85** (room temperature, CH_3CN) are centered at $g_0 = 1.977$–1.978 and display hfs $a_{Cr} = 19.35$–20 G (^{53}Cr, $I = 3/2$, 9.55% natural abundance), and shf splitting from two equivalent nitrogen atoms $a_N = 2.12$–2.17 G (^{14}N, $I = 1$, 99.63% natural abundance). The involvement of the oxochromium(V) complexes in the catalytic cycle is supported by the similarity of the products obtained under the catalytic conditions to those obtained from the reaction of alkenes with the isolated oxo complex. One oxochromium complex of the type **85**, obtained according to the following equation, was isolated as X-ray-quality single crystals [121]:

$$[Cr^{III}(Me_2salen)]^+ OTf^- (\mathbf{84}) + PhIO \rightarrow [Cr^V{=}O(Me_2salen)]^+ OTf^- (\mathbf{85}) + PhI$$

$$(3.3)$$

The oxygen transfer was proposed to occur via electrophilic attack by the oxochromium complex on the double bond of the alkene. The details of the oxygen atom transfer from the oxochromium(V) complex to the olefin were not elucidated.

Later, the intermediates formed in catalyst systems **86**/PhIO and **87**/PhIO (Figure 3.29) in various solvents were studied by EPR, ^1H and ^2H NMR, and UV–Vis/NIR techniques [122]. Two types of high-valent chromium complexes **A** and **B** were identified in those systems. The EPR parameters of complexes **A** noticeably differ from those of **B** (g-factors and a_N in the range of 1.970–1.974 and 1.6–2.0 G for **A**, and 1.976–1.980 and 2.0–2.3 G for **B**). As an example, the EPR spectra of oxochromium(V) intermediates $\mathbf{A_{86}}$ and $\mathbf{B_{86}}$ formed in the system **86**/PhIO in CH_3CN are presented in Figure 3.30. The subscript "86" is to highlight that species **A** and **B** originate from the starting complex **86**. The EPR parameters of complexes **85**, reported by Kochi, were close to those of complexes of the type **B**. Complex $\mathbf{A_{86}}$ (green solution, $g = 1.970$, $a_{Cr} = 19.3$ G) is unstable and almost

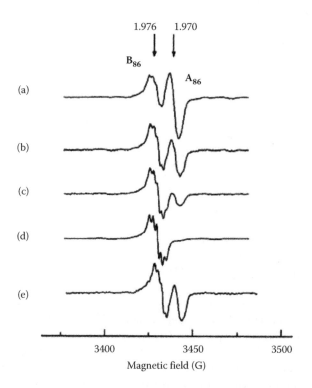

FIGURE 3.30 EPR spectra (CH$_3$CN, 17°C) of chromium(V) complexes formed after various times of stirring 16 μmol of complex **86** and 35 μmol of PhIO in 3 mL of CH$_3$CN: 10 min (a), 30 min (b), 50 min (c), 150 min (d), and 4.5 mg addition of fresh portion of PhIO followed by mixture stirring for 10 min (e). (Reproduced with permission from Bryliakov, K. P., Talsi, E. P. 2003. *Inorg. Chem.* 42: 7258–7265.)

completely decays within 3 h at room temperature, with concomitant formation of complex **B$_{86}$** (brown solution, $g = 1.976$, $a_N = 2.1$ G, $a_{Cr} = 19.3$ G); the latter is more stable, its concentration remaining nearly constant for several hours. When complex **87** was taken as the starting material for catalysis, the formation of **A$_{87}$** ($\tau_{1/2} = 5$ min, CH$_3$CN, 17°C) and **B$_{87}$** ($\tau_{1/2} = 25$ min, CH$_3$CN, 17°C) was detected. According to the NMR data, the self-decay of the chromium species **A** and **B** was accompanied by partial destruction of the salen ligand.

The EPR spectra of species of type **A** in CH$_3$CN displayed unresolved hyperfine structures. If tightly coordinating donor ligands D was added (DMSO, DMF, H$_2$O, pyridine), well-resolved shf structures from the nitrogen nuclei were observed due to slower monodentate ligand exchange (Figure 3.31, Equation 3.4).

$$O = Cr^V(Salen)L' + D \leftrightarrow Cr^V(Salen)D + L' \tag{3.4}$$

(The nature of L' is to be discussed later.)

NMR and EPR Spectroscopy

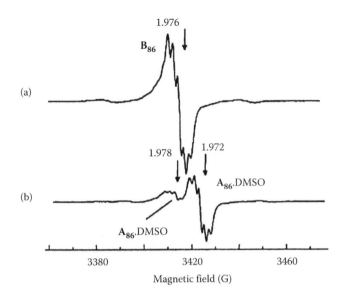

FIGURE 3.31 EPR spectra (CH$_3$CN, 17°C) of chromium(V) complexes formed upon stirring 16 μmol of complex **86** and 35 μmol of PhIO in 3 mL of CH$_3$CN: 150 min stirring (a) and addition of 140-fold excess of DMSO (b). (Reproduced with permission from Bryliakov, K. P., Talsi, E. P. 2003. *Inorg. Chem.* 42: 7258–7265.)

The reactivity of **A$_{87}$** in DMF toward *E*-β-methyl styrene was evaluated by EPR; the decay of **A$_{87}$** at differing *E*-β-methyl styrene concentrations demonstrated good accordance with a pseudo-first-order kinetics with a rate constant k_1, which was dependent on the substrate concentration (Figure 3.32):

$$k_1 = k_0 + k_2[\text{substrate}] \qquad (3.5)$$

For complex **A$_{87}$**, k_0 was $(6 \pm 2) \times 10^{-4}$ s^{-1} and $k_2 = 3.2 \pm 0.3 \times 10^{-3}$ L mol^{-1} s^{-1}.

On the basis of EPR, ^{1}H and ^{2}H NMR, and UV–Vis/NIR techniques, the intermediate **A** was assigned to a reactive complex Cr^V = O(salen)L (where L = Cl$^-$ or a solvent molecule), while intermediate **B** was concluded to be oxidation-inactive dinuclear chromium-salen complex, which acted as a reservoir for the active species. The latter complex demonstrated an EPR signal characteristic of oxochromium(V)-salen species, and ^{1}H NMR spectrum typical for chromium(III)-salen complexes, and was identified as mixed-valence binuclear complex L$_1$(salen)Cr^{III}OCr^V(salen)L$_2$ (L$_1$, L$_2$ = Cl$^-$ or solvent molecules) [122]. In support of this assignment, **B$_{86}$** was found to display a well-defined peak in the near-IR region (λ_{max} = 1075 nm), attributable to an intervalence charge transfer (ICT) transition in the bridged dinuclear complex. The proposed mechanism of chromium-salen-catalyzed epoxidation is presented in Scheme 3.7.

3.4.2 Oxocomplexes [Mn^V = O(Salen)]$^+$

The interest in manganese-based catalysts of selective oxidations increased drastically when the groups of Katsuki and of Jacobsen reported the manganese salen-type

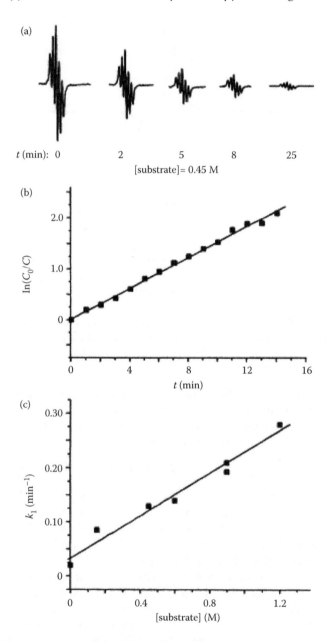

FIGURE 3.32 EPR spectra (17°C) of chromium(V) complex A_{87} in DMF at various moments of time after the addition of E-β-methyl styrene ([substrate] = 0.45 M) (a). Kinetic plot of A_{87} concentration in the above sample (b). Dependence of the first-order rate constant k_1 on the concentration of E-β-methyl styrene. (Reproduced with permission from Bryliakov, K. P., Talsi, E. P. 2003. *Inorg. Chem.* 42: 7258–7265.)

NMR and EPR Spectroscopy

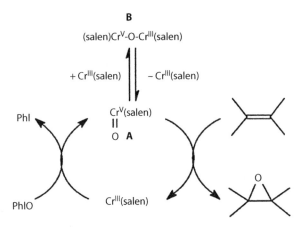

SCHEME 3.7 Proposed mechanism of chromium-salen-catalyzed epoxidation. (Reproduced with permission from Bryliakov, K. P., Talsi, E. P. 2003. *Inorg. Chem.* 42: 7258–7265.)

catalysts for the asymmetric epoxidation of unfunctionalized olefins with PhIO [123,124], followed by extensive mechanistic studies [125–131]. The major efforts were directed at Mn[III] complex **88a** (known as the Jacobsen's catalyst) (Figure 3.33).

The [(salen)Mn^V = O]$^+$ species was originally suggested by Kochi and coworkers as a possible active species of manganese-salen-catalyzed epoxidations by analogy to the well-characterized [(salen)Cr^V = O]$^+$ intermediate [132]. Later, direct evidence for the formation of [(salen)Mn^V = O]$^+$ was obtained by Plattner and coworkers by electrospray MS experiments [126]. Talsi and coworkers studied the mechanism of asymmetric epoxidation of styrene with various oxidants in the presence of catalyst **88a**, using EPR and NMR spectroscopy [128]. It was shown that EPR spectroscopy

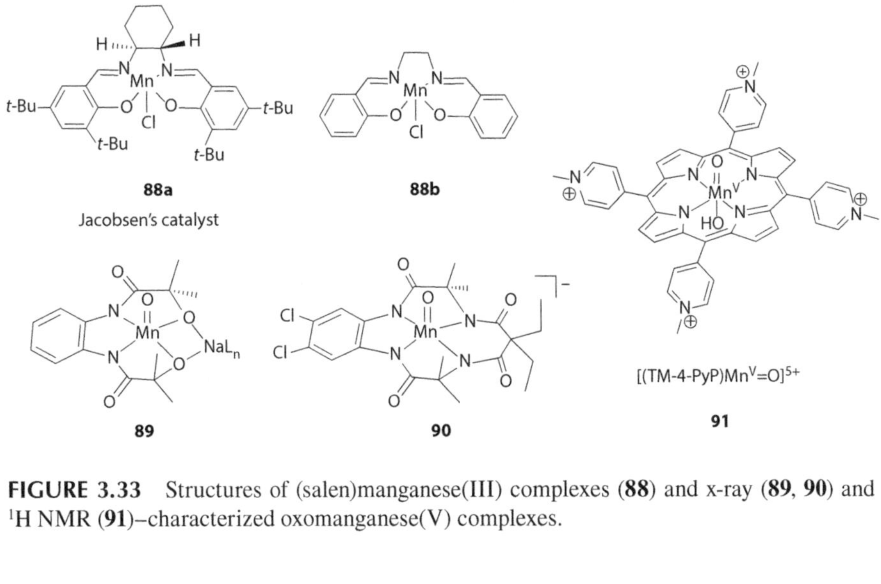

FIGURE 3.33 Structures of (salen)manganese(III) complexes (**88**) and x-ray (**89, 90**) and ^{1}H NMR (**91**)–characterized oxomanganese(V) complexes.

is suitable for the detection and identification of paramagnetic Mn^{II}, Mn^{III}, and Mn^{IV} species formed in the catalyst systems studied, whereas the ^1H NMR spectroscopy can be used for the detection of the low-spin ($S = 0$) oxomanganese(V) active species.

The X-band EPR spectrum of a frozen 0.1 M solution of Mn^{III} complex **88a** ($S = 2$) in CH_2Cl_2 at $-196°C$ is shown in Figure 3.34a. The weak signal at $g = 7.8 \pm 0.3$ belongs to the forbidden EPR transitions between the levels of the $|\pm 2\rangle$ non-Kramers doublet. Coordination of N-methylmorpholine-N-oxide (NMO) to **88a** changes the

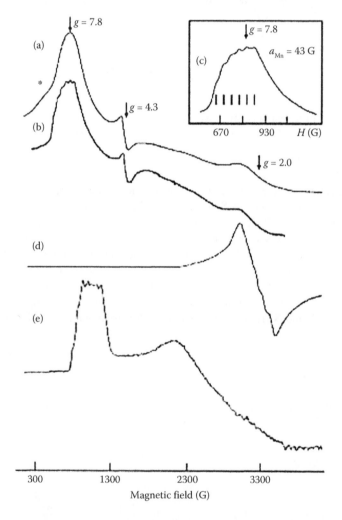

FIGURE 3.34 Frozen solution ($-196°C$) X-band EPR spectra of 0.05 M solutions of complex **88a** in CH_2Cl_2 (a) and in CH_2Cl_2 containing N-methylmorpholine-N-oxide ([NMO] = 1 M) (b, c); of the (salen)Mn^{II} precursor of complex **88a** in DMSO (d); and of the (salen)Mn^{IV} complexes, recorded 1 min after the onset of reaction of complex **88a** with one equivalent of m-chloroperoxybenzoic acid at 0°C (e). (Reproduced with permission from Bryliakov, K. P. et al. 2000. *J. Mol. Catal. A Chem.* 158: 19–35.)

NMR and EPR Spectroscopy

shape of its EPR spectrum so that the signal at $g = 7.8$ displays 6-line hyperfine structure A = 44 ± 3 G from one manganese nucleus ($I = 5/2$) (Figure 3.34b and c). The (salen)Mn^{II} ($S = 5/2$) precursor of **88a** exhibits an intense signal at $g \approx 2$, with partially resolved hfs from one manganese nucleus ($a_{Mn} = 87$ G, $I = 5/2$, 100% natural abundance) (Figure 3.34d). The amplification in the spectrum of Figure 3.34d is two orders of magnitude lower than that in Figure 3.34a–c [128].

The (salen)Mn^{IV} complex ($S = 3/2$) obtained via the reaction of **88a** with 1 equivalent of *meta*-chloroperoxybenzoic acid (*m*-CPBA) in CH_2Cl_2 at 0°C exhibits a resonance at $g = 5.4$, $a_{Mn} = 73$ G, typical for Mn^{IV} species with $D > h\nu$ [133] (Figure 3.34e).

Known structurally characterized Mn^V=O(L) complexes with tetra-anionic ligands **89** and **90** (Figure 3.33) are diamagnetic low-spin d^2 complexes and display sharp ^{1}H NMR resonances. These complexes are stable and inert toward the oxidation of organic substrates [134,135]. Groves and coworkers have detected by ^{1}H NMR unstable diamagnetic Mn intermediate in the catalyst system (TM-2-PyP)Mn^{III}Cl (**91**)/*m*-CPBA, and assigned it to the [(TM-2-PyP)Mn^V = O]$^{5+}$ complex (Figure 3.33) [136,137].

Inspired by those data, we undertook the search for the [(salen)Mn^V = O]$^+$ intermediate in the catalyst system **88a** + PhIO (at −20°C in CDCl$_3$) by means of the ^{1}H NMR spectroscopy. One could expect that this intermediate would be diamagnetic, like the structurally characterized isoelectronic complex [(salen)Mn^V≡N] [138]. Indeed, it was found that at the initial stage of interaction of **88a** with PhIO, unstable diamagnetic manganese complex with the proposed structure [(salen)Mn^V=O]$^+$ was formed, which displayed three resonances from the *t*Bu groups (1.68 (s, 9H), 1.64 (s, 9H), 1.42 (s, 18H), δ, CDCl$_3$, −20°C), similar to the [(salen)Mn^V≡N] (1.49 (s, 9H), 1.45 (s, 9H), 1.28 (s, 18H), δ, CDCl$_3$, −20°C) (other resonances of [(salen)Mn^V=O]$^+$ were masked by those of PhI) [128]. The concentration of the proposed [(salen)Mn^V=O]$^+$ species decreased rapidly in the presence of styrene at −20°C.

With *m*-CPBA as the oxidant, no species assignable to [(salen)Mn^V=O]$^+$ were detected; contrariwise, the major species at −70°C was assigned to an unstable (salen) Mn^{III}(OOCOAr) complex. The latter was suggested to act as the oxygen-transferring species [128]. The assumption that various oxidants can operate in (salen)Mn^{III}-based catalyst systems was confirmed by the study of chemo-, diastereo-, and enantioselective oxidation of olefins by various oxidants [129,139].

3.4.3 Oxocomplexes [(L)Fe^V=O]$^{3+}$ (L = Tetradentate N-Donor Ligand) as Proposed Active Species of Selective Epoxidation of Olefins

Selective oxidation of organic molecules is an important step of many biological and industrial processes. Nature employs a number of heme [140–144] and non-heme [145–151] iron enzymes to carry out such transformations. For heme enzymes such as cytochrome P450, formally, Fe^V=O species ((P$^{·+}$)Fe^{IV}=O, P$^{·+}$=porphyrin cation radical) has been accepted as the active intermediate of oxidation [140–144]. For non-heme iron enzymes such as α-KG-dependent dioxygenases, the high-spin ($S = 2$) Fe^{IV}=O intermediate is considered to be responsible for the selective oxidation [148].

96   Applications of EPR and NMR Spectroscopy in Homogeneous Catalysis

To date, significant efforts have been directed to the design of synthetic (biomimetic) catalyst systems, capable of mimicking the chemical reactivities of naturally occurring metalloenzymes. Iron complexes of the types **92–96** (Figure 3.35) with N_4-donor ligands are regarded as the best functional models of natural non-heme oxygenases [119,147,150,152–158]. One can expect that the catalytically active species, responsible for the oxidation of organic substrates by synthetic biomimetic catalyst systems, can be similar to those of the metalloenzyme-based systems (i.e., $Fe^{IV}=O$ or $Fe^{V}=O$ species). However, until now, for the existing synthetic non-heme catalyst systems, capable of preparative epoxidation of olefins with H_2O_2, only $Fe^{V}=O$ species have been considered as the active species of epoxidation, whereas $Fe^{IV}=O$ complexes were shown to be responsible for side reactions, deteriorating the epoxidation selectivity [120,159].

The first synthesis of an oxoiron(V) complex, $[(TAML)Fe^{V}=O]^-$ (**97**, see Figure 3.39), was reported in 2007 [160]. The reaction of $[(TAML)Fe^{III}(H_2O)]$ with *m*-CPBA in *n*-butyronitrile at −60°C yielded a deep green complex, identified as $[(TAML)Fe^{V}=O]^-$ by Mössbauer, EPR, EXAFS, and ESI-MS studies. This low-spin ($S = 1/2$) complex displayed a rhombically anisotropic EPR spectrum ($g_1 = 1.99$, $g_2 = 1.97$, $g_3 = 1.74$) and was capable of oxidizing a variety of substrates, such as thioanisole, styrene, cyclooctene, ethylbenzene, and 9,10-dihydroanthracene. Inspired by this discovery, we undertook the search for the $Fe^{V}=O$ intermediates in the catalyst systems **92–96**/H_2O_2. EPR spectroscopy was suitable for the detection of the

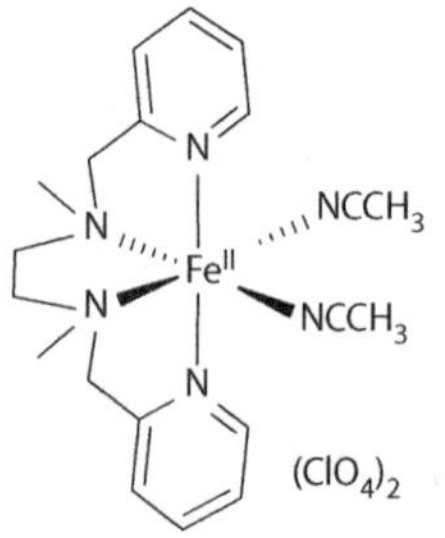

[(BPMEN)Fe^{II}(CH$_3$CN)$_2$](ClO$_4$)$_2$ [(TPA)Fe^{II}(CH$_3$CN)$_2$](ClO$_4$)$_2$ [(BQEN)Fe^{II}(OTf)$_2$]

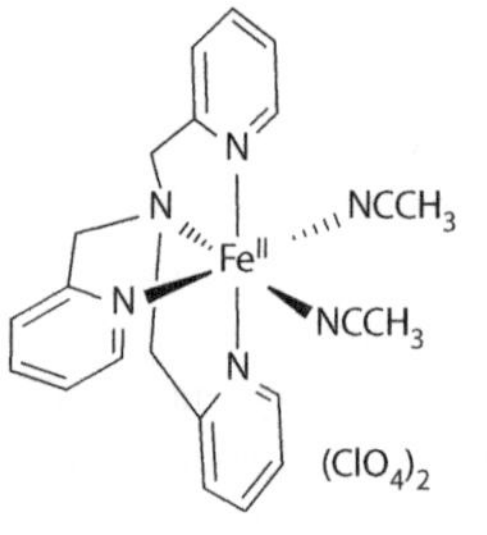

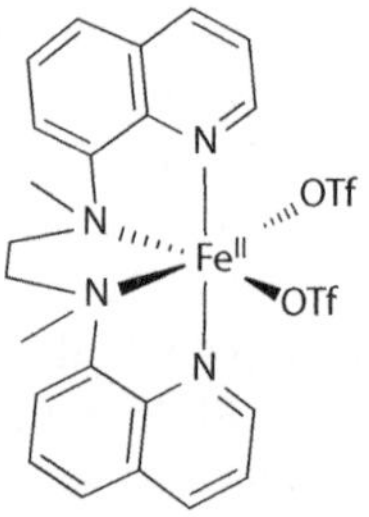

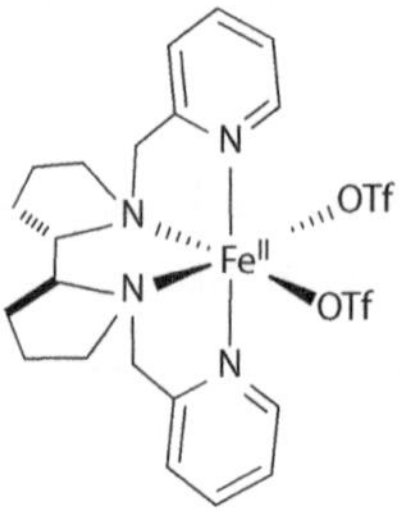

92   **93**   **94**

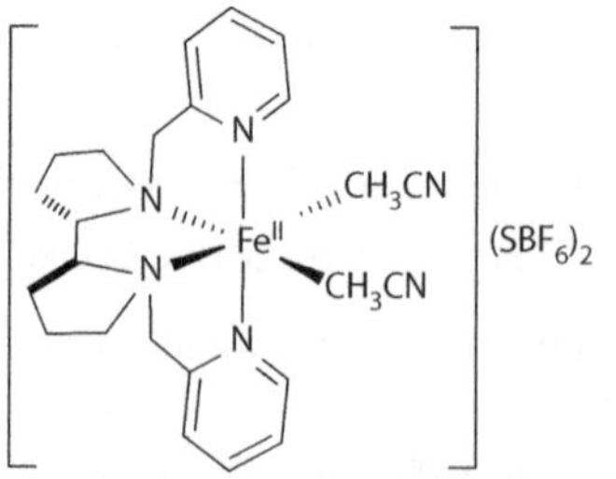

[((S,S)-PDP)Fe^{II}(CH$_3$CN)$_2$](SbF$_6$)$_2$ [(Me$_2$PyTACN)Fe^{II}(OTf)$_2$]

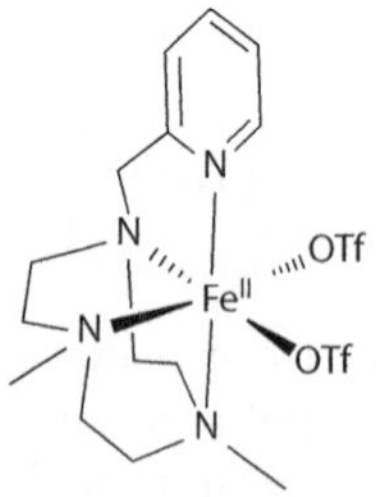

95   **95^{SbF6}**   **96**

FIGURE 3.35 Examples of nonheme iron complexes—efficient and selective catalysts of oxidation of alkenes and alkanes with H_2O_2 (**92–96**).

NMR and EPR Spectroscopy

$S = 1/2$ species formed in the catalyst systems **92–96**/H_2O_2 (e.g., Fe^{III}–OOH, Fe^{III}–OH and putative Fe^V=O intermediates), and ^1H and ^2H NMR spectroscopy was used to observe the Fe^{IV}=O intermediates in the catalyst systems studied [161–164].

The EPR spectrum of the sample frozen 30 s after the addition of 2 equivalents of H_2O_2 to the 0.027 M solution of **92** in a 1.7:1 CH_2Cl_2/CH_3CN mixture at −60°C displays resonances from the low-spin ($S = 1/2$) ferric species **92a** and **92b** (g values in the range of 1.9–2.4) and the resonance at $g = 4.2$ from an unidentified high-spin ($S = 5/2$) ferric species (Figure 3.36a). Storing this sample for 5 min at −70°C leads to the buildup of a new complex **92c** (Figure 3.36b) [162].

The EPR spectrum of the sample recorded 30 s after mixing **92** with 10 equivalents of H_2O_2 at −60°C predominantly displays resonances of complex **92b** (Figure 3.36c). Storing this sample for 5 min at −70°C results in an increase in the concentration of complexes **92c** and **92a** (Figure 3.36d). Overall, three types of low-spin iron species (**92a–c**) can be observed in the system **92**/H_2O_2 at low temperatures.

The expanded EPR spectrum of **92a** is presented in Figure 3.37a. This spectrum is a superposition of the spectra of two species: **92a-CH$_3$CN** ($g_1 = 2.218$, $g_2 = 2.175$, $g_3 = 1.966$) and **92a-H$_2$O** ($g_1 = 2.197$, $g_2 = 2.128$, $g_3 = 1.970$) (Table 3.5). The simulated spectrum (Figure 3.37b) is in excellent agreement with the experimental one (Figure 3.37a). Complexes **92a-CH$_3$CN** and **92a-H$_2$O** were assigned to hydroperoxo complexes [(BPMEN)Fe^{III}(η^1-OOH)(CH_3CN)]$^{2+}$ and [(BPMEN)Fe^{III}(η^1-OOH)(H_2O)]$^{2+}$, respectively [162]. ESI-MS data of Rybak-Akimova and Makhlynets have confirmed this assignment [165]. Complexes **92a-CH$_3$CN** and **92a-H$_2$O** do not directly react with cyclohexene. The rate of their self-decay at −70°C does not change in the presence of cyclohexene.

Complex **92b** ($g_1 = 2.43$, $g_2 = 2.21$, $g_3 = 1.91$) can be assigned to ferric hydroxo complex [(BPMEN)Fe^{III}(OH)(S)]$^{2+}$ ($S = H_2O$ or CH_3CN) since its EPR parameters are close to those of known ferric hydroxo complexes with N_4-donor ligand (Table 3.5) [166,167].

Importantly, in contrast to **92a** and **92b**, complex **92c** reacted with cyclohexene at −70°C to afford cyclohexene epoxide. A clear connection between the rate of decay of **92c** and the rate of cyclohexene epoxidation by the catalyst system **92**/H_2O_2 was documented [161]. Hence, it is reasonable to assign intermediate **92c** to the active species of epoxidation. Intermediate **93c** with similar EPR spectra was observed in the catalyst system **93**/H_2O_2/CH_3COOH (Table 3.5). The decay of **93c** accelerated with increasing cyclohexene concentration [161,162].

The catalyst system **92**/H_2O_2/CH_3COOH was the first example of non-heme iron-based systems that could be potentially useful for the preparative epoxidation of olefins [153]. Another milestone result was the synthesis and application of the catalyst [((S,S)-PDP)Fe^{II}(SbF$_6$)$_2$](**95^{SbF6}**) (Figure 3.35). The catalyst system **95^{SbF6}**/H_2O_2/CH_3COOH was capable of regioselective and stereospecific oxidation of aliphatic C–H groups in complex organic molecules, with selectivities in some cases approaching those of enzymatic reactions [154,155].

The EPR spectrum of the sample **95**/H_2O_2 = 1:2.5, frozen 1 min after the addition of H_2O_2 to the solution of **95** in a 1.7:1 CH_2Cl_2/CH_3CN mixture at −70°C, displays resonances from hydroperoxo complexes **95a-CH$_3$CN** and **95a-H$_2$O** (g-values in the range of 1.96–2.20), and resonances in the range of $g = 4.2$–8 from an unidentified high-spin

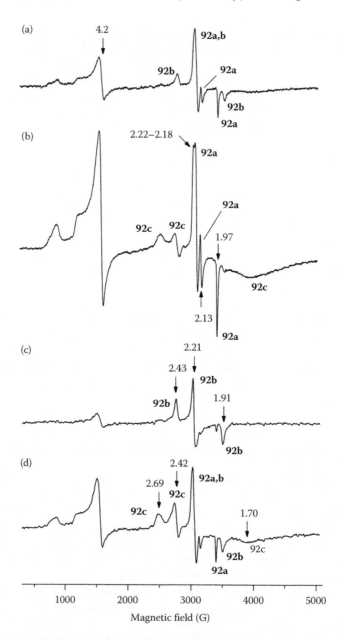

FIGURE 3.36 EPR spectra (−196°C) of the sample **92**/H$_2$O$_2$ ([H$_2$O$_2$]:[**92**] = 2, [**92**] = 0.027 M) frozen 30 s after mixing the reagents at −60°C (a) and 5 min after storing the sample in "A" at −70°C (b). EPR spectra (−196°C) of the sample **92**/H$_2$O$_2$ ([H$_2$O$_2$]:[**92**] = 10, [**92**] = 0.027 M) frozen 30 s after mixing the reagents at −60°C (c) and 5 min after storing the sample in "C" at −70°C (d). A 1:1.7 CH$_3$CN/CH$_2$Cl$_2$ mixture was used as a solvent. (Reproduced with permission from Lyakin, O. Y. et al. 2011. *Inorg. Chem.* 50: 5526–5538.)

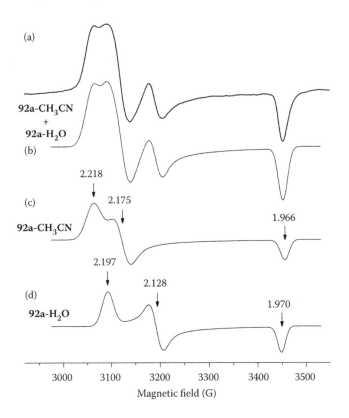

FIGURE 3.37 The expanded EPR spectrum of **92a** (a). Simulated superposition of spectra of **92a-CH₃CN** and **92a-H₂O** ([**92a-CH₃CN**]:[**92a-H₂O**] = 1.85) (b). Simulated spectrum of **92a-CH₃CN** ($g_1 = 2.218$, $g_2 = 2.175$, $g_3 = 1.966$; individual line widths $\Delta H_1 = 15$ G, $\Delta H_2 = 14$ G, $\Delta H_3 = 9$ G; Gaussian line shape) (c). Simulated spectrum of **92a-H₂O** ($g_1 = 2.197$, $g_2 = 2.128$, $g_3 = 1.970$; individual line widths $\Delta H_1 = 11.5$ G, $\Delta H_2 = 11.5$ G, $\Delta H_3 = 8$ G; Gaussian line shape) (d). (Reproduced with permission from Lyakin, O. Y. et al. 2011. *Inorg. Chem.* 50: 5526–5538.)

($S = 5/2$) ferric species (Figure 3.38a, Table 3.5). The increase in the [H₂O₂]:[**95**] ratio of the sample **95**/H₂O₂ results in the rise in the concentration of **95a** and in the appearance of resonances of a low-spin ($S = 1/2$) iron–oxygen species **95c** (Figure 3.38b, Table 3.5). The EPR parameters of **95c** are very close to those of the active intermediate of epoxidation **92c** (Table 3.5). The situation changes abruptly in the presence of acetic acid. The EPR spectrum of the sample **95**/H₂O₂/CH₃COOH = 1:2.5:15 (Figure 3.38c) shows predominantly the resonances of **95c**. Apparently, acetic acid promotes the conversion of hydroxo complex **95a** to the active species of epoxidation **95c**.

Chiral complex **95** catalyzed the chemo- and enantioselective epoxidation of chalcone (by the system **95**/H₂O₂/RCOOH) with up to 86% *ee* [163]. Remarkably, the epoxidation yield was improved by the addition of a carboxylic acid, and the epoxidation enantioselectivity increased monotonously in line with rising steric bulk of the carboxylic acid (Table 3.6). Apparently, the role of carboxylic acid additive

TABLE 3.5
EPR Spectroscopic Data for $S = 1/2$ Iron–Oxygen Species Formed in the Systems Discussed Herein in Comparison with Those for Related Complexes[a]

Complex	g_1	g_2	g_3	Reference
[(BPMEN)Fe^{III}(OOH)(CH$_3$CN)]$^{2+}$ (**92a-CH$_3$CN**)	2.218	2.175	1.966	[162]
[(BPMEN)Fe^{III}(OOH)(H$_2$O)]$^{2+}$ (**92a-H$_2$O**)	2.197	2.128	1.970	[162]
[(BPMEN)Fe^{III}(OH)(S)]$^{2+}$ (**92b**)[b]	2.43	2.21	1.91	[162]
[(N$_4$Py)Fe^{III}(OH)]$^{2+}$	2.41	2.15	1.92	[166]
[(TPEN)Fe^{III}(OH)]$^{2+}$	2.39	2.19	1.91	[167]
[(BPMEN)Fe^{V}=O(OC(O)CH$_3$)]$^{2+}$ (**92c**)	2.69	2.42	1.70	[161,162]
[(TPA)Fe^{III}(OOH)(CH$_3$CN)]$^{2+}$ (**93a-CH$_3$CN**)	2.194	2.152	1.970	[162]
[(TPA)Fe^{III}(OOH)(H$_2$O)]$^{2+}$ (**93a-H$_2$O**)	2.19	2.12	1.97	[162]
[(TPA)Fe^{V}=O(OC(O)CH$_3$)]$^{2+}$ (**93c**)	2.71	2.42	1.53	[161,162]
[((S,S)-PDP)Fe^{III}–OOH(CH$_3$CN)]$^{2+}$ (**95a-CH$_3$CN**)	2.206	2.171	1.955	[163]
[((S,S)-PDP)Fe^{III}–OOH(H$_2$O)]$^{2+}$ (**95a-H$_2$O**)	2.191	2.124	1.963	[163]
[((S,S)-PDP)Fe^{II}(OH)(S)]$^{2+}$ (**95b**)[b]	2.44	2.21	1.89	[163]
[((S,S-PDP)Fe^{V}=O(OC(O)CH$_3$)]$^{2+}$ (**95c**)	2.66	2.42	1.71	[163]
[(TMC)Fe^{V}=O(NC(O)CH$_3$)]$^{+}$	2.053	2.010	1.971	[172]
[(TAML)Fe^{V}=O]$^{-}$	1.99	1.97	1.74	[160]

[a] EPR spectra were recorded at $-196°C$ or lower. TAML = macrocyclic tetraamide ligand, N4Py = N,N-bis(2-pyridylmethyl)-N-(bis-2-pyridylmethyl)amine, TPEN = N,N,N',N'-tetrakis(2-pyridylmethyl)-ethylenediamine.

[b] S = CH$_3$CN or H$_2$O.

is dual: it assists the activation of hydrogen peroxide and affects the epoxidation enantioselectivity; obviously, the acid molecule is incorporated into the active species at the enantioselectivity-determining step, presumably acting as an auxiliary ligand. The active species is most plausibly represented as [((S,S)-PDP)Fe^{V}=O(OCOR)]$^{2+}$ (**95c**, Figure 3.39) [163]. A similar active species [(TPA)Fe^{V}=O(OCOCH$_3$)]$^{2+}$ (**93c**, Figure 3.39) was previously proposed for the catalyst system **93**/H$_2$O$_2$/CH$_3$COOH on the basis of the catalytic and ^{18}O labeling studies [168].

In 2011, the first successful trapping of the elusive HO–Fe^{V}=O intermediate in a practical catalyst system **96**/H$_2$O$_2$ was reported, using cryospray-assisted variable-temperature mass spectrometry (VT-MS) [169]. The presence of the oxoferryl moiety was confirmed by isotopic (^{18}O) labeling experiments, and the reactivity of the intermediate was confirmed by the *in situ* observation of the reaction with cyclooctene. The same system was subsequently studied by EPR spectroscopy, and the reactive species was found to be a low-spin ($S = 1/2$) complex **96c** with EPR parameters ($g_1 = 2.66$, $g_2 = 2.43$, $g_3 = 1.74$) close to those of the intermediates **92c**, **93c**, and **95c** (Figure 3.39, Table 3.5) [164]. These data support the assignment of intermediates **92c**, **93c**, **95c**, and **96c** to the oxoiron(V) species.

It is important to directly compare the reactivity of the putative Fe^{V}=O and Fe^{IV}=O intermediates in the catalyst systems studied. It was found that the catalyst system

NMR and EPR Spectroscopy

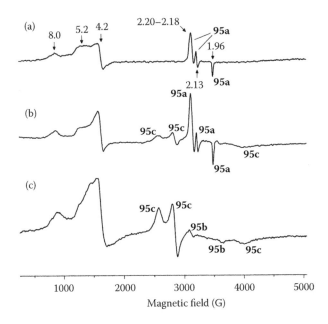

FIGURE 3.38 EPR spectra (1.7:1 CH$_2$Cl$_2$/CH$_3$CN, −196°C) of samples: (a) **95**/H$_2$O$_2$ (1:2.5), (b) **95**/H$_2$O$_2$ (1:10), and (c) **95**/H$_2$O$_2$/CH$_3$COOH (1:2.5:15), frozen 1 min after mixing the reagents at −70°C ([**95**] = 0.05 M).

TABLE 3.6
Asymmetric Epoxidation of Chalcone with H$_2$O$_2$ Catalyzed by Complex 95: Effect of Added Carboxylic Acid[a]

Entry	Catalyst (mol %)	Additive[b]	Conversion (%)/ Epoxide Yield (%)	ee (%)[c]
1	**95** (1.0)	–	13/13	61
2		AA	77/75	62
3		BA	65/63	64
4		CA	68/65	64
5		IBA	85/85	73
6		CHA	91/90	73
7		EBA	89/89	82
8		EHA	90/89	82
9		VPA	92/92	83

[a] The reactions were performed at 0°C; [H$_2$O$_2$]/[substrate]/[RCOOH] = 2.0:1.0:0.55; H$_2$O$_2$ added by a syringe pump over 30 min. Conversions and epoxide yields are given based on the substrate.

[b] AA = acetic acid, BA = n-butyric acid, CA = n-caproic acid, IBA = iso-butyric acid, CHA = cyclohexanecarboxylic acid, EBA = 2-ethylbutyric acid, EHA = 2-ethylhexanoic acid, VPA = 2-propylvaleric acid (Figure 3.44).

[c] Absolute configuration of chalcone epoxide was (2R,3S).

[(BPMEN)Fe^{V}=O(OAc)]$^{2+}$
92c

[(TPA)Fe^{V}=O(OAc)]$^{2+}$
93c

[((*S*,*S*)-PDP)Fe^{V}=O(OAc)]$^{2+}$
95c

[(Me$_2$PyTACN)Fe^{V}=O(OH)]$^{2+}$
96c

[(TAML)Fe^{V}=O]$^{-}$
97

[(TMC)Fe^{V}=O(NC(O)CH$_3$)]$^{+}$
98

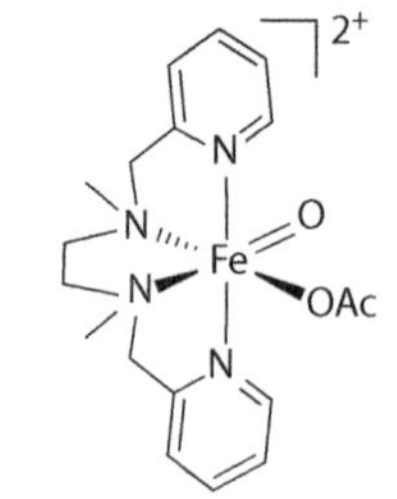

FIGURE 3.39 The putative oxoiron(V) intermediates **92c–96c**, and the spectroscopically characterized oxoiron(V) intermediates **97** and **98**.

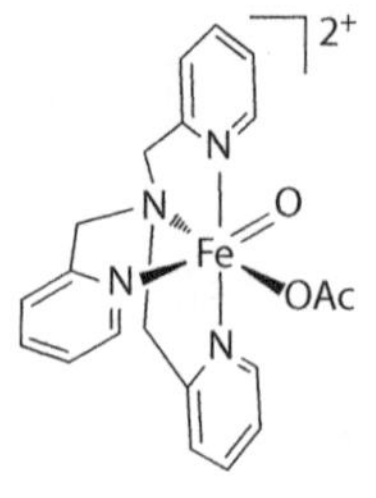

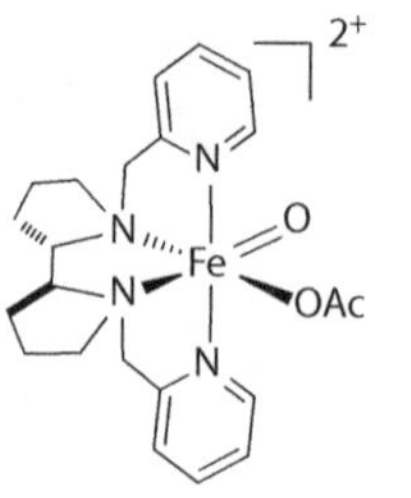

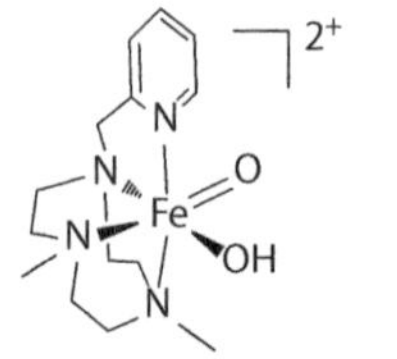

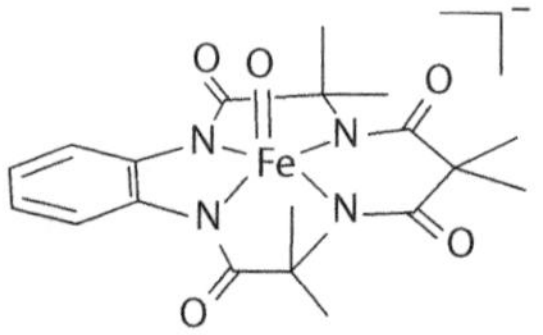

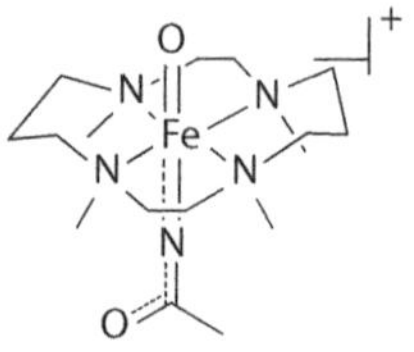

92/PhIO displays characteristic ^{1}H NMR peaks from oxoiron(IV) intermediate [(BPMEN)Fe^{IV}=O]$^{2+}$ (**92d**) (Figure 3.40), and no EPR resonances from the putative oxoiron(V) intermediate [(BPMEN)Fe^{V}=O]$^{3+}$ (**92c**). In contrast, catalyst system **92**/H_2O_2 shows resonances from **92c** (Figure 3.36b and d) but no signals from **92d** [162]. It was shown that **92c** is much more reactive toward cyclohexene at −70°C compared

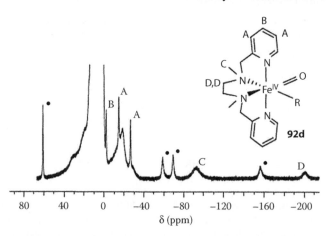

FIGURE 3.40 ^{1}H NMR spectrum (−70°C) of the sample **92**/PhIO ([Fe]:[PhIO] = 1:1) 5 min after mixing the reagents at −40°C in a 1.7:1 CD_2Cl_2/CD_3CN mixture. Assigned signals of **92d** are denoted by letters, and unassigned are marked by points. R = CH_3CN. (Reproduced with permission from Lyakin, O. Y. et al. 2011. *Inorg. Chem.* 50: 5526–5538.)

NMR and EPR Spectroscopy

with **92d**. The intensities of the ^{1}H NMR peaks of **92d** decreased twofold after 1 h storing the sample containing **92d** and 12 equivalents of cyclohexene, whereas the EPR peaks of **92c** completely disappeared after 10 min under the same conditions. Cyclohexene oxide was the major oxidation product of the catalyst system **92**/H_2O_2/C_6H_{10}, while mostly allylic oxidation products (i.e., 2-cyclohexen-1-one and 2-cyclohexen-1-ol) were observed in the catalysts system **92**/PhIO/C_6H_{10} [162]. Hence, **92c** is able to selectively epoxidize cyclohexene, whereas **92d** mainly drives its allylic oxidation. Reactivity studies of the well-characterized intermediates [(TPA)Fe^{IV}=O]$^{2+}$ (**93d**) [168], [(BQEN)Fe^{IV}=O]$^{2+}$ (**94d**) [170], and [(Me$_2$PyTACN)Fe^{IV}=O]$^{2+}$ (**96d**) [171] also showed that the Fe^{IV}=O species cannot drive the selective epoxidation of olefins. Therefore, oxoiron(V) intermediates are the likely active species of the catalyst systems based on complexes **92–96** and H_2O_2. We have assumed the highly unstable and reactive complexes **92c–96c** (Figure 3.39) to be these particular intermediates.

However, the EPR parameters of **92c–96c** drastically differ from those of known oxoiron(V) complexes [(TAML)Fe^V=O]$^-$ (**97**) [160] and [(TMC)Fe^V=O(NC(O)CH$_3$)]$^+$ (**98**) [172] (Figure 3.39, Table 3.5). Confused by this apparent discrepancy, Wang et al. hypothesized that **92c–96c** were ferric-peracetate complexes Fe^{III}–OOC(O)CH$_3$ rather than oxoiron(V) complexes [173]. In 2014, Oloo et al. reported the detection of the intermediate **X** ($g_1 = 2.58$, $g_2 = 2.38$, $g_3 = 1.72$) in the catalyst system **93***/H_2O_2/CH$_3$COOH in acetonitrile at $-40°C$. In contrast to complex **93** with unsubstituted TPA ligand, the TPA* ligand of **93*** contained donor substituents at the pyridine rings (Figure 3.41). By a combination of EPR, Mössbauer, resonance Raman, and ESI-MS

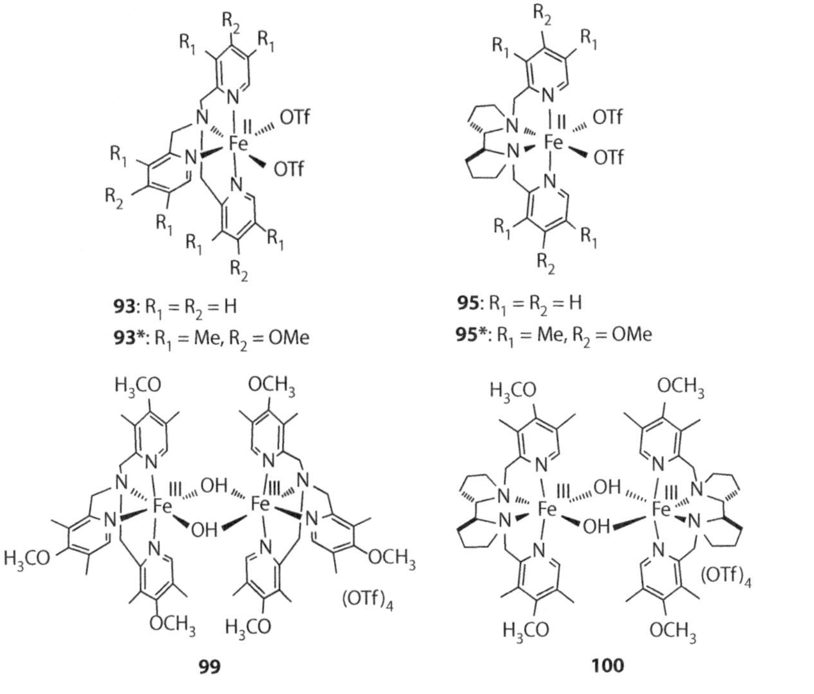

93: R$_1$ = R$_2$ = H
93*: R$_1$ = Me, R$_2$ = OMe

95: R$_1$ = R$_2$ = H
95*: R$_1$ = Me, R$_2$ = OMe

99

100

FIGURE 3.41 Iron complexes with aminopyridine ligands bearing electron-donating groups.

104   Applications of EPR and NMR Spectroscopy in Homogeneous Catalysis

methods, X was identified as a low-spin acylperoxoiron(III) species $[(TPA^*)-Fe^{III}(\eta^2-OOC(O)CH_3)]^{2+}$. The rate of decay of complex X at $-40°C$ was unaffected by the presence of cyclooctene, 1-octene, or 2-heptene in the reaction solution, ruling out the assignment of X to the actual oxygen-transferring species [174].

The g-values of X resemble those for complexes **92c–96c** (Table 3.5). However, the assignment of **92c–96c** to ferric-peracetate complexes $Fe^{III}-OOC(O)CH_3$ similar to X [173] would be incorrect. Species **92c**, **95c**, and **96c** can also be observed in the catalyst systems **92**, **95**, and **96**/H_2O_2, without any additives of CH_3COOH [161,163,164], and thus cannot be acylperoxoiron(III) species. Moreover, the decay of **92c** is accelerated in the presence of cyclohexene at $-70°C$ [161], whereas the rate of decay for complex X at $-40°C$ was unaffected by the presence of cyclohexene in the reaction solution [174]. To shed light on the possible reason of the sharp difference of the EPR parameters of **92c–96c** and of known $Fe^V{=}O$ species (Table 3.5), we undertook the search for the true active species in the catalyst systems **99**/H_2O_2/CH_3COOH and **100**/H_2O_2/CH_3COOH. Dinuclear ferric complexes **99** and **100** can be used instead of the corresponding mononuclear ferrous complexes **93*** and **95*** as the precatalysts (Figure 3.41) because it was shown that the first 0.5 equivalents of H_2O_2 rapidly convert **93*** and **95*** into **99** and **100**, respectively.

3.4.4 EPR SPECTROSCOPIC DETECTION OF THE ELUSIVE $Fe^V{=}O$ INTERMEDIATES IN SELECTIVE CATALYTIC EPOXIDATION OF OLEFINS MEDIATED BY FERRIC COMPLEXES WITH SUBSTITUTED AMINOPYRIDINE LIGANDS

The starting dinuclear complex **99** is EPR-silent. Just after the addition of 50 equivalents of CH_3COOH to the 0.05 M solution of **99** in CH_2Cl_2/$CH_3CN = 1.2:1$ (by volume) at room temperature and freezing the sample at $-196°C$, the EPR signal of a new ferric complex with the proposed structure $[(TPA^*)Fe^{III}(\eta^2-OC(O)CH_3)]^{2+}$ (**101**) is observed (Figure 3.42). This complex displays resonances at $g_1 = 2.58$, $g_2 = 2.38$, and $g_3 = 1.72$ [175]. We note that related complexes $[(TPA)Fe^{III}(acac)]^{2+}$ and $[(5-Me_3TPA)Fe^{III}(acac)]^{2+}$ (Hacac = acetylacetone) show very similar EPR spectra ($g_1 = 2.57$, $g_2 = 2.35$, and $g_3 = 1.71$) [176]. The EPR spectrum of **101** disappears with the half-life time $\tau_{1/2} = 2$ min at room temperature. Besides the resonances of **101**, the EPR spectrum of Figure 3.42a exhibits the intense resonances at $g = 5.6$ and $g = 4.3$ from unidentified stable high-spin ($S = 5/2$) ferric species. The EPR spectrum ($-196°C$) of the sample **99**/H_2O_2/$CH_3COOH = 1:6:20$ recorded 1.5 min after mixing the reagents at $-80°C$ exhibits resonances of residual **101** and a rhombic spectrum $g_1 = 2.070$, $g_2 = 2.005$, and $g_3 = 1.956$ from a new species **99c** (Figure 3.42b, Table 3.5) [176]. The latter is much less temperature stable ($\tau_{1/2} = 5$ min at $-85°C$) as compared to intermediate X ($\tau_{1/2} = 5$ min at $-40°C$ [174]). The evaluated maximum steady-state concentration of **99c** approaches ca. 2–3% of the total iron concentration.

To directly assess the reactivity of **99c** toward olefin epoxidation, the rates of decay of **99c** in the absence and in the presence of various olefins were compared. It was found that the addition of relatively small amounts of cyclohexene or *cis*-β-methylstyrene ([alkene]/[Fe] = 0.5) to the sample **99**/H_2O_2/$CH_3COOH = 1:6:20$ ([**99**] = 0.02 M) results in the drop of the half-life time of **99c** from $\tau_{1/2} = 5$ min to $\tau_{1/2} < 0.5$ min at $-85°C$, whereas electron-deficient alkenes (1-acetyl-1-cyclohexene

NMR and EPR Spectroscopy

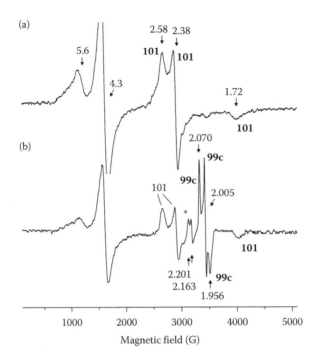

FIGURE 3.42 EPR spectra (−196°C) of the sample (a) **99**/CH$_3$COOH ([CH$_3$COOH]/[**99**] = 50, [**99**] = 0.05 M) frozen 1 min after mixing the reagents at room temperature in a 1.2:1 CH$_2$Cl$_2$/CH$_3$CN mixture, and (b) **99**/H$_2$O$_2$/CH$_3$COOH ([**99**]:[H$_2$O$_2$]:[CH$_3$COOH] = 1:6:20, [**99**] = 0.02 M) frozen 1.5 min after mixing the reagents at −80°C in a 1.8:1 CH$_2$Cl$_2$/CH$_3$CN mixture. (Reproduced with permission from Lyakin, O. Y. et al. 2015. *ACS Catal.* 5: 2702–2707.)

and cyclohexene-1-carbonitrile with an [alkene]/[Fe] ratio up to 10) do not appreciably affect the rate of decay of **99c**. In the presence of 1-octene as a substrate, the pseudo-first-order rate constant versus the amount of added 1-octene was measured. A linear correlation was observed, revealing the second-order rate constant $k_2 = 3.2 \times 10^{-3}$ M^{-1} s^{-1}. Hence, **99c** is highly reactive toward electron-rich alkenes even at −85°C.

The *g*-values of **99c** drastically differ from those of intermediates **92c–96c** (Table 3.5). Apparently, this difference is associated with the presence of electron-donating substituents at the pyridine rings of **99**. To verify this assumption, we undertook the search for the intermediate similar to **99c** in a structurally related catalyst system **100**/H$_2$O$_2$/CH$_3$COOH. It was found that EPR spectrum (−196°C) recorded just after the addition of 20 equivalents of CH$_3$COOH to the solution of **100** in CH$_2$Cl$_2$/CH$_3$CN = 1.8:1 at room temperature exhibits a rhombic EPR spectrum of a new complex **102** (Figure 3.43a, Table 3.7). Complex **102** is relatively stable and can be observed even after 30 min of storing the sample at room temperature. This species can be assigned to the ferric complex [((S,S)-PDP*)Fe^{III}(η^2-OC(O)CH$_3$)]$^{2+}$ by analogy with complex **101**. After the addition of 6 equivalents of H$_2$O$_2$ to the

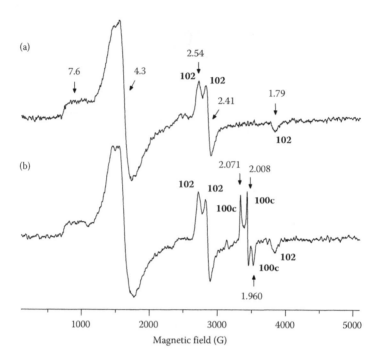

FIGURE 3.43 EPR spectra (1.8:1 CH$_2$Cl$_2$/CH$_3$CN, −196°C) (A) of the sample **100**/CH$_3$COOH ([CH$_3$COOH]/[**100**] = 20, [**100**] = 0.02 M) frozen 1 min after mixing the reagents at room temperature and (B) frozen after addition of 6 equivalents of H$_2$O$_2$ to the sample in "A" at −75°C and mixing the reagents during 1 min. (Reproduced with permission from Lyakin, O. Y. et al. 2015. *ACS Catal.* 5: 2702–2707.)

TABLE 3.7
EPR Spectroscopic Data for $S = 1/2$ Iron Species Formed in the Catalyst Systems Based on Iron Complexes with Tetradentate N_4-Donor Ligands

Number	Compound	g_1	g_2	g_3	Reference
1	[(TMC)Fe^V=O(NC(O)CH$_3$)]$^+$	2.053	2.010	1.971	[172]
2	[(TAML)Fe^V=O]$^−$	1.99	1.97	1.74	[160]
3	[(TPA)Fe^V=O(OC(O)CH$_3$)]$^{2+}$ (**93c**)	2.71	2.42	1.53	[161]
4	[(BPMEN)Fe^V=O(OC(O)CH$_3$)]$^{2+}$ (**92c**)	2.69	2.42	1.70	[161]
5	[((S,S)-PDP)Fe^V=O(OC(O)CH$_3$)]$^{2+}$ (**95c**)	2.66	2.42	1.71	[163]
6	[(Me$_2$PyTACN)Fe^V=O(OH)]$^{2+}$ (**96c**)	2.66	2.43	1.74	[164]
7	[(TPA*)Fe^V=O(OC(O)CH$_3$)]$^{2+}$ (**99c**)	2.070	2.005	1.956	[175]
8	[((S,S)-PDP*)Fe^V=O(OC(O)CH$_3$)]$^{2+}$ (**100c**)	2.071	2.008	1.960	[175]
9	[(TPA*)Fe^{III}(κ^2-OC(O)CH$_3$)]$^+$ (**101**)	2.58	2.38	1.72	[175]
10	[((S,S)-PDP*)Fe^{III}(κ^2-OC(O)CH$_3$)]$^{2+}$ (**102**)	2.54	2.41	1.79	[175]

TMC = tetramethylcyclam, TAML = macrocyclic tetraamide ligand.

NMR and EPR Spectroscopy

sample of Figure 3.43a at $-75°C$ and mixing the reagents during 1 min, the EPR resonances of a new intermediate **100c** appear (Figure 3.43b, Table 3.7). The g values of **100c** ($g_1 = 2.071$, $g_2 = 2.008$, $g_3 = 1.960$) are very close to those of **99c** ($g_1 = 2.070$, $g_2 = 2.005$, $g_3 = 1.956$). The half-life time of **100c** (2 min at $-80°C$) is rather close to that for **99c** ($\tau_{1/2} = 5$ min at $-80°C$). The evaluated maximum concentration of **100c** did not exceed 1% of the total iron concentration.

To evaluate the reactivity of **100c** toward olefins, the sample (**100**/H_2O_2/ $CH_3COOH = 1:6:20$, **[100]** = 0.02 M) containing the preliminarily generated **100c** was prepared. An olefin was added to this sample at $-90°C$ within 1 min, and decay of **100c** was monitored at $-85°C$. In the absence of olefins or in the presence of 1-acetyl-1-cyclohexene and cyclohexene-1-carbonitrile, **92** decayed with $\tau_{1/2} = 4$ min at $-85°C$. At the same time, the EPR resonances of **100c** disappeared within <0.5 min in the presence of cyclohexene, cis-β-methylstyrene, and 1-octene at the same temperature ([alkene]/[Fe] = 0.5), demonstrating the direct reactivity of **100c** toward electron-rich alkenes.

Both **99c** and **100c** are electrophilic oxidants. To quantify the products of the interaction of **99c** and **100c** with cyclohexene, catalytic oxidation experiments at $-85°C$ were performed. Fifteen minutes after the reaction onset with catalysts **99** and **100**, 4.5 and 5.3 TN cyclohexene oxide were obtained (TN = moles of product per mole of Fe) [176]. The total amount of other cyclohexene oxidation products was $<1\%$ with respect to the epoxide. Thus, the true active species **99c** and **100c** oxidize cyclohexene even at $-85°C$, demonstrating $>99\%$ epoxide selectivity. The reactivity of **99c** and **100c** resembles that of **92c**, which epoxidizes cyclohexene at $-70°C$ [161]. One can assume that **99c**, **100c**, and **92c** are chemically similar intermediates. However, the EPR spectra of **99c** and **100c** demonstrate a relatively small g-factor anisotropy ($g_{max}-g_{min} = 0.1$) and sharp resonances ($\Delta H_{1/2} = 15–40$ G), whereas **92c–96c** display highly anisotropic EPR spectra ($g_{max}-g_{min} = 1.0–1.2$) with much broader resonances ($\Delta H_{1/2} = 100–350$ G) (Table 3.7).

The g-values and line widths of **99c** and **100c** are very similar to those of the spectroscopically well-characterized oxoiron complex [(TMC)Fe^V=O(NC(O) CH$_3$)]$^+$ (Table 3.7). The great difference in the EPR spectra of **99c**, **100c**, and of the species **92c–96c** (Table 3.7) implies that the electronic structures of those intermediates differ significantly. It is well known that paramagnetic metal complexes demonstrate larger g-factor anisotropy and broader EPR resonances than organic radicals, owing to the occurrence of relatively low-lying excited states and stronger spin–orbit coupling [177]. One can expect that iron complexes with the unpaired spin mainly resting at the metal should display broader resonances and larger g-factor anisotropy than complexes with the unpaired spin mostly delocalized over the ligand. As an example, the EPR parameters of complex with the proposed structure [(TBP$_8$Cz$^{+\bullet}$)Fe^{IV}=O] ($g_1 = 2.09$, $g_2 = 2.05$, $g_3 = 2.02$, and $g_{max}-g_{min} = 0.07$; antiferromagnetic coupling between Fe^{IV}=O ($S = 1$) and TBP$_8$Cz$^{+\bullet}$ ($S = 1/2$) is suggested) drastically differ from those of the $S = 1/2$ complex [(TBP$_8$Cz)Fe^{III}(Py)]: $g_1 = 2.39$, $g_2 = 2.20$, $g_3 = 1.90$, and $g_{max}-g_{min} = 0.49$ (TBP$_8$Cz = octakis(4-$tert$-butylphenyl) corrolazinato) [178]. It is also worth mentioning that antiferromagnetic coupling between Fe^{IV}=O ($S = 1$) and a ligand cation radical ($S = 1/2$) is typical for compounds I of heme enzymes, for example, CYP119, the thermophilic cytochrome

108   Applications of EPR and NMR Spectroscopy in Homogeneous Catalysis

P450 from *Sulfolobus acidocaldarius*. The reaction of ferric CYP119 with an excess of *m*-chloroperoxybenzoic acid produces $S = 1/2$ CYP119 compound I with EPR parameters $g_1 = 2.00$, $g_2 = 1.96$, and $g_3 = 1.86$ [179]. Very recently, complex $[(Me_3tacn)Fe^{III}(Cl-acac)Cl]^+$ (Me_3tacn = 1,4,7-trimethyl-1,4,7-triazacyclononane, Cl-acac = 3-chloro-acetylacetonate) has been used for the oxidation of hydrocarbons with oxone, including cyclohexane, propane, and ethane. ESI-MS, EPR, and UV–Vis spectroscopy, ^{18}O labeling experiments, and DFT studies pointed to $[(Me_3tacn)Fe^{IV}=O(\{Cl-acac\}^{+\bullet})]^{2+}$ as the catalytically active species [180]. The EPR parameters of this complex ($g_1 = 1.97$, $g_2 = 1.93$, $g_3 = 1.91$) are rather close to those of CYP119 compound I.

On the basis of these data, it seems reasonable to assign the intermediates **99c** and **100c** to species of the type $[(L^{+\bullet})Fe^{IV}=O(OC(O)CH_3)]^{2+}$. We believe that the presence of strong electron donors in the structures of **99** and **100** may favor the intramolecular redistribution of electron density, leading to the reduction of iron to the formal +4 oxidation state, the positive charge being spread over the substituted aromatic rings in structures of the type $[(L^{+\bullet})Fe^{IV}=O(OC(O)CH_3)]^{2+}$. In these structures, the unpaired spin is mostly delocalized over the organic ligand framework, thus accounting for the observed low *g*-tensor anisotropy. On the contrary, this scenario is unlikely for intermediates **92c–96c** due to the lack of electron density at their unsubstituted pyridine rings. We conclude that **99c**, **100c**, and $[(TMC)Fe^V=O(NC(O)CH_3)]^+$ may be better represented as $(L^{+\bullet})Fe^{IV}=O$ rather than $(L)Fe^V=O$; in contrast, **92c–96c** are most likely $(L)Fe^V=O$ species. Apparently, further studies on a broader set of complexes and carboxylic acids are needed to verify this assumption.

The obtained data clearly show that active iron–oxygen intermediates with EPR parameters close to those of known $Fe^V=O$ species can be detected in bioinspired catalyst systems for chemo- and stereoselective oxidation of organic substrates with hydrogen peroxide. On the basis of our data and milestone results of other research groups [181,182], the following mechanism of alkene epoxidation by the catalyst systems studied was proposed (Scheme 3.8). At the initial stage, dimeric iron complexes

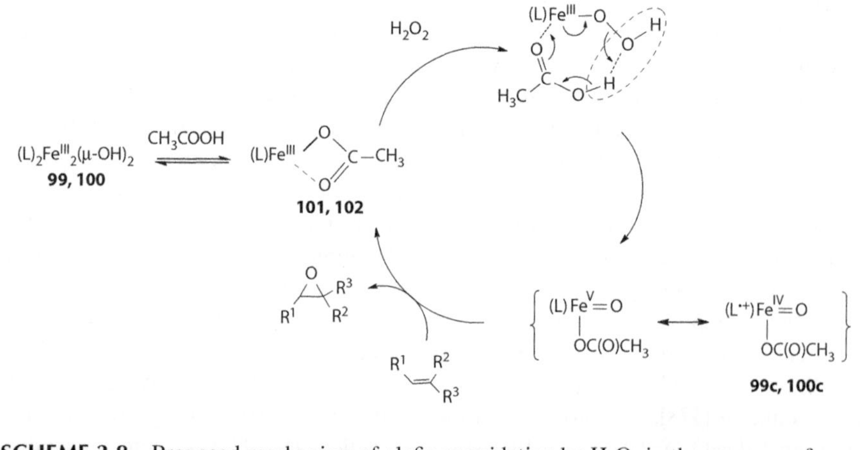

SCHEME 3.8   Proposed mechanism of olefin epoxidation by H_2O_2 in the presence of complexes **99** and **100**.

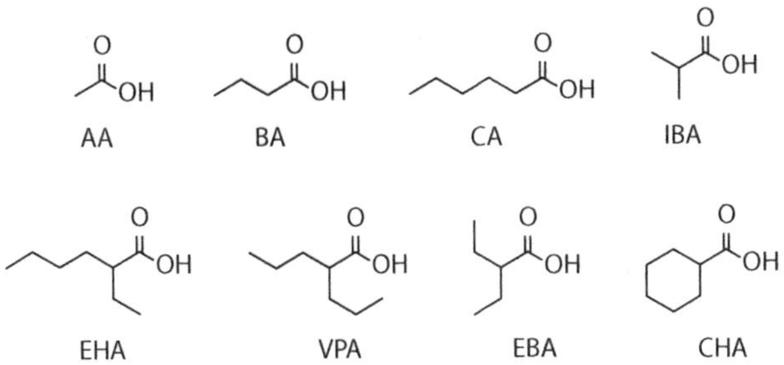

NMR and EPR Spectroscopy

FIGURE 3.44 Structures and abbreviations of the carboxylic acids discussed.

reversibly transform into monomeric ferric-acetate complexes (**101, 102**) observed by EPR even at room temperature. Subsequent interaction of **101, 102** with H_2O_2, followed by heterolytic O—O bond cleavage with removal of a water molecule, generates the active species **99c** and **100c** that directly epoxidize alkenes even at $-85°C$.

Another important parameter, capable of affecting the structure and properties of the catalytically active species, is the nature of the carboxylic acid since the use of different carboxylic acids has been shown to significantly affect the epoxidation enantioselectivity (Table 3.6). To get direct insight into the effect of the structure of the carboxylic acid on the catalytic reactivity of Fe catalysts, undertook the search for the active species in the catalyst systems $[((S,S)\text{-PDP})Fe(CF_3SO_3)_2]$ (**95**)/H_2O_2/ RCOOH with eight different carboxylic acids RCOOH (Figure 3.44). Spectacularly, it was found that the use of carboxylic acids with primary (AA) and secondary α-carbons (BA, CA) leads to the intermediates with large g-factor anisotropy of the type **95c^{AA}** (Table 3.8), while in the presence of some acids with tertiary α-carbon, intermediates with small g-factor anisotropy were formed (Table 3.8). In particular, the intermediates with small g-factor anisotropy (of the type **95c^{EHA}**) were observed for EHA, EBA, and VPA (Figure 3.45), while IBA and CHA led to the intermediates with large g-factor anisotropy (Table 3.8).

Taking into account the above explanation of the reason of the small g-factor anisotropy, it is logical to expect that in the intermediates of the type **95c^{EHA}**, the unpaired spin mostly resides on the coordinated carboxylic moiety (Scheme 3.9) [183]. Apparently, the active oxidizing species formed in the presence of EHA, VPA, and EBA may be better represented as $[((S,S)\text{-PDP})Fe^{IV}=O(^\bullet OC(O)R)]^{2+}$. The much higher stereoselectivity (Table 3.6) of the corresponding catalyst systems **95**/H_2O_2/ RCOOH (RCOOH = EHA, VPA, EBA) is readily explained by the additional stabilization of the corresponding active species $[((S,S)\text{-PDP})Fe^{IV}=O(^\bullet OC(O)R)]^{2+}$ due to the unpaired electron delocalization over the carboxylic moiety, which leads to more product-like transition states and thus ensures tighter interactions between the substrate and the oxoferryl species within the transition state, eventually ensuring higher enantioselectivity [183].

For iron complexes with aminopyridine ligands bearing electron-donating substituents (**99, 100**), the oxoferryl intermediates with small g-factor anisotropy are formed even with acetic acid [175]. In effect, the use of all carboxylic acids (Figure 3.44) in the catalyst system **100**/H_2O_2/RCOOH afforded active species with

110 Applications of EPR and NMR Spectroscopy in Homogeneous Catalysis

TABLE 3.8
EPR Data for the Oxoferryl Intermediates Originated from 95 and 100 in the Presence of Various Carboxylic Acids

Intermediate	g_1	g_2	g_3
$[((S,S)\text{-PDP})Fe^V=O(OC(O)R]^{2+}$, type **95c**[AA]:			
95c[AAa]	2.66	2.42	1.71
95c[BA]	2.70	2.42	1.67
95c[CA]	2.70	2.42	1.66
95c[IBA]	2.72	2.42	1.66
95c[CHA]	2.69	2.42	1.67
$[((S,S)\text{-PDP})Fe^{IV}=O(\cdot OC(O)R)]^{2+}$, type **95a**[EHA]:			
95c[EHA]	2.069	2.007	1.963
95c[EBA]	2.069	2.007	1.961
95c[VPA]	2.069	2.007	1.962
$[((S,S)\text{-PDP}^*)^{\cdot+}Fe^{IV}=O(OC(O)R)]^{2+}$, type **100c**:			
100c[AAb]	2.071	2.008	1.960
100c[BA]	2.070	2.008	1.958
100c[CA]	2.069	2.008	1.957
100c[IBA]	2.069	2.007	1.957
100c[CHA]	2.070	2.008	1.957
100c[EHA]	2.070	2.008	1.958
100c[EBA]	2.069	2.008	1.957
100c[VPA]	2.069	2.007	1.958

[a] From Reference [163].
[b] From Reference [175].

virtually the same g-factor values (Table 3.8). However, the use of carboxylic acid with tertiary α-carbon still causes smaller but positive effect on the chalcone epoxidation enantioselectivity (Table 3.9). Remarkably, the enantioselectivities correlated well with the observed maximum steady-state concentrations of the active species as evaluated by EPR (Table 3.9). This provides a spectacular illustration for the widely used Hammond–Leffler principle, which for the present catalyst system may be expressed as follows: *the less reactive (i.e., those approaching higher steady-state concentration) oxygen-transferring species leads to more product-like transition states*, resulting in higher epoxidation stereoselectivity [183].

These studies provide the most plausible explanation of the influence of (1) electron-donating substituents and (2) branched carboxylic acids on the enantioselectivity of biomimetic catalysts systems Fe(PDP)/H$_2$O$_2$/RCOOH and, probably, of the structurally related manganese-based catalyst systems [163], and delineates the directions for rational tuning their reactivity and stereoselectivity. We note that the data obtained fit well within the heme paradigm, which assumes that the oxygen-transferring species in biological, CYP-mediated oxidations are (P$^{\cdot+}$)Fe$^{IV}=$O species (where P = porphyrin) [120,179–181].

NMR and EPR Spectroscopy

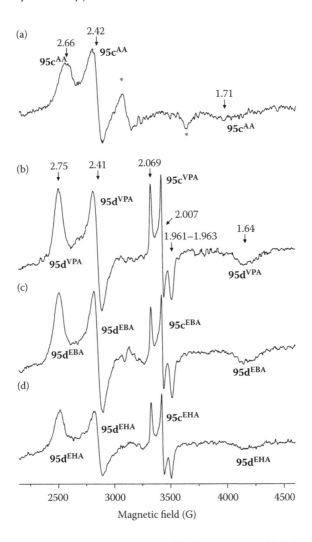

FIGURE 3.45 EPR spectrum (1.8:1 CH$_2$Cl$_2$/CH$_3$CN mixture, −196°C) of the sample **95**/H$_2$O$_2$/CH$_3$COOH ([**95**]:[H$_2$O$_2$]:[CH$_3$COOH] = 1:3:10, [**95**] = 0.04 M) frozen 1 min after mixing the reagents at −75°C and storing the sample at −85°C for 1 min (a). Signals denoted by asterisks belong to ferric hydroxo complex [((*S,S*)-PDP)Fe^{III}–OH(CH$_3$CN)]$^{2+}$ (g_1 = 2.44, g_2 = 2.21, g_3 = 1.89 [163]). (b) EPR spectra of the samples **95**/H$_2$O$_2$/RCOOH ([**95**]:[H$_2$O$_2$]:[RCOOH] = 1:3:10, [**95**] = 0.04 M) frozen 2.5 min after mixing the reagents at −65°C, (b) RCOOH = VPA, (c) RCOOH = EBA, and (d) RCOOH = EHA. Signals marked as **95d^{VPA}**, **95d^{EBA}**, and **95d^{EHA}** belong to stable low-spin (S = 1/2) ferric species.

The transition metal complexes considered above mainly operate in model biomimetic catalyst systems. In industrial catalyst systems, information on the nature of the active oxidizing species in most cases is lacking. Some well-known exceptions are the alkylperoxo complexes of molybdenum and titanium, conducting the Halcon epoxidation of propene, and asymmetric Sharpless epoxidation of allylic alcohols,

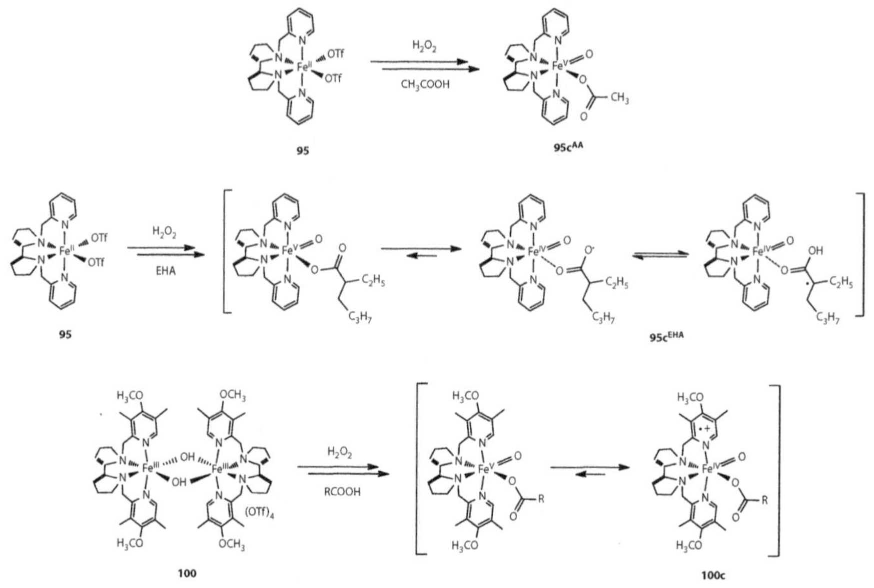

112 Applications of EPR and NMR Spectroscopy in Homogeneous Catalysis

SCHEME 3.9 Plausible mechanistic scenarios for the formation of the active species of the type **95c^{AA}** (with large g-factor anisotropy), **95c^{EHA}** (small g-factor anisotropy), and **100c** (small g-factor anisotropy). For **95c^{EHA}**, possible tautomeric forms are presented to illustrate the additional stabilization caused by the carboxylic acid with tertiary α-carbon.

TABLE 3.9
Asymmetric Epoxidation of Chalcone with H_2O_2 Catalyzed by Complex 100: Enantioselectivity versus Concentration of the Active Species[a]

Number	Carboxylic Acid[b]	Conversion/Epoxide Yield (%)	ee (%)[c]	Concentration of Active Species[d]
1	AA	74/73	69	1.0
2	CA	48/46	70	1.0
3	BA	61/59	70	1.4
4	IBA	92/92	78	2.2
5	CHA	72/71	78	2.2
6	EBA	54/54	81	2.0
7	VPA	69/69	82	2.6
8	EHA	73/73	83	3.0

[a] At 0°C, [substrate]:[H_2O_2]:[carboxylic acid] = 100:200:55 µmol, catalyst load 0.5 mol.% (1 mol.% Fe), oxidant was added by a syringe pump over 30 min, and the mixture was stirred for additional 2.5 h, followed by HPLC analysis.

[b] For structures and abbreviations of carboxylic acids, see Figure 3.44.

[c] Chalcone epoxide absolute configuration (2R,3S).

[d] Maximum concentration of the active species calculated from EPR spectra, expressed in % of total iron concentration in the sample.

NMR and EPR Spectroscopy

respectively. Another example is the conversion of *p*-xylene to terephthalic acid; below, we consider the NMR spectroscopic characterization of metal complexes operating in this industrial process.

3.5 STRUCTURE OF Co(III) ACETATE IN SOLUTION

Most part of the large-scale homogeneous oxidation processes proceed via the classical autoxidation mechanism (e.g., conversion of *p*-xylene to terephthalic acid, or cyclohexane to adipic acid). Simple alkanes or alkylbenzenes can be oxidized by molecular oxygen in AcOH in the presence of $Co^{II}(OAc)_2 \cdot 4H_2O$ as the precatalyst. It is generally agreed that *in situ*–formed Co^{III} complexes, the so-called Co(III) acetate, abstracts an electron from the substrate and thus initiates the autoxidation (Equations 3.6 and 3.7) [184,185].

$$ArCH_3 + Co^{III} \leftrightarrow [ArCH_3]^{\bullet+} + Co^{II} \tag{3.6}$$

$$[ArCH_3]^{\bullet+} \rightarrow ArCH_2^{\bullet} + H^+ \tag{3.7}$$

In spite of numerous studies, the structure of Co(III) acetate in the solid state or in solution was not entirely clear. Crystal structures of compounds containing oxo-centered trinuclear cations $[Co_3^{III}O(OAc)_5(\mu\text{-}OH)(py)_3]^+$ (**103**), $[Co_3^{III}O(OAc)_5(\mu\text{-}OMe)(py)_3]^+$ (**104**), and the binuclear hydroxo-bridged cation $[Co_2^{III}(OAc)_2(\mu\text{-}OAc)(\mu\text{-}OH)_2(Py)_4]^+$ (**105**) were reported (Figure 3.46) [186–188]. On the basis of those data, three main complexes observed in the solution of Co(III) acetate in water/acetic acid mixtures, were tentatively assigned to $[Co_3^{III}O(OAc)_6(AcOH)_3]^+$, $[Co_2^{III}Co^{II}O(OAc)_6(AcOH)_3]$, and $[Co_2^{III}(OAc)_4(\mu\text{-}OH)_2]$ [189,190]. Apparently, more structural and/or spectroscopic data on these complexes are needed.

For a long time, it was thought that NMR spectroscopy is not applicable for the characterization of Co(III) acetate because of the presence of paramagnetic Co(II)

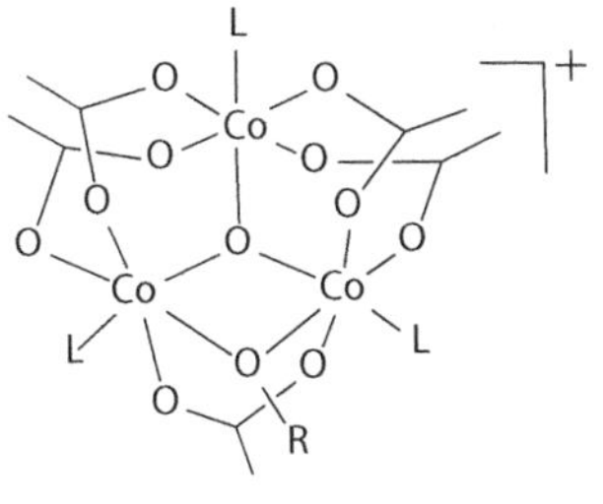

103 R = H, L = Py
104 R = Me, L = Py
 A R = Me, L = MeOH
 B R = H, L = MeOH
 I R = Ac, L = AcOH
 II R = H, L = AcOH

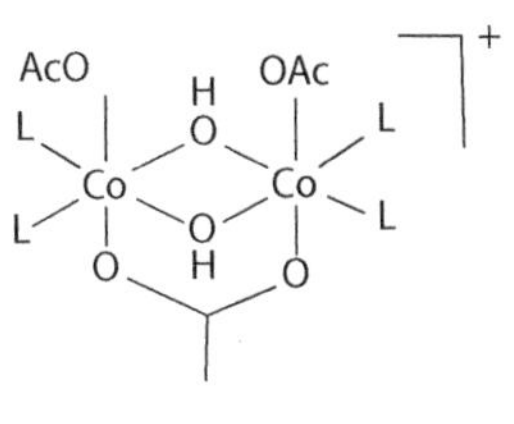

105 L = Py
 C L = MeOH
 III L = AcOH

FIGURE 3.46 The structure of Co(III) trinuclear and dinuclear cations formed in solutions of "Co(III) acetate" in various solvents.

species in the reaction solution. However, in 1998, we showed that ^{59}Co and ^{1}H NMR spectroscopy could be very suitable for this goal. The solid sample of Co(III) acetate was prepared upon the interaction of Co(OAc)$_2$·4H$_2$O with 40% peracetic acid in AcOH, followed by solvent removal in a vacuum. The green solid obtained was used for further investigations [191].

By means of ^{59}Co and ^{1}H NMR spectroscopy, three main types of diamagnetic Co(III) complexes (**A**, **B**, **C**) were observed in solutions of Co(III) acetate in methanol. Five minutes after dissolving Co(III) acetate in MeOH, predominantly complexes **A** and **B** were observed in ^{59}Co and ^{1}H NMR spectra. **B** slowly converted with time into complex **A** ($\tau_{1/2}$ = 15 min at 20°C) (Figure 3.47). The concentration of **C** was low in the freshly prepared solution of Co(III) acetate (Figure 3.47). Heating this sample for 5 min at 50°C resulted in the disappearance of ^{59}Co resonances of **A** and **B**, and the increase of that of **C**. In effect, **C** became the predominant diamagnetic Co(III) species in the solution. Complex **C** displayed ^{59}Co resonance at δ 1250 ($\Delta\nu_{1/2}$ = 3 kHz). A number of paramagnetically shifted ^{1}H NMR resonances (assigned to unidentified mixed-valence Co(III)–Co(II) species) were also observed.

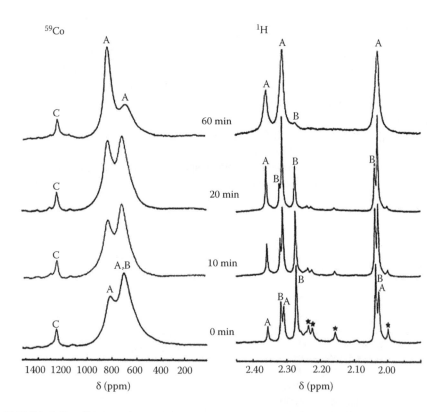

FIGURE 3.47 ^{59}Co and ^{1}H NMR spectra (methanol-d_4, 20°C) of Co(III) acetate at various moments of time after dissolving 30 mg of Co(III) acetate in 1 mL of CD$_3$OD at room temperature. Asterisks denote resonances of unidentified cobalt species. (Reproduced with permission from Babushkin, D. E., Talsi, E. P. 1998. *J. Mol. Catal. A* 130: 131–137.)

The analysis of NMR spectra allows the unambiguous identification of **A** and **B** (Figure 3.47). Complex **A** displays two [59]Co resonances at 850 ppm ($\Delta\nu_{1/2} = 5$ kHz) and 680 ppm ($\Delta\nu_{1/2} = 5$ kHz) with relative integral intensities 2:1. These resonances belong to two equivalent and one unique cobalt atoms. The ^{1}H NMR pattern of **A** (CD$_3$OD, 20°C, δ: 2.38 (s, 3H), 2.33 (s, 6H), 2.05 (s, 6H)) is similar to that of the acetate ligands of the trinuclear cation [Co$_3$O(OAc)$_5$(OMe)(Py)$_3$]$^+$ [189]. Three additional resonances appear in the ^{1}H NMR spectrum of **A**, when CH$_3$OH is used instead of CD$_3$OD as a solvent. They belong to the μ-OMe group (1.33 (s, 3H)) and axial MeOH molecules (3.52 (s, 6H), 3.76 (s, 3H)) of complex **A** ([Co$_3$O(OAc)$_5$(μ-OMe)(MeOH)$_3$]$^+$). The ^{1}H NMR patterns of the acetate ligands of complexes **A** and **B** are similar. In both cases, three resonances with relative intensities 1:2:2 are observed (Figure 3.47). Apparently, **B** is a cation [Co$_3$O(OAc)$_5$(μ-OH)(MeOH)$_3$]$^+$. The substitution of μ-OH by μ-OR in ROH was reported for [Co$_3$O(OAc)$_5$(μ-OMe)(Py)$_3$]$^+$ [190]. We could not observe two separate [59]Co resonances for **B**. Probably, at given line widths, the 2:1 [59]Co resonances of **B** overlap. The structure of **C** is less clear than that of **A** and **B**. By analogy with isolated complex **105** [186], we proposed that **C** is a dinuclear complex of the type [Co$_2^{III}$(OAc)$_3$(μ-OH)$_2$(MeOH)$_4$]$^+$.

[59]Co, ^{13}C, and ^{1}H NMR spectra of the freshly prepared sample of Co(III) acetate in CD$_2$Cl$_2$ and CDCl$_3$ display resonances of diamagnetic complexes **I** and **II** (Figure 3.48). The analysis of the NMR spectra suggested the assignment of **I** and

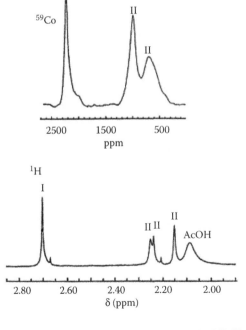

FIGURE 3.48 [59]Co and ^{1}H NMR spectra of Co(III) acetate (in CH$_2$Cl$_2$ and CDCl$_3$, respectively). (Reproduced with permission from Babushkin, D. E., Talsi, E. P. 1998. *J. Mol. Catal. A* 130: 131–137.)

116   Applications of EPR and NMR Spectroscopy in Homogeneous Catalysis

II to the cations $[Co_3O(OAc)_6(AcOH)_3]^+$ and $[Co_3O(OAc)_5(\mu\text{-}OH)(AcOH)_3]^+$, respectively [191]. The ^{59}Co NMR spectra of one and the same sample of Co(III) acetate in CH_2Cl_2 and MeOH shows that **I** converts to **A** and **II** to **B** immediately upon dissolving of Co(III) acetate in MeOH.

The sample of Co(III) acetate, dissolved in AcOH, exhibits broad ^{59}Co resonances of complexes **I** and **II** (30–50 kHz in AcOH vs. 5–10 kHz in CH_2Cl_2 or MeOH) [190]. Because of the high line widths, **II** displays only one ^{59}Co resonance at ca. 900 ppm. The increase in water concentration of AcOH leads to the conversion of **I** to **II**. Besides, the intensity of ^{59}Co resonance of complex **III** at 1200 ppm ($\Delta v_{1/2} = 15$ kHz) increases. **III** converts into **C** upon the addition of great excess of MeOH to the sample. Overall, ^{59}Co, ^{13}C, and ^1H NMR spectroscopic data clearly show that the freshly prepared sample of Co(III) acetate consists mainly of the trinuclear complexes $[Co_3O(OAc)_6(AcOH)_3]^+$(**I**) and $[Co_3O(OAc)_5(\mu\text{-}OH)(AcOH)_3]^+$(**II**). The increase in temperature or water concentration of AcOH promotes the conversion of **I** and **II** into complex **III**, which is probably dinuclear complex of the type $[Co_2(\mu\text{-}OH)_2(OAc)_3(AcOH)_4]^+$ (Figure 3.46).

In 2011, a similar NMR spectroscopic approach was successfully used to establish the structure of complexes formed by dissolution of palladium diacetate in MeOH and $CDCl_3$ [192]. By means of ^1H and ^{13}C NMR spectroscopy (including 2D-HSQC and 2-DOSY techniques), it was shown that dissolution of diacetate cyclic trimer $[Pd(OAc)_2]_3$ in MeOH at $-18°C$ leads to a methoxo complex $Pd_3(\mu\text{-}OMe)(OAc)_5$, which is partially converted to the dimethoxo complex $Pd_3(\mu\text{-}OMe)_2(OAc)_4$. In chloroform, $[Pd(OAc)_2]_3$ is reversibly converted to the hydroxo complex $Pd_3(\mu\text{-}OH)(OAc)_5$.

REFERENCES

1. Sawyer, D. T., Valentine, J. S. 1981. How super is superoxide? *Acc. Chem. Res.* 14: 393–400.
2. Lunsford, J. H. 1973. ESR of adsorbed oxygen species. *Catal. Rev.* 8: 135–157.
3. Jones, R. D., Summerville, D. A., Basolo, F. 1979. Synthetic oxygen carriers related to biological systems. *Chem. Rev.* 79: 139–179.
4. Nishinaga, A., Nishizawa, K., Tomita, H., Matsuura, T. 1977. Novel peroxycobalt(III) complexes derived from 4-aryl-2,6-di-*tert*-butylphenols. A model intermediate of dioxygenase reaction. *J. Am. Chem. Soc.* 99: 1287–1288.
5. Nishinaga, A., Tomita, H. 1980. Model catalytic oxygenations with Co(III)-Schiff base complexes and the role of cobalt-oxygen complexes in oxygenation process. *J. Mol. Catal.* 7: 179–199.
6. Nishinaga, A., Tomita, H., Nishizawa, K., Matsuura, T. 1981. Regioselective formation of peroxyquinolatocobalt(III) complexes in the oxygenation of 2,6-di-*t*-butylphenols with cobalt(II) Schiff-base complexes. *J. Chem. Soc. Dalton Trans.* 1501–1514.
7. Zombeck, A., Drago, R. S., Corden, B. B., Gaul, J. H. 1981. Activation of molecular oxygen. Mechanistic studies of the oxidation of hindered phenols with cobalt-dioxygen complexes. *J. Am. Chem. Soc.* 103: 7580–7585.
8. Corden, B. B., Drago, R. S., Perito, R. P. 1985. Steric and electronic effects of ligand variation on cobalt dioxygen catalysts. *J. Am. Chem. Soc.* 107: 2903–2907.
9. Hoffman, B. M., Diemente, D. L., Basolo, F. 1970. Electron paramagnetic resonance studies of some cobalt(II) Schiff base compounds and their monomeric oxygen adducts. *J. Am. Chem. Soc.* 92: 61–65.
10. Rodley, G. A., Robinson, W. T. 1972. Structure of a monomeric oxygen-carrying complex. *Nature* 235: 438–439.

NMR and EPR Spectroscopy

11. Talsi, E. P., Babenko, V. P., Likholobov, V. A., Nekipelov, V. M., Chinakov, V. D. 1985. A new superoxo complex of palladium that oxidizes alkenes to epoxides. *J. Chem. Soc. Chem. Commun.* 1768–1769.
12. Talsi, E. P., Babenko, V. P., Shubin, A. A., Chinakov, V. D., Nekipelov, V. M., Zamaraev, K. I. 1987. Formation, structure, and reactivity of palladium superoxo complexes. *Inorg. Chem.* 26: 3871–3878.
13. Stephenson, T. A., Morehouse, S. M., Powell, A. R., Heffer, J. P., Wilkinson, G. 1965. Carboxylates of palladium, platinum, and rhodium, and their adducts. *J. Chem. Soc.* 3632–3640.
14. Selke, M., Valentine, J. S. 1998. Switching on the nucleophilic reactivity of a ferric porphyrin peroxo complex. *J. Am. Chem. Soc.* 120: 2652–2653.
15. Cai, X., Majumdar, S., Fortman, G. C., Cazin, C. S. J., Slawin, A. M. Z., Lhermitte, C., Prabhakar, R. et al. 2011. Oxygen binding to [Pd(L)(L')] (L = NHC, L' = NHC or PR_3, NHC = N-heterocyclic carbine). Synthesis and structure of paramagnetic *trans*-[Pd(NHC)$_2$ (η^1-O$_2$)$_2$] complex. *J. Am. Chem. Soc.* 133: 1290–1293.
16. Huacuja, R., Graham, D. J., Fafard, C. M., Chen, C.-H., Foxman, B. M., Herbert, D. E., Alliger, G., Thomas, C. M., Ozerov, O. V. 2011. Reactivity of a Pd(I)-Pd(I) dimer with O_2: Monohapto Pd superoxide and dipalladium peroxide in equilibrium. *J. Am. Chem. Soc.* 133: 3820–3823.
17. Fujita, K., Schenker, R., Gu, W., Brunold, T. C., Cramer, S. P., Riordan, C. G. 2004. A monomeric nickel-dioxygen adduct derived from a nickel(I) complex and O_2. *Inorg. Chem.* 43: 3324–3326.
18. Kieber-Emmons, M. T., Annaraj, J., Seo, M. S., Van Heuvelen, K. M., Tosha, T., Kitagawa, T., Brunold, T. C., Nam, W., Riordan, C. G. 2006. Identification of an "end-on" nickel-superoxo adduct, [Ni(tmc)(O$_2$)]$^+$. *J. Am. Chem. Soc.* 128: 14230–14231.
19. Kieber-Emmons, M. T., Riordan, C. G. 2007. Dioxygen activation at monovalent nickel. *Acc. Chem. Res.* 40: 618–625.
20. Yao, S., Bill, E., Milsmann, C., Wieghardt, K., Driess, M. 2008. A "side-on" superoxonickel complex [LNi(O$_2$)] with a square-planar tetracoordinate nickel(II) center and its conversion into [LNi(μ-OH)$_2$NiL]. *Angew. Chem. Int. Ed.* 47: 7110–7113.
21. Company, A., Yao, S., Ray, K., Driess, M. 2010. Dioxygenase-like reactivity of an isolable superoxo-nickel(II) complex. *Chem. Eur. J.* 16: 9669–9675.
22. Cho, J., Kang, H. Y., Liu, L. V., Sarangi, R., Solomon, E. I., Nam, W. 2013. Mononuclear nickel(II)-superoxo and nickel(III)-peroxo complexes bearing a common macrocyclic TMC ligand. *Chem. Sci.* 4: 1502–1508.
23. Mirica, L. M., Ottenwaelder, X., Stack, T. D. P. 2004. Structure and spectroscopy of copper-dioxygen complexes. *Chem. Rev.* 104: 1013–1045.
24. Lewis, E. A., Tolman, W. B. 2004. Reactivity of dioxygen-copper systems. *Chem. Rev.* 104: 1047–1076.
25. Klinman, J. P. 2006. The copper enzyme family of dopamine β-monooxygenase and peptidylcycline α-hydroxylating monooxygenase: Resolving the chemical pathway for substrate hydroxylation. *J. Biol. Chem.* 281: 3013–3016.
26. Humphreys, K. J., Mirica, L. M., Wang, Y., Klinman, J. P. 2009. Galactose oxidase as a model for reactivity of a copper superoxide center. *J. Am. Chem. Soc.* 131: 4657–4663.
27. Maiti, D., Fry, H. C., Woertink, J. S., Vance, M. A., Solomon, E. I., Karlin, K. D. 2007. A 1:1 copper adduct is an end-on bound superoxo copper(II) complex which undergoes oxygenation reactions with phenols. *J. Am. Chem. Soc.* 129: 264–265.
28. Maiti, D., Lee, D. H., Gaoutchenova, K., Würtele, C., Holthausen, M. C., Sarjeant, A. A. N., Sundermeyer, J., Schindler, S., Karlin, K. D. 2008. Reactions of a copper(II) superoxo complex lead to C-H and O-H substrate oxygenation: Modeling coppermonooxygenase C-H hydroxylation. *Angew. Chem. Int. Ed.* 47: 82–85.

118   Applications of EPR and NMR Spectroscopy in Homogeneous Catalysis

29. Donoghue, P. J., Gupta, A. K., Boyce, D. W., Cramer, C. J., Tolman, W. B. 2010. An anionic, tetragonal copper(II) superoxide complex. *J. Am. Chem. Soc.* 132: 15869–15871.

30. Kunishita, A., Kubo, M., Sugimoto, H., Ogura, T., Sato, K., Takui, T., Itoh, S. 2009. Mononuclear copper(II)-superoxo complexes that mimic the structure and reactivity of the active centers of PHM and DβM. *J. Am. Chem. Soc.* 131: 2788–2789.

31. Peterson, R. L., Himes, R. A., Kotani, H., Suenobu, T., Tian, L., Siegler, M. A., Solomon, E. I., Fukuzumi, S., Karlin, K. D. 2011. Cupric superoxo-mediated intermolecular C-H activation chemistry. *J. Am. Chem. Soc.* 133: 1702–1705.

32. Peterson, R. L., Ginsbach, J. W., Cowley, R. E., Qayyum, M. F., Himes, R. A., Siegler, M. A., Moore, C. D. et al. 2013. Stepwise protonation and electron-transfer reduction of a primary copper-dioxygen adduct. *J. Am. Chem. Soc.* 135: 16454–16467.

33. Liu, Y., Mukherjee, A., Nahumi, N., Ozbil, M., Brown, D., Angeles-Boza, A. M., Dooley, D. M., Prabhakar, R., Roth, J. P. 2013. Experimental and computational evidence of metal-O_2 activation and rate-limiting proton-coupled electron transfer in a copper amine oxidase. *J. Phys. Chem. B* 117: 218–229.

34. Ginsbach, J. W., Peterson, R. L., Cowley, R. E., Karlin, K. D., Solomon, E. I. 2013. Correlation of the electronic and geometric structures in mononuclear copper(II) superoxide complexes. *Inorg. Chem.* 52: 12872–12874.

35. Lee, J. Y., Peterson, R. L., Ohkubo, K., Garcia-Bosch, I., Himes, R. A., Woertink, J., Moore, C. D., Solomon, E. I., Fukuzumi, S. 2014. Mechanistic insights into the oxidation of substituted phenols via hydrogen atom abstraction by a cupric-superoxo complex. *J. Am. Chem. Soc.* 136: 9925–9937.

36. Solomon, E. I., Heppner, D. E., Johnston, E. M., Ginsbach, J. W., Cirera, J., Qayyum, M., Kieber-Emmons, M. T., Kjaergaard, C. H., Hadt, R. G., Tian, L. 2014. Copper active sites in biology. *Chem. Rev.* 114: 3659–3853.

37. Prigge, S. T., Eipper, B. A., Mains, R. E., Amzel, L. M. 2004. Dioxygen binds end-on to mononuclear copper in a precatalytic enzyme complex. *Science* 304: 864–867.

38. Fujisawa, K., Tanaka, M., Moro-oka, Y., Kitajima, N. 1994. A monomeric side-on superoxocopper(II) complex: $Cu(O_2)(HB(3\text{-}tBu\text{-}5\text{-}iPrpz)_3)$. *J. Am. Chem. Soc.* 116: 12079–12080.

39. Würtele, C., Gaoutchenova, E., Harms, K., Holthausen, M. C., Sundermeyer, J., Schindler, S. 2006. Crystallographic characterization of a synthetic 1:1 end-on copper dioxygen adduct complex. *Angew. Chem. Int. Ed.* 45: 3867–3869.

40. Woertink, J. S., Tian, L., Maiti, D., Lucas, H. R., Himes, R. A., Karlin, K. D., Neese, F. et al. 2010. Spectroscopic and computational studies of an end-on bond superoxo-Cu(II) complex: Geometric and electronic factors that determine the ground state. *Inorg. Chem.* 49: 9450–9459.

41. Hong, S., Huber, S. M., Gagliardi, L., Cramer, C. C., Tolman, W. B. 2007. Copper(I)-α-ketocarboxylate complexes: Characterization and O_2 reactions that yield copper-oxygen intermediates capable of hydroxylating arenes. *J. Am Chem. Soc.* 129: 14190–14192.

42. Itoh, S. 2006. Mononuclear copper active-oxygen complexes. *Curr. Opin. Chem. Biol.* 10: 115–122.

43. Schlichting, I., Berendzen, J., Chu, K., Stock, A. M., Maves, S. A., Benson, D. E., Sweet, R. M., Ringe, D., Petsko, G. A., Sligar, S. G. 2000. The catalytic pathway of cytochrome P450$_{cam}$ at atomic resolution. *Science* 287: 1615–1622.

44. Ortiz de Montellano, P. R. 2010. Hydrocarbon hydroxylation by cytochrome P450 enzymes. *Chem. Rev.* 110: 932–948.

45. Ray, K., Pfaff, F. F., Wang, B., Nam, W. 2014. Status of reactive non-heme metal-oxygen intermediates in chemical and enzymatic reactions. *J. Am. Chem. Soc.* 136: 13942–13958.

NMR and EPR Spectroscopy

46. Xing, G., Diao, Y., Hoffart, L. M., Barr, E. W., Prabhu, K. S., Arner, R. J., Reddy, C. C., Krebs, C., Bollinger, J. M., Jr. 2006. Evidence for C-H cleavage by an iron-superoxide complex in the glycol cleavage reaction catalyzed by *myo*-inositol oxygenase. *Proc. Natl. Acad. Sci. U. S. A.* 103: 6130–6135.

47. Kovaleva, E. G., Lipscomb, J. D. 2007. Crystal structures of Fe^{2+} dioxygenase superoxo, alkylperoxo, and bond product intermediates. *Science* 316: 453–456.

48. Mbughuni, M. M., Chakrabarti, M., Hayden, J. A., Bominaar, E. L., Hendrich, M. P., Münck, E., Lipscomb, J. D. 2010. Trapping and spectroscopic characterization of an Fe^{III}-superoxo intermediate from a nonheme mononuclear iron-containing enzyme. *Proc. Natl. Acad. Sci. U. S. A.* 107: 16788–16793.

49. Shan, X., Que, L. Jr. 2005. Intermediates in the oxygenation of a nonheme diiron(II) complex, including the first evidence for a bond superoxo species. *Proc. Natl. Acad. Sci. U. S. A.* 102: 5340–5345.

50. Zhao, M., Helms, B., Slonkina, E., Friedle, S., Lee, D., DuBois, J., Hedman, B., Hodgson, K. O., Fréchet, J. M. J., Lippard, S. J. 2008. Iron complexes of dendrimer-appended carboxylates for activating dioxygen and oxidizing hydrocarbons. *J. Am. Chem. Soc.* 130: 4352–4363.

51. Chiang, C. W., Kleespies, S. T., Stout, H. D., Meier, K. K., Li, P. Y., Bominaar, E. L., Que, L. Jr., Münck, E., Lee, W. Z. 2014. Characterization of a paramagnetic mononuclear nonheme iron-superoxo complex. *J. Am. Chem. Soc.* 36: 10846–10849.

52. Landau, R., Sullivan, G. A., Brown, D. 1979. Propylene oxide by the co-product process. *CHEMTECH* 9: 602–607.

53. Sheldon, R. A., Van Doorn, J. A. 1973. Metal-catalyzed epoxidation of olefins with organic hydroperoxides: A comparison of various metal catalysts. *J. Catal.* 31: 427–437.

54. Talsi, E. P., Klimov, O. V., Zamaraev, K. I. 1993. Characterization with ^{95}Mo, ^{17}O, ^{1}H NMR and EPR of alkylperoxo, alkoxo, peroxo and diolo molybdenum(VI) complexes formed in the course of catalytic epoxidation of cyclohexene with organic hydroperoxides. *J. Mol. Catal.* 83: 329–346.

55. Dorosheva, T. S., Bryliakov, K. P., Talsi, E. P. 2003. Multinuclear NMR-spectroscopic characterization of alkylperoxo complexes of molybdenum(VI). *Mendeleev Commun.* 13: 8–9.

56. Kühn, F. E., Groarke, M., Bencze, E., Herdtweck, E., Prazeres, A., Santos, A. M., Calhorda, M. J. et al. 2002. Octahedral bipyridine and bipyrimidine dioxomolybdenum(VI) complexes: Characterization, application in catalytic epoxidation, and density functional mechanistic study. *Chem. Eur. J.* 8: 2370–2383.

57. Groarke, M., Gonçalves, I. S., Herrmann, W. A., Kühn, F. E. 2002. New insights into the reaction of t-butylhydroperoxide with dichloro- and dimethyl(dioxo)molybdenum(VI). *J. Organomet. Chem.* 649:108–112.

58. Katsuki, T., Sharpless, K. B. 1980. The first practical method for asymmetric epoxidation. *J. Am. Chem. Soc.* 102: 5974–5976.

59. Woodard, S. S., Finn, M. G., Sharpless, K. B. 1991. Mechanism of asymmetric epoxidation. 1. Kinetics. *J. Am. Chem. Soc.* 113: 106–113.

60. Finn, M. G., Sharpless, K. B. 1991. Mechanism of asymmetric epoxidation. 2. Catalyst structure. *J. Am. Chem. Soc.* 113: 113–126.

61. Berrisford, D. J., Bolm, C., Sharpless, K. B. 1995. Ligand-accelerated catalysis. *Angew. Chem. Int. Ed.* 34: 1059–1070.

62. Bellussi, G., Carati, A., Clerici, M. G., Maddinelli, G., Millini, R. 1992. Reactions of titanium silicalite with protic molecules and hydrogen peroxide. *J. Catal.* 133: 220–230.

63. Clerici, M. G., Ingallina, P. 1993. Epoxidation of lower olefins with hydrogen peroxide and titanium silicalite. *J. Catal.* 140: 71–83.

64. Sheldon, R. A. 1973. Molybdenum-catalyzed epoxidation of olefins with alkyl hydroperoxides. II Isolation and structure of the catalyst. *Recl. Trav. Chim. Pays-Bas.* 92: 253–373.

120 Applications of EPR and NMR Spectroscopy in Homogeneous Catalysis

65. Sheldon, R. A. 1980. Synthetic and mechanistic aspects of metal-catalyzed epoxidations with hydroperoxides. *J. Mol. Catal.* 7: 107–126.
66. Boche, G., Möbus, K., Harms, K., Marsch, M. 1996. [((η^2-tert-butylperoxo)titanatrane)$_2$·3 dichloromethane]: X-ray structure and oxidation reactions. *J. Am. Chem. Soc.* 118: 2770–2771.
67. Al-Ajlouni, A., Valente, A. A., Nunes, C. D., Pillinger, M., Santos, A. M., Zhao, J., Romão, C. C., Gonçalves, I. S., Kühn, F. E. 2005. Kinetics of cyclooctene epoxidation with tert-butyl hydroperoxide in the presence of [MoO$_2$X$_2$L]-type catalysts (L = bidentate Lewis base). *Eur. J. Inorg. Chem.* 1716–1723.
68. Valente, A. A., Moreira, J., Lopes, A. D., Pillinger, M., Nunes, C. D., Romão, C. C., Kühn, F. E., Gonçalves, I. S. 2004. Dichloro and dimethyl dioxomolybdenum(VI)-diazabutadiene complexes as catalysts for the epoxidation of olefins. *New. J. Chem.* 28: 308–313.
69. Bruno, S. M., Balula, S. S., Valente, A. A., Almeida Paz, F. A., Pillinger, M., Sousa, C., Klinowski, J., Freire, C., Ribeiro-Claro, P., Gonçalves, I. S. 2007. Synthesis and catalytic properties in olefin epoxidation of dioxomolybdenum(VI) complexes bearing a bidentate or tetradentate salen-type ligand. *J. Mol. Catal. A Chem.* 270: 185–194.
70. Mimoun, H., Chaumette, P., Mignard, M., Saussine, L., Fischer, J., Weiss, R. 1983. 1st D-degrees metal alkylperoxidic complexes-synthesis, x-ray structure and hydroxylating properties of vanadium(V) dipicolinato alkylperoxides. *Nouv. J. Chim.* 7: 467–475.
71. Hanson, R. M., Sharpless, K. B. 1986. Procedure for the catalytic asymmetric epoxidation of allylic alcohols in the presence of molecular sieves. *J. Org. Chem.* 51: 1922–1925.
72. Gao, Y., Hanson, R. M., Klunder, J. M., Ko, S. Y., Masamune, H., Sharpless, K. B. 1987. Catalytic asymmetric epoxidation and kinetic resolution – modified procedures including *in situ* derivatization. *J. Am. Chem. Soc.* 109: 5765–5976.
73. Johnson, R. A., Sharpless, K. B. 2000. Catalytic asymmetric epoxidation of allylic alcohols. In *Catalytic Asymmetric Synthesis*, ed. Ojima, I. pp. 231–281. Wiley-VCH, New York.
74. Modena, G., Furia, F. D., Seraglia, R. 1984. Synthesis of chiral sulfoxides by metal-catalyzed oxidation with tert-butylhydroperoxide. *Synthesis* 325–326.
75. Pitchen, P., Dunach, E., Deshmukh, M. N., Kagan, H. B. 1984. An efficient asymmetric oxidation of sulfides. *J. Am. Chem. Soc.* 106: 8188–8193.
76. Bonchio, M., Licini, G., Modena, G., Bortolini, O., Moro, S., Nugent, W. A. 1999. Enantioselective Ti(IV) sulfoxidation catalysts bearing C$_3$-symmetric trialkanolamine ligands: Solution speciation by ^{1}H NMR and ESI-MS analysis. *J. Am. Chem. Soc.* 121: 6258–6268.
77. Fujiwara, M., Wessel, H., Hyang-Suh, P., Roesky, H. W. 2002. Formation of titanium tert-butylperoxo intermediate from cubic silicon-titanium complex with tert-butyl hydroperoxide and its reactivity for olefin epoxidation. *Tetrahedron* 58: 239–243.
78. Babushkin, D. E., Talsi, E. P. 1999. ^{13}C and ^{1}H NMR spectroscopic characterization of titanium(IV) alkylperoxo complexes Ti(OOtBu)$_n$(OiPr)$_{4-n}$ with n = 1, 2, 3, 4. *React. Kinet. Catal. Lett.* 67: 359–364.
79. Babushkin, D. E., Talsi, E. P. 2003. Formation, solution structure and reactivity of alkylperoxo complexes of titanium. *J. Mol. Catal. A Chem.* 200: 165–175.
80. Indictor, N., Brill, W. F. 1965. Metal acetylacetonate catalyzed epoxidation of olefins with t-butyl hydroperoxide. *J. Org. Chem.* 30: 2074–2075.
81. Mimoun, H., Mignard, M., Brechot, P., Saussine, L. 1986. Selective epoxidation of olefins by oxo[N-(2-oxidophenyl)salicyldenaminato]vanadium(V) alkylperoxides. On the mechanism of the Halcon epoxidation process. *J. Am. Chem. Soc.* 108: 3711–3718.
82. Talsi, E. P., Chinakov, V. D., Babenko, V. P., Zamaraev, K. I. 1993. Role of vanadium alkylperoxo complexes in epoxidation of cyclohexene by organic hydroperoxides in the presence of bis(acetylacetonato)vanadyl. *J. Mol. Catal.* 81: 235–254.

NMR and EPR Spectroscopy

83. Babushkin, D. E., Talsi, E. P. 2000. Multinuclear NMR spectroscopic characterization of vanadium(V) alkylperoxo complexes $VO(OO\text{-}t\text{Bu})_k(OnBu)_{3-k}$, where k = 1, 2, 3. *React. Kinet. Catal. Lett.* 71: 115–120.

84. Sharpless, K. B., Michaelson, R. C. 1973. High stereo- and regioselectivities in the transition metal catalyzed epoxidations of olefinic alcohols by tert-butyl hydroperoxide. *J. Am. Chem. Soc.* 95: 6136–6137.

85. Michaelson, R. C., Palermo, R. E., Sharpless, K. B. 1977. Chiral hydroxamic acids as ligands in the vanadium catalyzed asymmetric epoxidation of allylic alcohols by tert-butyl hydroperoxide. *J. Am. Chem. Soc.* 99: 1990–1992.

86. Murase, N., Hoshino, Y., Oishi, M., Yamamoto, H. 1999. Chiral vanadium-based catalysts for asymmetric epoxidation of allylic alcohols. *J. Org. Chem.* 64: 338–339.

87. Hoshino, Y., Yamamoto, H. 2000. Novel α-amino acid-based hydroxamic acid ligands for vanadium-catalyzed asymmetric epoxidation of allylic alcohols. *J. Am. Chem. Soc.* 122: 10452–10453.

88. Bolm, C., Kühn, T. 2000. Asymmetric epoxidation of allylic alcohols using vanadium complexes of N-hydroxy-[2.2]paracyclophane-4-carboxylic amids. *SYNLETT* 899–901.

89. Bolm, C. 2003. Vanadium-catalyzed asymmetric oxidations. *Coord. Chem. Rev.* 237: 245–256.

90. Bryliakov, K. P., Talsi, E. P., Kühn, T., Bolm, C. 2003. Multinuclear NMR study of the reactive intermediates in enantioselective epoxidation of allylic alcohols catalyzed by a vanadium complex derived from a planar-chiral hydroxamic acid. *New J. Chem.* 27: 609–614.

91. Ji, Y., Benkovics, T., Beutner, G. L., Sfouggatakis, C., Eastgate, M. D., Blackmond, D. G. 2015. Mechanistic insights into the vanadium-catalyzed Achmatovicz rearrangement of furfurol. *J. Org. Chem.* 80: 1696–1702.

92. Mimoun, H., Seree de Roch, I., Sajus, L. 1970. Epoxidation des olefins par les complexes peroxydiques covalents du molybden-VI. *Tetrahedron* 26: 37–50.

93. Sharpless, K. B., Townsend, J. M., Williams, D. R. 1972. On the mechanism of epoxidation of olefins by covalent peroxides of molybdenum(VI). *J. Am. Chem. Soc.* 94: 295–296.

94. Deubel, D. V., Sundermeyer, J., Frenking, G. 2000. Mechanism of the olefin epoxidation catalyzed by molybdenum diperoxo complexes: Quantum chemical calculations give an answer to a long-standing question. *J. Am. Chem. Soc.* 122: 10101–10108.

95. Gisdakis, P., Yudanov, I. V., Rösch, N. 2001. Olefin epoxidation by molybdenum and rhenium peroxo and hydroperoxo compounds: A density functional study of energetics and mechanisms. *Inorg. Chem.* 40: 3755–3765.

96. Talsi, E. P., Shalyaev, K. V., Zamaraev, K. I. 1993. ^{17}O, ^{95}Mo and ^1H NMR study of the mechanism of epoxidation of alkenes with hydrogen peroxide in the presence of molybdenum complexes. *J. Mol. Catal. A Chem.* 83: 347–366.

97. Conte, V., Floris, B. 2011. Vanadium and molybdenum peroxides: Synthesis and catalytic activity in oxidation reactions. *Dalton Trans.* 40: 1419–1436.

98. Conte, V., Coletti, A., Floris, B., Licini, G., Zonta, C. 2011. Mechanistic aspects of vanadium catalyzed oxidations with peroxides. *Coord. Chem. Rev.* 255: 2165–2177.

99. Conte, V., Di Furia, F., Licini, G. 1997. Liquid phase oxidation reactions by peroxides in the presence of vanadium complexes. *Appl. Catal. A General* 157: 335–361.

100. Mimoun, H., Saussine, L., Daire, E., Postel, M., Fischer, J., Weiss, R. 1983. Vanadium(V) peroxo complexes. New versatile biomimetic reagents for epoxidation of olefins and hydroxylation of alkanes and aromatic hydrocarbons. *J. Am. Chem. Soc.* 105: 3101–3110.

101. Ballistreri, F. P., Tomaselli, G. A., Toscano, R. M., Conte, V., Di Furia, F. 1991. Application of the thianthrene 5-oxide mechanistic probe to peroxometal complexes. *J. Am. Chem. Soc.* 113: 6209–6212.

102. Bonchio, M., Di Furia, F., Modena, G. 1989. Metal catalysis in oxidations by peroxides. 31. The hydroxylation of benzene by $VO(O_2)(Pic)(H_2O)_2$: Mechanistic and synthetic aspects. *J. Org. Chem.* 54: 4368–4371.

103. Talsi, E. P., Shalyaev, K. V. 1994. ^{51}V NMR and EPR study of the mechanistic details of oxidation with $VO(O_2)(Pic)(H_2O)_2$. *J. Mol. Catal.* 92: 245–255.
104. Kirillova, M. V., Kuznetsov, M. L., Romakh, V. B., Shul'pina, L. S., da Silva, J. J. R. F., Pombeiro, A. J. L., Shul'pin, G. B. 2009. Mechanism of oxidations with H_2O_2 catalyzed by vanadate anion or oxovanadium(V)triethanolaminate(vanadatrene) in combination with pyrazine-2-carboxylic acid (RCA): Kinetic and DFT studies. *J. Catal.* 267: 140–157.
105. Nakagawa, Y., Kamata, K., Kotani, M., Yamaguchi, K., Mizuno, N. 2005. Polyoxovanadometalate-catalyzed selective epoxidation of alkenes with hydrogen peroxide. *Angew. Chem. Int. Ed.* 44: 5136–5141.
106. Kamata, K., Sugahara, K., Yonehara, K., Ishimoto, R., Mizuno, N. 2011. Efficient epoxidation of electron-deficient alkenes with hydrogen peroxide catalyzed by $[\gamma\text{-}PW_{10}O_{38}V_2(\mu\text{-}OH)_2]^{3-}$. *Chem. Eur. J.* 17: 7549–7559.
107. Kamata, K., Yamaura, T., Mizuno, N. 2012. Chemo- and regioselective direct hydroxylation of arenes with hydrogen peroxide catalyzed by a divanadium-substituted phosphotungstate. *Angew. Chem. Int. Ed.* 51: 7275–7278.
108. Kamata, K., Yonehara, K., Nakagawa, Y., Uehara, K., Mizuno, N. 2010. Efficient stereo- and regioselective hydroxylation of alkanes catalyzed by a bulky polyoxometalate. *Nat. Chem.* 2: 478–483.
109. Ivanchikova, I. D., Maksimchuk, N. V., Maksimovskaya, R. I., Maksimov, G. M., Kholdeeva, O. A. 2014. Highly selective oxidation of alkylphenols to p-benzoquinones with aqueous hydrogen peroxide catalyzed by divanadium-substituted polyoxotungstates. *ACS Catal.* 4: 2706–2713.
110. Bolm, C., Bienewald, F. 1996. Asymmetric sulfide oxidation with vanadium catalysts and H_2O_2. *Angew. Chem. Int. Ed.* 34: 2640–2642.
111. Bryliakov, K. P., Karpyshev, N. N., Fominsky, S. A., Tolstikov, A. G., Talsi, E. P. 2001. ^{51}V and ^{13}C NMR spectroscopic study of the peroxovanadium intermediates in vanadium catalyzed enantioselective oxidation of sulfides. *J. Mol. Catal. A Chem.* 171: 73–80.
112. Mimoun, H., Postel, M., Casabianca, F., Fischer, J., Mitschler, A. 1982. Novel unusually stable peroxotitanium(IV) compounds. Molecular and crystal structure of per oxobis(picolinato)(hexamethylphosphoric triamide)titanium(IV). *Inorg. Chem.* 21: 1303–1306.
113. Ledon, H. J., Varescon, F. 1984. Pole of peroxo vs. alkylperoxo titanium porphyrin complexes in the epoxidation of olefins. *Inorg. Chem.* 23: 2735–2737.
114. Yamase, T., Ishikawa, E., Asai, Y., Kanai, S. 1996. Alkene epoxidation by hydrogen peroxide in the presence of titanium-substituted Keggin-type polyoxotungstates $[PTi_xW_{12-x}O_{40}]^{(3+2x)-}$ and $[PTi_xW_{12-x}O_{40-x}(O_2)_x]^{(3+2x)-}$ (x = 1 and 2). *J. Mol. Catal. A* 114: 237–245.
115. Kholdeeva, O. A., Maksimov, G. M., Maksimovskaya, R. I., Kovaleva, L. A., Fedotov, M. A., Grigoriev, V. A., Hill, C. L. 2000. A dimeric titanium-containing polyoxometalate. Synthesis, characterization, and catalysis of H_2O_2-based thioether oxidation. *Inorg. Chem.* 39: 3828–3837.
116. Kholdeeva, O. A., Trubitsina, T. A., Maksimovskaya, R. I., Golovin, A. V., Neiwert, W. A., Kolesov, B. A., López, X., Poblet, J. M. 2004. First isolated active titanium peroxo complex: Characterization and theoretical study. *Inorg. Chem.* 43: 2284–2292.
117. Kholdeeva, O. A., Trubitsina, T. A., Timofeeva, M. N., Maksimov, G. M., Maksimovskaya, R. I., Rogov, V. A. 2005. The role of protons in cyclohexene oxidation with H_2O_2 catalyzed by Ti(IV)-monosubstituted Keggin polyoxymetalate. *J. Mol. Catal. A* 232, 173–178.
118. Antonova, N. S., Carbó, J. J., Kortz, U., Kholdeeva, O. A., Poblet, J. M. 2010. Mechanistic insights into alkene epoxidation with H_2O_2 by Ti- and other TM-containing polyoxometalates: Role of the metal nature and coordination environment. *J. Am. Chem. Soc.* 132: 7488–7497.

NMR and EPR Spectroscopy

119. Talsi, E. P., Bryliakov, K. P. 2012. Chemo- and stereoselective C-H oxidations and epoxidations/cis-dihydroxylations with H_2O_2, catalyzed by non-heme iron and manganese complexes. *Coord. Chem. Rev.* 256: 1418–1434.

120. Bryliakov, K. P., Talsi, E. P. 2014. Active sites and mechanisms of bioinspired oxidation with H_2O_2, catalyzed by non-heme iron and manganese complexes. *Coord. Chem. Rev.* 256: 1418–1434.

121. Samsel, E. G., Srinivasan, K., Kochi, J. K. 1985. Mechanism of the chromium-catalyzed epoxidation of olefins. Role of chromium(V) cations. *J. Am. Chem. Soc.* 107: 7606–7617.

122. Bryliakov, K. P., Talsi, E. P. 2003. Cr^{III}(salen)Cl catalyzed asymmetric epoxidations: Insight into the catalytic cycle. *Inorg. Chem.* 42: 7258–7265.

123. Zhang, W., Loebach, J. L., Wilson, S. R., Jacobsen, E. N. 1990. Enantioselective epoxidation of unfunctionalized olefins catalyzed by salen manganese complexes. *J. Am. Chem. Soc.* 112: 2801–2803.

124. Irie, R., Noda, K., Ito, Y., Matsumoto, N., Katsuki, T. 1990. Catalytic asymmetric epoxidation of unfunctionalized olefins. *Tetrahedron Lett.* 31: 7345–7348.

125. Katsuki, T. 1995. Catalytic asymmetric oxidations using optically active (salen) manganese(III) complexes as catalysts. *Coord. Chem. Rev.* 140: 189–214.

126. Feichtinger, D., Plattner, D. A. 1997. Direct proof for O = Mn^V(salen) complexes. *Angew. Chem.* 36:1718–1719.

127. Palucki, M., Finney, N. S., Pospisil, P. J., Güler, M. L., Ishida, T., Jacobsen, E. N. 1998. The mechanistic basis for effects on enantioselectivity in the salenMn(III)-catalyzed epoxidation reaction. *J. Am. Chem. Soc.* 120: 948–954.

128. Bryliakov, K. P., Babushkin, D. E., Talsi, E. P. 2000. ^{1}H NMR and EPR spectroscopic monitoring of the reactive intermediates of (Salen)Mn(III) catalyzed olefin epoxidation. *J. Mol. Catal. A Chem.* 158: 19–35.

129. Adam, W., Roschmann, K. J., Saha-Möller, C. R., Seebach, D. 2002. *cis*-Stilbene and (1α, 2α, 3α)-(2-ethenyl-3-methoxycyclopropyl) benzene as mechanistic probes in Mn^{III}(salen)-catalyzed epoxidation: Influence of the oxygen source and the counter ion on the diastereoselectivity of the competitive concerted and radical-type oxidation transfer. *J. Am. Chem. Soc.* 124: 5068–5073.

130. Palucki, M., Pospisil, P. J., Zhang, W., Jacobsen, E. N. 1994. Highly enantioselective, low-temperature epoxidation of styrene. *J. Am. Chem. Soc.* 116: 9333–9334.

131. McGarrigle, E. M., Gilheany, D. G. 2005. Chromium- and manganese-salen promoted epoxidation of alkenes. *Chem. Rev.* 105: 1564–1598.

132. Srinivasan, K., Michaud, P., Kochi, J. K. 1986. Epoxidation of olefins with cationic (salen)manganese(III) complexes. The modulation of catalytic activity by substituents. *J. Am. Chem. Soc.* 108: 2309–2320.

133. Smegal, J. A., Schardt, B. C., Hill, C. L. 1983. Isolation, purification, and characterization of intermediate (iodosylbenzene)metalloporphyrin complexes from the (tetraphenylporphinato)manganese(III) – iodosylbenzene catalytic hydrocarbon functionalization system. *J. Am. Chem. Soc.* 105: 3510–3515.

134. Collins, T. J., Gordon-Wylie, S. W. 1989. A manganese(V)-oxo complex. *J. Am. Chem. Soc.* 111: 4511–4513.

135. Collins, T. J., Powell, R. D., Slebodnick, S., Uffelman, E. S. 1990. A water stable manganese(V)-oxo complex: Definitive assignment of a Mn^V°O triple bond infrared vibration. *J. Am. Chem. Soc.* 112: 899–901.

136. Jin, N., Groves, J. T. 1999. Unusual kinetic stability of a ground-state singlet oxomanganese(V) porphyrin. Evidence for a spin crossing effect. *J. Am. Chem. Soc.* 121: 2923–2924.

137. Groves, J. T., Lee, J., Marla, S. S. 1997. Detection and characterization of an oxomanganese(V) porphyrin complex by rapid-mixing stopped-flow spectrophotometry. *J. Am. Chem. Soc.* 119: 6269–6273.

138. Du Bois, J., Hong, J., Carreira, E. M., Day, M. W. 1996. Nitrogen transfer from a nitridomanganese(V) complex: Amination of silyl enol ethers. *J. Am. Chem. Soc.* 118: 915–916.

139. Collman, J. P., Zeng, L., Brauman, J. I. 2004. Donor ligand effect on the nature of the oxygenating species in Mn^{III}(salen)-catalyzed epoxidation of olefins: Experimental evidence for multiple active oxidants. *Inorg. Chem.* 43: 2672–2679.

140. Groves, J. T. 2006. High-valent iron in chemical and biological oxidations. *J. Inorg. Biochem.* 100: 434–447.

141. Meunier, B., de Visser, S. R., Shaik, S. 2004. Mechanism of oxidation reactions catalyzed by cytochrome P450 enzymes. *Chem. Rev.* 104: 3947–3980.

142. Mansuy, D. 1993. Activation of alkanes: The biomimetic approach. *Coord. Chem. Rev.* 125: 129–141.

143. Nam, W. 2007. High-valent iron(IV)-oxo complexes of heme and non-heme ligands in oxygenation reactions. *Acc. Chem. Res.* 40: 522–531.

144. Costas, M. 2011. Selective C-H oxidation catalyzed by metalloporphyrins. *Coord. Chem. Rev.* 255: 2912–2932.

145. Solomon, E. I., Brunold, T. C., Davis, M. J., Kemsley, J. N., Lee, S. K., Lehnert, N., Neese, F., Skulan, A. J., Yang, Y. S., Zhou, J. 2000. Geometric and electronic structure/function correlations in non-heme iron enzymes. *Chem. Rev.* 100: 235–349.

146. Costas, M., Mehn, M. P., Jensen, M. P., Que, L., Jr. 2004. Dioxygen activation of mononuclear nonheme iron active sites: Enzymes, models, and intermediates. *Chem. Rev.* 104: 939–986.

147. Que, L. Jr. 2007. The road to non-heme oxoferryls and beyond. *Acc. Chem. Res.* 40: 493–500.

148. Krebs, C., Fujimori, D. G., Walsh, C. T., Bollinger, M. 2007. Non-heme Fe(IV)-oxo intermediates. *Acc. Chem. Res.* 40: 484–492.

149. McDonald, A. R., Que, L. Jr. 2013. High-valent nonheme iron-oxo complexes: Synthesis, structure, and spectroscopy. *Coord. Chem. Rev.* 257: 414–428.

150. Que, L. Jr., Tolman, W. B. 2008. Biologically inspired oxidation catalysis. *Nature* 455: 333–340.

151. Bruijnincx, P. C. A., van Koten, G., Klein Gebbink, R. J. M. 2008. Mononuclear nonheme iron enzymes with the 2-His-1-carboxylate facial triad: Recent developments in enzymology and modeling studies. *Chem. Soc. Rev.* 37: 2716–2744.

152. Prat, I., Gómez, L., Canta, M., Ribas, X., Costas, M. 2013. An iron catalyst for oxidation of alkyl C-H bonds showing enhanced selectivity for methylene sites. *Chem. Eur. J.* 19: 1908–1913.

153. White, M. C., Doyle, A. G., Jacobsen, E. N. 2001. A synthetically useful, self-assembling MMO mimic system for catalytic alkene epoxidation with aqueous H_2O_2. *J. Am. Chem. Soc.* 123: 7194–7195.

154. Chen, M. S., White, M. C. 2007. A predictably selective aliphatic C-H oxidation reaction for complex molecule synthesis. *Science* 318: 783–787.

155. Chen, M. S., White, M. C. 2010. Combined effects on selectivity in Fe-catalyzed methylene oxidation. *Science* 327: 566–571.

156. Company, A., Gómez, L., Fontrodona, X., Ribas, X., Costas, M. 2008. A novel platform for modeling oxidative catalysis in non-heme iron oxygenases with unprecedented efficiency. *Chem. Eur. J.* 14: 5727–5731.

157. Gomez, L., Garcia-Bosch, I., Company, A. A., Benet-Buchholz, J., Polo, A., Sala, X., Ribas, X., Costas, M. 2009. Stereospecific C-H oxidation with H_2O_2 catalyzed by a chemically robust site isolated iron catalyst. *Angew. Chem. Int. Ed.* 48: 5720–5723.

158. Gussó, O., Garcia-Bosch, I., Ribas, X., Lloret-Fillol, J., Costas, M. 2013. Asymmetric epoxidation with H_2O_2 by manipulating the electronic properties of non-heme iron catalysts. *J. Am. Chem. Soc.* 135: 14871–14878.

NMR and EPR Spectroscopy

159. Lyakin, O. Y., Ottenbacher, R. V., Bryliakov, K. P., Talsi, E. P. 2013. Active species of nonheme iron and manganese-catalyzed oxidations. *Top. Catal.* 56: 939–949.

160. de Oliveira, F. T., Chanda, A., Banerjee, D., Shan, X., Mondal, S., Que, L., Jr., Bominaar, E. L., Münck, E., Collins, T. J. 2007. Chemical and spectroscopic evidence for an Fe(V) oxo complex. *Science* 315: 835–838.

161. Lyakin, O. Y., Bryliakov, K. P., Britovsek, G. J. P., Talsi, E. P. 2009. EPR spectroscopic trapping of the active species of nonheme iron-catalyzed oxidation. *J. Am. Chem. Soc.* 131: 10798–10799.

162. Lyakin, O. Y., Bryliakov, K. P., Talsi, E. P. 2011. EPR, ^{1}H and ^{2}H NMR, and reactivity studies of the iron-oxygen intermediates in bioinspired catalyst systems. *Inorg. Chem.* 50: 5526–5538.

163. Lyakin, O. Y., Ottenbacher, R. V., Bryliakov, K. P., Talsi, E. P. 2012. Asymmetric epoxidations with H_2O_2 on Fe and Mn aminopyridine catalysts: probing the nature of active species by combined electron paramagnetic resonance and enantioselectivity study. *ACS Catal.* 2: 1196–1202.

164. Lyakin, O. Y., Prat, I., Bryliakov, K. P., Costas, M., Talsi, E. P. 2012. EPR detection of Fe^V=O active species in nonheme iron-catalyzed oxidations. *Catal. Commun.* 29: 105–108.

165. Makhlynets, O. V., Rybak-Akimova, E. V. 2010. Aromatic hydroxylation at a non-heme iron center: Observed intermediates and insights into the nature of the active species. *Chem. Eur. J.* 16: 13995–14006.

166. Roelfes, G., Lubben, M., Chen, K., Ho, R. Y. N., Meetsma, A., Genseberger, S. et al. 1999. Iron chemistry of a pentadentate ligand that generates a metastable Fe^{III}-OOH intermediate. *Inorg. Chem.* 38: 1929–1936.

167. Duelund, L., Hazell, R., McKenzie, C. L., Nielsen, L. P., Tofflund, H. J. 2001. Solid and solution state structures of mono- and di-nuclear iron(III) complexes of related hexadentate and pentadentate aminopyridyl ligands. *J. Chem. Soc. Dalton Trans.* 152–156.

168. Mas-Balleste, R., Que, L., Jr. 2007. Iron catalyzed olefin epoxidation in the presence of acetic acid: Insights into the nature of the metal-based oxidant. *J. Am. Chem. Soc.* 129: 15964–15972.

169. Prat, I., Mathieson, J. S., Güell, M., Ribas, X., Luis, J. M., Cronin, L., Costas, M. 2011. Observation of Fe(V)=O using variable-temperature mass spectrometry and its enzyme-like C-H and C=C oxidation reactions. *Nat. Chem.* 3: 788–793.

170. Yoon, J., Wilson, S. A., Jang, Y. K., Seo, M. S., Nehru, K., Hedman, B., Hodson, K. O., Bill, E., Solomon, E. I., Nam, W. 2009. Reactive intermediates in oxygenation reactions with mononuclear nonheme iron catalysts. *Angew. Chem. Int. Ed.* 48: 1257–1260.

171. Company, A., Prat, I., Frisch, J. R., Mas-Balesté, R., Güell, M., Juhász, G. et al. 2011. Modeling the *cis*-oxo-labile binding site of non-heme iron oxygenases: Water exchange and oxidation reactivity of a non-heme iron(IV)-oxo compound bearing a tripodal tetradentate ligand. *Chem. Eur. J.* 17: 1622–1634.

172. Van Heuvelen, K. M., Fiedler, A. T., Shan, X., De Hont, R. F., Meier, K. K., Bominaar, E. L., Münck, E., Que, L., Jr. 2012. One-electron oxidation of an oxoiron(IV) complex to form an [O=Fe(V)=NR]$^+$ centre. *PNAS* 109: 11933–11938.

173. Wang, Y., Janardanan, D., Usharani, D., Han, K., Que, L., Jr., Shaik, S. 2013. The nonheme iron oxidant formed in the presence of H_2O_2 and acetic acid is the cyclic ferricperacetate complex and not a perferryl-oxo complex. *ACS Catal.* 3: 1334–1341.

174. Oloo, W. N., Meier, K. K., Wang, Y., Shaik, S., Münck, E., Que, L., Jr. 2014. Identification of a low-spin acylperoxoiron(III) intermediate in bio-inspired non-heme iron-catalyzed oxidations. *Nat. Commun.* 5, Article no. 3046.

175. Lyakin, O. Y., Zima, A. M., Samsonenko, D. G., Bryliakov, K. P., Talsi, E. P. 2015. EPR spectroscopic detection of the elusive Fe^V=O intermediates in selective catalytic oxofunctionalizations of hydrocarbons by biomimetic ferric complexes. *ACS Catal.* 5: 2702–2707.

176. Zang, X., Kim, J., Dong, Y., Wilkinson, E. C., Appelman, E. H., Que, L., Jr. 1997. Stereospecific alkane hydroxylation with H_2O_2 catalyzed by an iron(II)-tris(2-pyridylmethyl)amine complex. *J. Am. Chem. Soc.* 119, 5964–5965.
177. Weil, J. A., Bolton, J. R. 2007. Electron paramagnetic resonance. In *Elementary Theory and Practical Applications*, 2nd ed. pp. 85–117. John Wiley & Sons, Inc., Hoboken, NJ.
178. McGown, A. J., Kerber, W. D., Fujii, H., Goldberg, D. P. 2009. Catalytic reactivity of a *meso-N*-substituted corrole and evidence for a high-valent iron-oxo species. *J. Am. Chem. Soc.* 131: 8040–8048.
179. Rittle, J., Green, M. T. 2010. Cytochrome P450 compound I: Capture, characterization, and C-H bond activation kinetics. *Science* 330: 933–937.
180. Tse, C.-W., Show, T. W.-S., Guo, Z., Lee, H. K., Huang, J.-S., Che, C.-M. 2014. Nonheme iron mediated oxidation of light alkanes with oxone: Characterization of reactive oxoiron(IV) ligand cation radical intermediates by spectroscopic studies and DFT calculations. *Angew. Chem. Int. Ed.* 53: 798–803.
181. Oloo, W. N., Que, L., Jr. 2015. Bioinspired nonheme iron catalysts for C-H and C=C bond oxidation: Insights into the nature of the metal-based oxidants. *Acc. Chem. Res.* 48: 2612–2621.
182. Cussó, O., Ribas, X., Costas, M. 2015. Biologically inspired non-heme iron catalysts for asymmetric epoxidation; design principles and perspectives. *Chem. Commun.* 51: 14285–14298.
183. Zima, A. M., Lyakin, O. Y., Ottenbacher, R. V., Bryliakov, K. P., Talsi, E. P. 2016. Dramatic effect of carboxylic acid on the electronic structure of the active oxoferryl species in Fe(PDP)-catalyzed biomimetic oxidation. *ACS Catal.* 6: 5399–5404.
184. Sheldon, R. A., Kochi, J. K. 1981. *Metal-Catalyzed Oxidations of Organic Compounds*. Academic Press, New York.
185. Parshall, G. W., Ittel, D. I. 1993. *Homogeneous Catalysis*, 2nd ed. Wiley, New York.
186. Sumner, C. E., Jr. 1988. Interconversion of dinuclear and oxo-centered trinuclear cobaltic acetates. *Inorg. Chem.* 27: 1320–1327.
187. Sumner, C. E., Jr., Steinmetz, G. R. 1985. Isolation of oxo-centered cobalt(III) clusters and their role in the cobalt bromide catalyzed autoxidation of aromatic hydrocarbons. *J. Am. Chem. Soc.* 107: 6124–6126.
188. Sumner, C. E., Jr., Steinmetz, G. R. 1989. Synthesis and reactivity of alkoxide-bridged cobaltic acetates. *Inorg. Chem.* 28: 4290–4294.
189. Blake, A. B., Chipperfield, J. R., Lau, S., Webster, D. E. 1990. Studies on the nature of cobalt(III) acetate. *J. Chem. Soc. Dalton Trans.* 3719–3724.
190. Chipperfield, J. R., Lau, S., Webster, D. E. 1992. Speciation in cobalt(III)-catalyzed autoxidation: Studies using thin-layer chromatography. *J. Mol. Catal. A* 75: 123–128.
191. Babushkin, D. E., Talsi, E. P. 1998. Multinuclear NMR spectroscopic characterization of Co(III) species: Key intermediates of cobalt catalyzed autoxidation. *J. Mol. Catal. A* 130: 131–137.
192. Nosova, V. M., Ustynyuk, Y. A., Bruk, L. G., Temkin, O. N., Kisin, A. V., Storozhenko, P. A. 2011. Structure of complexes formed by dissolution of palladium diacetate in methanol and chloroform. *Inorg. Chem.* 50: 9300–9310.

4 NMR and EPR Spectroscopy in the Study of the Mechanisms of Metallocene and Post-Metallocene Polymerization and Oligomerization of α-Olefins

4.1 METALLOCENE CATALYSTS

4.1.1 INTRODUCTION

The discovery of Ziegler–Natta polymerization catalysts has been one of the most important achievements of modern chemistry [1,2]. More than 110 million tons of polyolefins are produced annually with the use of these catalysts [3]. Traditional Ziegler–Natta catalysts are heterogeneous systems generated from titanium tetrachloride, aluminum alkyl activators, for example, $AlEt_3$, and a solid support such as $MgCl_2$. To control the morphology of polymer particles, the catalyst is often placed into the pores of spherical granules of silica gel [4]. Conclusive characterization of the active centers of heterogeneous Ziegler–Natta catalysts remains a still unsolved problem because the applicability of spectroscopic techniques for the detailed studies of these complex, nonuniform (multisite) systems is rather limited.

Fortunately, in 1980, homogeneous, metallocene-based single-site analogs of Ziegler–Natta catalysts were discovered [5]. For these catalysts, the detailed data on the structure of the active centers can be obtained [6–13]. The interest in single-site polymerization largely increased after the discovery that *ansa*-metallocenes can produce stereoregular polypropylene, providing a precise control of tacticity [14,15]. Polyolefins produced with single-site catalysts exhibited well-defined microstructures, controlled molecular weight distributions, and superior mechanical and optical properties. Nowadays, metallocene-based polymers constitute about 10% of the total polyolefin production on Ziegler–Natta catalysts. This proportion is constantly

128 Applications of EPR and NMR Spectroscopy in Homogeneous Catalysis

increasing since the demand for special varieties of polymers, produced on single-site catalysts, increases faster than the demand for traditional polyolefins [3].

Modern homogeneous single-site catalysts are usually generated *in situ*, by reacting a metallocene or post-metallocene complex with an activator. In industry, the most widely used activators are the so-called methylalumoxane (MAO, a product of a controlled partial hydrolysis of trimethylaluminum [TMA]) or modified methylalumoxane (MMAO, which contains other alkyl groups, C_2 and higher). The latter has an advantage of being well soluble in nonpolar solvents, typically saturated hydrocarbons. The well-known activators for single-site catalysts also include tris(pentafluorophenyl)borate, $B(C_6F_5)_3$, and trityl tetrakis(pentafluorophenyl)borate, $[CPh_3][B(C_6F_5)_4]$.

The discovery of unique cocatalytic activity of MAO opened the avenue for single-site catalysis [16]. Now, MAO is produced commercially; however, rather little is known about its structure. Limited information on the structure of MAO can be obtained by means of NMR and EPR spectroscopy. Much more fortunate is the situation with the characterization of cationic parts of ion pair intermediates formed upon the activation of metallocenes and post-metallocenes with MAO. In this case, conclusive NMR data on the structure of the active centers of polymerization can be obtained. Below, we will present some results of NMR and EPR spectroscopic studies of MAO and ion pairs formed upon the activation of metallocenes and post-metallocenes with MAO. Recent data on the structure of MAO, obtained by other techniques, will also be briefly summarized.

4.1.2 SIZE OF MAO OLIGOMERS

MAO is a product of partial hydrolysis of TMA. The composition of MAO was found to be approximately $Me_{1.5}AlO_{0.75}$ [17]. Commercially available solutions of MAO consist of a polymeric MAO fraction and residual TMA. The ^{27}Al and ^{17}O NMR spectroscopic data can provide valuable information on the coordination geometries of aluminum and oxygen atoms in structurally unknown compounds because the corresponding chemical shifts are rather sensitive to these geometries. ^{27}Al NMR chemical shifts for three- (210–280 ppm), four- (125–180 ppm), and five- (100–125 ppm) coordinated Al atoms in organoaluminum compounds are different [18]. The ^{27}Al NMR chemical shifts reported for MAO in some publications (δ 149–152, $\Delta v_{1/2} = 1750$ Hz) [19,20] were nearly the same as that for TMA (δ 153, $\Delta v_{1/2} = 850$ Hz) [21]. One can assume that the ^{27}Al resonance of residual TMA was erroneously ascribed to MAO, while the very broad resonance of MAO is unobservable at room temperature. Babushkin, Talsi, and coworkers have undertaken a multinuclear NMR spectroscopic study of TMA-depleted samples of MAO (DMAO). It was shown that toluene solution of DMAO displays extremely broad ^{27}Al resonance at room temperature ($\Delta v_{1/2} \sim 50$ kHz), which cannot be observed without special probehead, and particular spectrometer settings. At elevated temperatures, the ^{27}Al resonance at δ 110 ± 10 ($\Delta v_{1/2} = 10$ kHz) attributable to DMAO can be clearly seen (Figure 4.1) [22]. After cooling the sample back to room temperature, the ^{27}Al resonance at δ 110 disappeared. This behavior was reproducible in several repeated heating/cooling cycles. Apparently, with

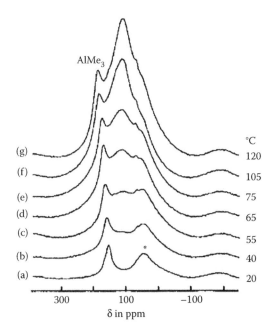

FIGURE 4.1 ^{27}Al NMR spectra of DMAO solution in toluene (2 M) at various temperatures: 20°C (a), 40°C (b), 55°C (c), 65°C (d), 75°C (e), 105°C (f), and 120°C (g). Asterisk (*) marks the signal of the NMR probehead. (Reproduced with permission from Babushkin, D. E. et al. 1997. *Macromol. Chem. Phys.* 198: 3845–3854.)

decreasing temperature, oligomeric molecules of MAO tend to associate to form larger aggregates, invisible by ^{27}Al NMR spectroscopy due to dramatic line broadening. The chemical shift of the observed resonance of MAO was within the range reported for aluminoxane clusters [(*t*Bu)Al(μ$_3$-O)]$_6$ (δ 110) and [(*t*Bu)Al(μ$_3$-O)]$_9$ (δ 120), with cage structure and RAlO$_3$ environment (Figure 4.2) [23]. On the basis of this similarity, it was proposed that MAO oligomers observed at elevated temperatures also have cage structure (Figure 4.2). The ^{17}O NMR resonance of MAO at 50°C (δ 65, Δv$_{1/2}$ = 1.7 kHz) fitted within the range typical for three-coordinated oxo ligands [22].

The size of MAO cages at 120°C was evaluated by comparison of the width of the observed ^{27}Al resonance of MAO (Δv$_{1/2}$ = 10 kHz, Figure 4.1g) and those for aluminoxane clusters [(*t*Bu)Al(μ$_3$-O)]$_6$ (Δv$_{1/2}$ = 7 kHz) and [(*t*Bu)Al(μ$_3$-O)]$_9$ (Δv$_{1/2}$ = 6.5 kHz) [23]. The evaluated radius of MAO oligomers was found to be about 5 Å at 120°C and about 7 Å at room temperature. This corresponds to (MeAlO)$_n$ oligomers with n = 9–14 and n = 20–30, respectively [22].

Hansen and coworkers have evaluated the size of MAO oligomers from the ^{1}H NMR spin–lattice relaxation time data (T_1) [24]. Their model provided an estimate of the molecular diameter of MAO oligomers in toluene at 20°C: d = 19 Å at 20°C. This value is larger than that obtained by Babushkin et al. (d = 14 Å) [22]. Recently, Macchioni and coworkers have evaluated the size of MAO and DMAO oligomers by diffusion NMR spectroscopy [25]. The average hydrodynamic volume

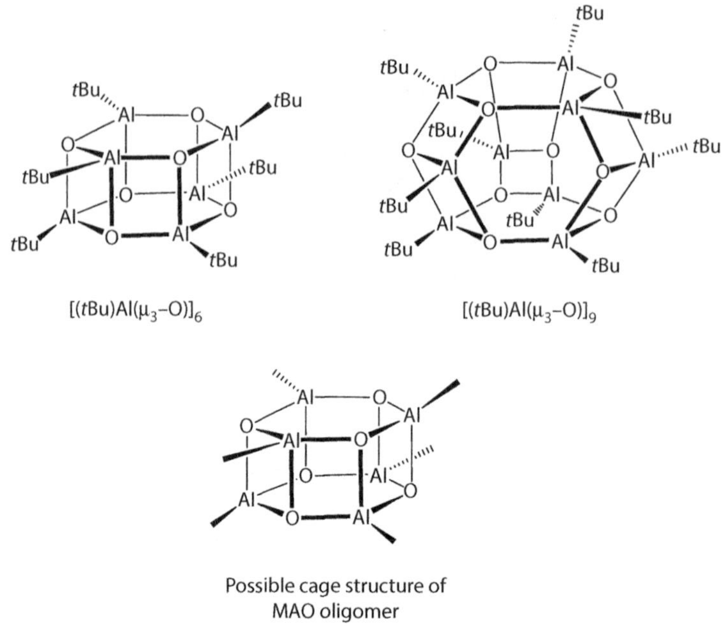

130 Applications of EPR and NMR Spectroscopy in Homogeneous Catalysis

[(tBu)Al(µ₃–O)]₆   [(tBu)Al(µ₃–O)]₉

Possible cage structure of
MAO oligomer

FIGURE 4.2 (tBuAlO)$_n$ cage structures as reported by Barron and coworkers [23], and possible structure of one of the MAO oligomers.

(V_H) depended on the concentration, and varied from 2720 Å^3 at [Al] = 17.4 mM to 3490 Å^3 at [Al] = 299.3 mM for MAO and from 3610 Å^3 at [Al] = 10 mM to 4700 Å^3 at [Al] = 127 mM for DMAO. This estimation was in nice agreement with that obtained by Hansen and coworkers.

Bochmann, Linnolahti, Stellbrink, and coworkers reported combined chemical, spectroscopic, neutron scattering, and computational studies of MAO and its interaction with TMA [26]. According to the PFG-NMR experiments, 75% of TMA molecules diffuse quickly, whereas 25% of TMA molecules are bound to MAO. The hydrodynamic radius R_H of MAO oligomers in toluene solutions at 25°C was found to be 12 ± 0.3 Å, which corresponds to the average size of MAO oligomers of about 50–60 Al atoms. Small-angle neutron scattering (SANS) gave similar size of MAO oligomers.

Overall, we can conclude that different methods provide various values for the radius of MAO oligomers, falling in the range of 7–12 Å at room temperature, that corresponds to 30–60 Al atoms per MAO oligomer.

4.1.3 On the Active Centers of MAO

One of the main roles of MAO was suggested to be cationization of metallocene or post-metallocene precatalysts, to form ion pairs of the type [LMCH$_3$]$^+$[Me–MAO]$^-$ (M = metal, L = metallocene or post-metallocene ligand). This implies that MAO should contain strong Lewis acidic sites. In heterogeneous catalysis, one of the

NMR and EPR Spectroscopy of the Mechanisms of Metallocene

widespread methods for the characterization of acidic sites is the application of spin probes (usually stable nitroxyl radicals). The EPR parameters of stable nitroxyl radical coordinated to the acidic site of a heterogeneous catalyst noticeably differ from those of the free radical and depend on the strength of the acidic sites [27]. We applied the EPR spin-probe technique for the characterization of Lewis acidic sites of MAO, using stable radical 2,2′,6,6′-tetramethylpiperidine-N-oxyl (TEMPO) as the spin probe [28].

First, the spin-probe technique was "calibrated" using a series of Lewis acidic aluminum compounds, $AlCl_3$, $AlEtCl_2$, and $AlEt_2Cl$, in toluene solutions. It was found that the value of hfs a_{Al} from ^{27}Al ($I = 5/2$) in the EPR spectrum of the adducts $AlCl_xEt_y$·TEMPO decreased in the following sequence: $AlCl_3$ ($a_{Al} = 8.8$ G) > $AlCl_2Et$ ($a_{Al} = 6.9$ G) > $AlClEt_2$ ($a_{Al} = 3.7$ G), which correlates with decreasing electron deficiency of the Al center and hence with its decreasing Lewis acidity.

For the EPR spin-probe study of MAO, solutions of TEMPO and DMAO in toluene were mixed at $-20°C$ in an EPR tube ([TEMPO]/[Al]$_{DMAO}$ = 1:20, [Al]$_{DMAO}$ = 0.2 M), and immediately frozen by immersion into liquid nitrogen. Then, the tube was placed into the resonator of the EPR spectrometer, and the sample was allowed to melt and to warm to room temperature over a period of ca. 30 min, with the EPR spectra being periodically recorded. The interaction of TEMPO with DMAO in toluene results in the formation of two types of adducts. The EPR spectrum measured 5 min after melting in the resonator of the EPR spectrometer displays a triplet ($a_N = 18.6$ G) with characteristic inflection points, which unambiguously documents the existence of unresolved hfs from aluminum ($a_{Al} = 1.0 \pm 0.2$ G) (Figure 4.3a). This signal disappears at room temperature, and after 15 min, a new signal (triplet of sextets) with $a_N = 19.6$ G and $a_{Al} = 1.9$ G is observed (Figure 4.3b). The intensity of the triplet of sextets decreases with time ($\tau_{1/2} = 3$ h at 20°C). So, DMAO contains at least two types of Lewis acidic sites, I_L and II_L. The hyperfine structure from the aluminum nucleus for sites II_L ($a_{Al} = 1.9$ G) is larger than that for sites I_L ($a_{Al} = 1.0 \pm 0.1$ G). On the basis of the presented data for $AlCl_xEt_y$·TEMPO adducts, we assumed that sites I_L and II_L are coordinatively unsaturated aluminum atoms in the $AlOMe_2$ and AlO_2Me environment [28].

Collins, McIndoe, and coworkers contributed systematic, negative, and positive ion studies of MAO by ESI-MS in fluorobenzene [29]. They suggested that MAO partially ionizes in sufficiently polar media to form ion pairs of the general formula $[Me_2Al][(MeAlO)_x(AlMe_3)_yMe]$. In the presence of octamethyltrisiloxane (OMTS), the most abundant ion pair had $x = 23$ and $y = 7$. The cation $[Me_2Al(OMTS)]^+$ (with m/z 293) and the anion (with m/z 1853) were detected [29]. On the basis of these data, it was assumed that MAO can be the source of $AlMe_2^+$ cations, and they can be responsible for the ionization of metallocenes.

In agreement with this assumption, the ^1H NMR studies of the interaction of MAO with donor molecules such as THF, pyridine, and diphenylphosphinopropane show that, besides TMA–donor ligand complexes, the cations $[Me_2AlL_2]^+$ are formed (L = donor molecule) [26]. The presented results suggest that MAO not only contains Lewis acidic sites but may also act as a source of strongly Lewis acidic $[AlMe_2]^+$ cations.

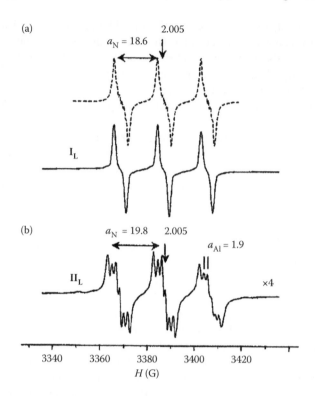

FIGURE 4.3 EPR spectra (toluene) of MAO + TEMPO mixtures ([TEMPO]/[Al]$_{MAO}$ = 1:20, [Al]$_{MAO}$ = 0.2 M), recorded 5 min after melting in the resonator (a, solid line) and after 20 min at 20°C (b). Simulation of the experimental spectrum with a_N = 18.6 G, a_{Al} = 1.0 G, line width = 1.0 G (a, dashed line). (Reproduced with permission from Bryliakov, K. P. et al. 2006. *Macromol. Chem. Phys.* 207: 327–335.)

4.1.4 STRUCTURE OF ION PAIRS FORMED UPON THE INTERACTION OF Cp$_2$ZrMe$_2$ WITH MAO

Cationic group 4 metalalkyl complexes of the general formula [Cp$_2$'M-R]$^+$ (R = alkyl, M = Ti, Zr, Hf) are highly active catalysts of the polymerization of olefins [30–35]. The first detailed ^{1}H, ^{13}C NMR spectroscopic study of cationic species formed in the catalyst systems [(Cp$_2$')M(CH$_3$)$_2$]/AlMe$_3$/[CPh$_3$][B(C$_6$F$_5$)$_4$] (Cp' = C$_5$H$_5$, Cp$_2$' = Me$_2$Si(Ind)$_2$, C$_2$H$_4$(Ind)$_2$, Me$_2$C(Cp)(Flu); M = Zr, Hf) was contributed by Bochmann and Lancaster [36]. Ion pairs [Cp$_2$'M-Me]$^+$[B(C$_6$F$_5$)$_4$]$^-$ (**I$_B$**), {[Cp$_2$'M-Me]$_2$(μ-Me)}$^+$[B(C$_6$F$_5$)$_4$]$^-$ (**II$_B$**), and [Cp$_2$'M(μ-Me)$_2$AlMe$_2$]$^+$[B(C$_6$F$_5$)$_4$]$^-$ (**III$_B$**) were detected and NMR characterized (Figure 4.4).

Inspired by those findings, Tritto and coworkers undertook the ^{13}C NMR study of cationic intermediates formed in the catalyst system Cp$_2$Zr(^{13}CH$_3$)$_2$/MAO at Al/Zr ratios 1–40 (isotopically ^{13}C-enriched precatalyst was used) [37]. Outer-sphere ion pairs {[Cp$_2$Zr-Me]$_2$(μ-Me)}$^+$[Me-MAO]$^-$ (**II**) and [Cp$_2$Zr(μ-Me)$_2$AlMe$_2$]$^+$[Me-MAO]$^-$ (**III**) were characterized by analogy with the ion pairs **II$_B$** and **III$_B$** (Figure 4.4).

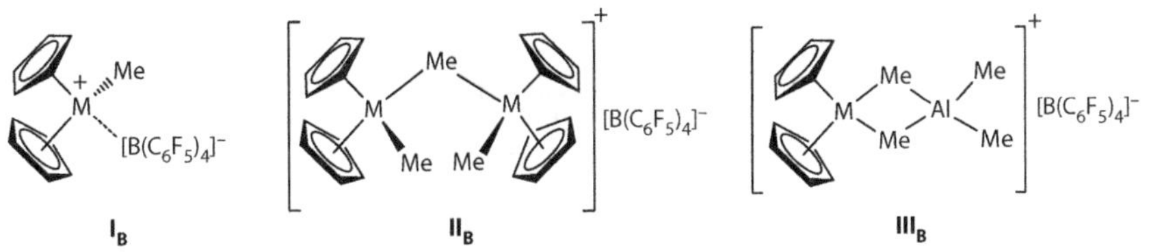

NMR and EPR Spectroscopy of the Mechanisms of Metallocene

FIGURE 4.4 The ion pairs formed upon the activation of Cp_2MMe_2 with $[Ph_3C][B(C_6F_5)_4]$ and $AlMe_3/[Ph_3C][B(C_6F_5)_4]$. M = Zr, Hf.

Besides sharp ^{13}C resonances of ion pairs **II** and **III**, very broad Cp and Zr-Me resonances of a new complex **IV** were observed [37]. The latter complex was identified by Babushkin and coworkers [38]. Using isotopically ^{13}C-enriched MAO, very broad Zr-Me and Zr-Me-Al resonances of complex **IV** were observed at δ 42 and 9 (Figure 4.5, Table 4.1). The resonance at δ 42 exhibited J^1_{CH} coupling of 120 ± 5 Hz that corresponds to terminal Zr-Me group. It was assumed that **IV** is contact ion pair $[Cp_2ZrMe^+...Me\text{-}MAO^-]$. The ^{13}C NMR chemical shifts of a structurally similar methyl-bridged complex $Cp_2ZrMe(\mu\text{-}Me)Al(C_6F_5)_3$ (δ 40.5 for the terminal methyl group and δ 7.9 ppm for the bridging one) [39] are close to those of **IV**.

In contrast to **II** and **III**, species **IV** can be considered as inner-sphere ion pair. The observed broadening of its resonances is caused by nonuniformity and big size of the $[Me\text{-}MAO]^-$ counteranions, tightly bound to the zirconocene. Contrariwise, outer-sphere ion pairs **II** and **III** display sharp resonances (Figure 4.5). The structures of zirconium species found in the system $Cp_2Zr(CH_3)_2/^{13}C\text{-}MAO$ are shown in Figure 4.6. Besides the aforementioned complexes **II–IV**, the adduct $Cp_2MeZr\text{-}Me{\rightarrow}Al{\equiv}MAO$ (**I**) is depicted in this figure. The existence of such adduct is evident from the broadening of the corresponding Zr-Me resonance (Figure 4.5). This resonance is a weighted average of sharp resonances of the free Cp_2ZrMe_2 and a broader resonance of **I**. Owing to rapid exchange between Cp_2ZrMe_2 and **I**, the chemical shift of the resulting resonance depends on the Al/Zr ratio (Figure 4.7).

The application of isotopically (^{13}C) enriched MAO allowed the monitoring of ion pairs formed in the system $Cp_2ZrMe_2/^{13}C\text{-}MAO$ at an Al/Zr ratio up to 4000. It was found that at high Al/Zr ratios, only ion pairs **III** and **IV** are present in the reaction solution (Figures 4.7 and 4.8) [38].

Babushkin and Brintzinger have estimated the size of the ion pair $[(C_5H_5)_2Zr(\mu\text{-}Me)_2]^+[Me\text{-}MAO]^-$ (**III**) by pulsed field-gradient (PFG) NMR spectroscopy. An effective hydrodynamic radius of 12.2–12.5 Å was obtained for **III** in benzene solution at different zirconocene and MAO concentrations [40]. The volume of the $Me\text{-}MAO^-$ anion suggested that it should contain 150–200 Al atoms. Macchioni and coworkers have confirmed this conclusion by the same technique showing that $[Me\text{-}MAO]^-$ counterions have large hydrodynamic radius >8.5 Å (>40 Al units per cluster) [25].

The detailed data on the structure of "zwitterion-like" intermediates formed upon the activation of Cp_2ZrMe_2 with MAO in toluene have stimulated the characterization of related intermediates in various zirconocene/MAO systems. In some cases, clear correlations between the spectroscopic and ethylene polymerization data were

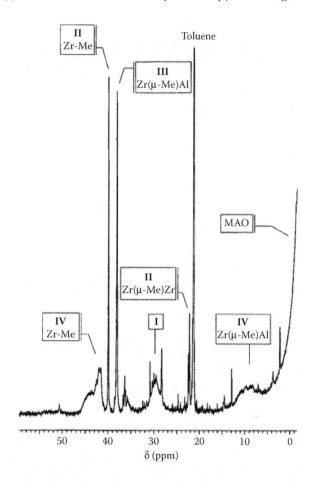

FIGURE 4.5 ^{13}C NMR spectrum (toluene, –25°C) of Cp$_2$ZrMe$_2$ + ^{13}C-MAO, [Al] = 1.8 M, Al/Zr = 100, in the range of Me-carbon signals. (Reproduced with permission from Babushkin, D. E. et al. 2000. *Macromol. Chem. Phys.* 201: 558–567.)

observed, confirming the key role of the zwitterion-like intermediates as the active sites of polymerization or their direct precursors.

4.1.5 Detection of Ion Pairs Formed upon Activation of (Cp-R)$_2$ZrCl$_2$ (R = nBu, tBu) with MAO

The ratio between ion pairs **III** and **IV** in the catalyst systems (Cp-R)$_2$ZrCl$_2$ (R = *n*Bu, *t*Bu) at various Al/Zr ratios was monitored by ^{1}H NMR spectroscopy, observing the resonances of the Cp ring. The ^{1}H NMR spectrum in the range of Cp hydrogen atoms of the system (Cp-*n*Bu)$_2$ZrCl$_2$/MAO (Al/Zr = 50) in toluene-*d*$_8$ displays sharp multiplets from **III** and several broad resonances denoted as **IV$_1$**, **IV$_2$**, and **IV$_3$**, which can be attributed to various types of inner-sphere ion pairs **IV** (Figure 4.9a) [41]. Similar resonances were observed in the catalyst system (Cp-*n*Bu)$_2$ZrMe$_2$/MAO

TABLE 4.1
^{13}C and ^{1}H NMR Chemical Shifts (δ), Line Widths $\Delta v_{1/2}$ (Hz), and J^1_{CH} Coupling Constants (Hz) for Complexes I–IV

Complex	^{13}C (Cp)	^{1}H (Cp)	^{13}C (Zr-Me)	^{1}H (Zr-Me)	^{13}C (μ-Me)	^{1}H (μ-Me)	^{13}C (Al-Me)	^{1}H (Al-Me)
Cp$_2$ZrMe$_2$[a]	109.11	5.64	29.26	–0.15	–	–	–	–
I[b]	112	5.7[c]	29.5[d]	–	29.5[d]	–	–	–
	$\Delta v_{1/2} = 40$							
II[e,f]	113.26	5.73	39.98	0.03	22.22	–1.30	–	–
	$J^1_{CH} = 176$		$J^1_{CH} = 121$		$J^1_{CH} = 133$			
III[b,f,g]	115.73	5.5[h]	–	–	38.07	–0.27	–6.00	–0.58
	$J^1_{CH} = 178$				$J^1_{CH} = 114$		$J^1_{CH} = 114$	
IV[b]	113.9	5.7[c]	42[i]	–	9	–	–	–
	$\Delta v_{1/2} = 70$		$J^1_{CH} = 120$		$\Delta v_{1/2} = 500$			

[a] At –30°C.
[b] At –25°C.
[c] Broad unresolved.
[d] Only one resonance ($\Delta v_{1/2} \geq 240$ Hz) with double intensity is observed.
[e] At –5°C.
[f] Sharp resonances.
[g] Chemical shifts at high total Zr concentrations.
[h] Chemical shifts depend on total Zr concentration, down to δ 5.44 at low [Zr].
[i] $\Delta v_{1/2}$ = approximately 200 Hz.

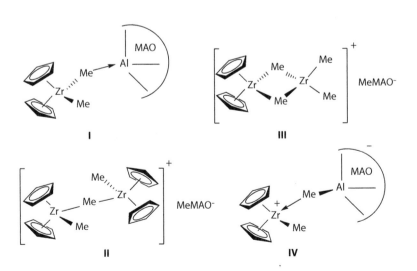

FIGURE 4.6 Structures proposed for intermediates I–IV.

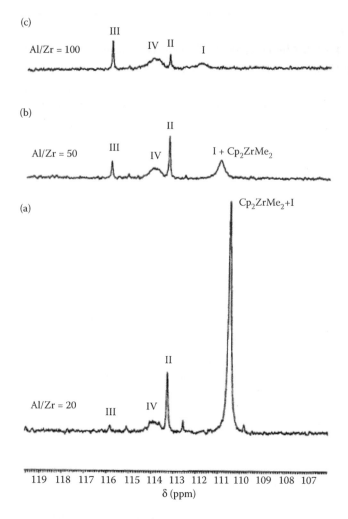

FIGURE 4.7 ^{13}C NMR spectra (toluene, −25°C) of Cp$_2$ZrMe$_2$ + MAO, [Al] = 1.8 M, in the range of Cp carbon signals. Al/Zr ratios are 20 (a), 50 (b), and 100 (c). (Reproduced with permission from Babushkin, D. E. et al. 2000. *Macromol. Chem. Phys.* 201: 558–567.)

(Al/Zr = 60). With the increase in the Al/Zr ratio, ion pair **III** becomes the predominant species in the solution (Figure 4.9b and c). No significant difference assignable to the presence of different Me-MAO$^-$ anions at different Al/Zr ratios could be detected in the ^{1}H NMR spectra of the outer-sphere ion pair **III**. The polymerization data show that the catalyst system (Cp-*n*Bu)$_2$ZrCl$_2$/MAO is virtually inactive at Al/Zr ratios below 200. The activity at an Al/Zr ratio of 1000 is higher by a factor of 13 than that at Al/Zr = 200 (Table 4.2). Although the concentration of zirconocene under the conditions of polymerization experiment is 1–2 orders of magnitude lower than under the NMR conditions, it is natural to expect that the proportion of **III** in the polymerization experiments should also increase with the increase in the

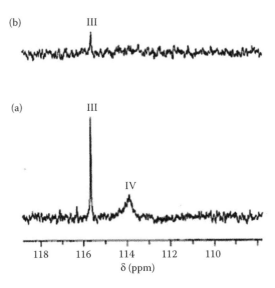

FIGURE 4.8 ^{13}C NMR spectra (toluene, −25°C) of Cp$_2$ZrMe$_2$ + MAO, [Al] = 1.8 M, in the range of Cp signals. Al/Zr ratios are 1000 (a) and 4000 (b). Total Zr concentrations are 1.8 × 10^{-3} M (a) and 4.5 × 10^{-4} M (b). Numbers of transients are 10^4 in both spectra. Intensity scales of the spectra (a) and coincide (b). (Reproduced with permission from Babushkin, D. E. et al. 2000. *Macromol. Chem. Phys.* 201: 558–567.)

Al/Zr ratio. Therefore, the increase in the catalytic activity of the (Cp-nBu)$_2$ZrMe$_2$/MAO system at high Al/Zr ratios can be associated with the increase in the portion of the outer-sphere ion pairs **III**.

The ^{1}H NMR spectra of the catalyst system (Cp-tBu)$_2$ZrCl$_2$/MAO (Al/Zr = 50) show that in this case, complexes (Cp-tBu)$_2$ZrCl$_2$ and (Cp-tBu)$_2$ZrMeCl are the major species in the solution (Figure 4.10a). At an Al/Zr ratio of up to 600–1000, as in the previous case, only species of the type **III** are observed (Figure 4.10b–d) [41]. This result differs from that for the Cp$_2$ZrMe$_2$/MAO system where complexes **III** and **IV** were present in comparable concentrations even at an Al/Zr ratio of 1000 (Figure 4.8) [38]. Probably, the bulky tBu-substituents hamper the anion–cation contacts in **IV**, which favors the formation of outer-sphere ion pairs **III**. As for R = nBu, the polymerization activity of the (Cp-tBu)$_2$ZrCl$_2$/MAO system at an Al/Zr ratio of 1000 is much higher than that at Al/Zr = 200 (Table 4.2). Summarizing these considerations, one can conclude that ion pairs **III** (or their polymeryl homologs) are the immediate precursors of the crucial alkyl zirconocene olefin cations required for the polymer chain growth.

4.1.6 Detection of Ion Pairs Formed in the Catalyst Systems (Cp-R)$_2$ZrCl$_2$/MAO (R = Me, 1,2-Me$_2$, 1,2,3-Me$_3$, 1,2,4-Me$_3$, Me$_4$)

The ^{1}H NMR spectrum of the system (Cp-1,2,3-Me$_3$)$_2$ZrCl$_2$/MAO at Al/Zr = 50 (in the range of Cp hydrogen atoms) displays the signal of **III** and several weaker signals (Figure 4.11a). In contrast to the systems (Cp-R)$_2$ZrCl$_2$/MAO (R = H, Me, nBu, tBu),

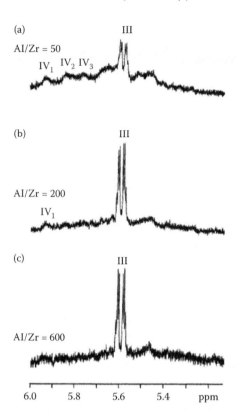

FIGURE 4.9 ^{1}H NMR spectra (toluene-d_8, 20°C, the range of Cp hydrogen atoms) of the system (Cp-nBu)$_2$ZrCl$_2$/MAO at Al/Zr ratios of 50 (a), 200 (b), and 600 (c). [(Cp-nBu)$_2$ZrCl$_2$] = 10^{-3} M; [Al$_{MAO}$] = 0.05 M (a), 0.2 M (b), and 0.6 M (c). (Reproduced with permission from Bryliakov, K. P. et al. 2003. *J. Organomet. Chem.* 683: 92–102.)

complex **III** dominated in the system (Cp-1,2,3-Me$_3$)$_2$ZrCl$_2$/MAO even at an Al/Zr ratio of 50 [41]. Moreover, this is the case at higher Al/Zr ratios (Figure 4.11b–d). As distinct from the systems (Cp-R)$_2$ZrCl$_2$/MAO (R = H, Me, nBu, tBu), the polymerization activity of the system (Cp-1,2,3-Me$_3$)$_2$ZrCl$_2$/MAO was virtually constant in the range Al/Zr = 200–1000 (Table 4.2). Ion pair **III** (R = 1,2,3-Me$_3$) appears to be stabilized relative to the contact ion pair **IV** by the more highly substituted Cp ligands.

According to the ^{1}H NMR spectroscopic data, only species **III** is present in the catalyst systems (Cp-R)$_2$ZrCl$_2$/MAO (R = 1,2,3-Me$_3$, 1,2,4-Me$_3$, Me$_4$) at Al/Zr = 200. One could expect then that their catalytic activities should be close at Al/Zr ratios of 200 and 1000. Indeed, the polymerization data showed that the initial polymerization activity of the systems (Cp-R)$_2$ZrCl$_2$/MAO (R = 1,2,3-Me$_3$, 1,2,4-Me$_3$, Me$_4$) in the range Al/Zr = 200–1000 increases less significantly with the increase of the Al/Zr ratio than the corresponding activities for the systems (Cp-R)$_2$ZrCl$_2$/MAO (R = Me, nBu, tBu) (Table 4.2).

For the most part of the (Cp-R)$_2$ZrCl$_2$/MAO systems studied, intermediate **III** seems to be the main precursor of the active centers of polymerization. These

NMR and EPR Spectroscopy of the Mechanisms of Metallocene

TABLE 4.2
Ethylene Polymerization over $(Cp-R)_2ZrCl_2$/MAO Catalysts

R	Al_{MAO}/Zr (mol/mol)	C_2H_4 Pressure (atm)	PE Yield[a] g	kg of PE/mol of Zr	Initial Activity[b] (kg of PE/mol Zr min atm C_2H_4)
H	200	2	7.7	3850	255
Me	200	5	5.2	2600	60
nBu	200	2	0.7	350	35
tBu	200	5	0.1	50	–
1,2-Me$_2$	200	2	13	6500	470
1,2,3-Me$_3$	200	5	10	5000	102
1,2,4-Me$_3$	200	5	13.9	6950	134
Me$_4$	200	5	17.1	8550	190
Me$_2$Si(Ind)$_2$	200	2	15.6	7800	420
H	1000	2	17.1	8550	610
Me	1000	2	14.3	7150	320
nBu	1000	2	23.4	11,700	460
tBu	1000	5	7.5	3750	92
1,2-Me$_2$	1000	2	19	9500	520
1,2,3-Me$_3$	1000	5	8.6	4300	115
1,2,4-Me$_3$	1000	2	8.7	4350	232
Me$_4$	1000	2	11.6	5800	300
Me$_2$Si(Ind)$_2$	1000	2	16.0	8000	450

[a] PE yield within 15 min.
[b] Initial activity calculated from the PE yield within 5 min.

assumptions are in agreement with the spectroscopic and polymerization data for the highly active catalyst systems $(Cp-R)_2ZrCl_2/AlMe_3/[Ph_3C][B(C_6F_5)_4]$ (R = H, Me, 1,2-Me$_2$, 1,2,3-Me$_3$, 1,2,4-Me$_3$, Me$_4$, nBu, tBu, Zr:Al:B = 1:100:1). ^{1}H NMR spectroscopy indicates the presence of only ion pairs $[(Cp-R)_2Zr(\mu-Me)_2AlMe_2]^+[B(C_6F_5)_4]^-$, thus confirming their key role as the precursors of the active species of polymerization [41].

4.1.7 Ion Pairs Formed upon Activation of Ansa-Zirconocenes with MAO

The analysis of the ^{1}H NMR spectra of the L_2ZrCl_2/MAO systems witnesses that for all catalysts studied (rac-C$_2$H$_4$(Ind)$_2$ZrCl$_2$, rac-Me$_2$Si(Ind)$_2$ZrCl$_2$, rac-Me$_2$Si(1-Ind-2-Me)$_2$ZrCl$_2$, rac-C$_2$H$_4$(1-Ind-4,5,6,7-H$_4$)ZrCl$_2$, rac-Me$_2$Si(2-Me-Benzind)$_2$ZrCl$_2$, rac-Me$_2$Si(1-Ind-2-Me-4-Ph)$_2$, (1-Ind-2-Me)$_2$ZrCl$_2$, Me$_2$C(Cp)(Flu)ZrCl$_2$, Me$_2$C(Cp-3-Me)(Flu)ZrCl$_2$, Me$_2$Si(Flu)$_2$ZrCl$_2$), broad resonances of **IV** were not detected at all or were observed only at low Al/Zr ratios (50–100), and sharp resonances of heterobinuclear ion pairs **III** strongly dominated in the solution [41]. As an example,

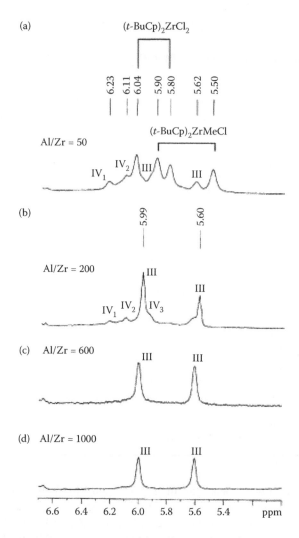

FIGURE 4.10 ^{1}H NMR spectra (toluene-d_8, 20°C, the range of Cp hydrogen atoms) of the system (Cp-tBu)$_2$ZrCl$_2$/MAO at Al/Zr ratios of 50 (a), 200 (b), 600 (c), and 1000 (d). [Al$_{MAO}$] = 0.5 M; [(Cp-tBu)$_2$ZrCl$_2$] = 10^{-2} M (a), 2.5 × 10^{-3} M (b), 8 × 10^{-4} M (c), and 5 × 10^{-4} M (d). (Reproduced with permission from Bryliakov, K. P. et al. 2003. *J. Organomet. Chem.* 683: 92–102.)

Figure 4.12 shows the ^{1}H NMR spectrum of [*rac*-C$_2$H$_4$(Ind)$_2$Zr(μ-Me)$_2$AlMe$_2$]$^+$[Me-MAO]$^-$ in toluene at 20°C. Probably, the substituents and bulkiness of the ligands disfavor the anion–cation contacts in the intermediates **IV**, complex **III** predominating in the solution.

The first example of a structurally characterized group IV metallocene AlMe$_3$ adduct [{Me$_2$Si-(2-Me-4-Ph-Ind)$_2$}Zr(μ-Me)$_2$AlMe$_2$]$^+$[B(C$_6$F$_5$)$_4$]$^-$ was reported by Carpentier, Kirillov, and coworkers. In the presence of excess of AlMe$_3$, this adduct

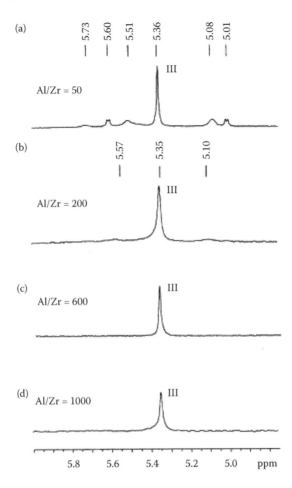

FIGURE 4.11 ^{1}H NMR spectra (toluene-d_8, 20°C, the range of Cp hydrogen atoms) of the system (Cp-1,2,3-Me$_3$)$_2$ZrCl$_2$/MAO at various Al/Zr ratios: 50 (a), 200 (b), 600 (c), and 1000 (d). [Al$_{MAO}$] = /0.5 M; [(Cp-1,2,3-Me$_3$)$_2$ZrCl$_2$] = /10^{-2} M (a), 2.5 × 10^{-3} M (b), 8 × 10^{-4} M (c), and 5 × 10^{-4} M (d). (Reproduced with permission from Bryliakov, K. P. et al. 2003. *J. Organomet. Chem.* 683: 92–102.)

decomposed via C-H activation of the bridging methyl unit to yield a new species with trimetallic {Zr(μ-CH$_2$)(μ-Me)AlMe(μ-Me)AlMe$_2$}core (Figure 4.13) [42].

More recently, detailed comparative studies on different metallocenium ion pairs derived from the parent metallocenes **1a–c** and **2a–c** (Figure 4.14) and different cationizing agents, including B(C$_6$F$_5$)$_3$, Al(C$_6$F$_5$)$_3$, [PhNMe$_2$H][B(C$_6$F$_5$)$_4$], [Ph$_3$C][B(C$_6$F$_5$)$_4$], and commercial-grade MAO have been reported [43]. The following ion pairs were characterized by X-ray diffraction (when appropriate), ^{1}H and ^{19}F NMR and NOESY nmr: [(L)ZrMe(μ-Me)E(C$_6$F$_5$)$_3$] (**3a–c**-MeE(C$_6$F$_5$)$_3$) (E = B, Al), [(L)ZrMe(NMe$_2$Ph)]$^+$[B(C$_6$F$_5$)$_4$]$^-$ (**4a–c**), [(L)ZrMe]$^+$[B(C$_6$F$_5$)$_4$]$^-$ (**5a–c**), [(L)Zr(μ-Me)$_2$AlMe$_2$]$^+$[B(C$_6$F$_5$)$_4$]$^-$ (**6a–c**), [(L)Zr(μ-Me)$_2$AlMe$_2$]$^+$[Me-MAO]$^-$ (**7a–c**) (L = L$_a$, L$_b$,

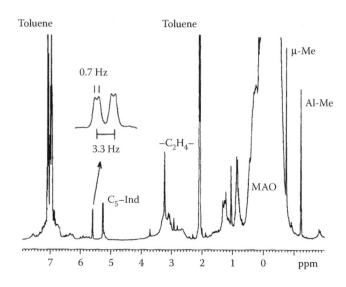

FIGURE 4.12 ^{1}H NMR spectrum (toluene, 20°C) of [ethanediyl(Ind)$_2$Zr(μ-Me)$_2$AlMe$_2$]$^+$[Me-MAO]$^-$. [Zr] = 3 × 10^{-3} M, Al/Zr = 300. (Reproduced with permission from Bryliakov, K. P. et al. 2003. *J. Organomet. Chem.* 683: 92–102.)

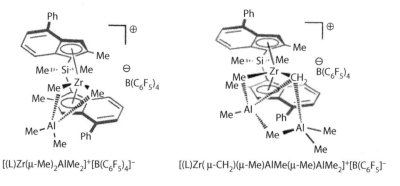

FIGURE 4.13 The first structurally characterized metallocene AlMe$_3$ adduct [(L)Zr(μ-Me)$_2$AlMe$_2$]$^+$[B(C$_6$F$_5$)$_4$]$^-$ and its decomposition product [(L)Zr(μ-CH$_2$)(μ-Me)AlMe(μ-Me) AlMe$_2$]$^+$[B(C$_6$F$_5$)]$^-$ (L = *rac*-{Me$_2$Si-(2-Me-4-Ph-Ind)$_2$}). (From Theurkauff, G. et al. 2015. *Angew. Chem. Int. Ed.* 54: 6343–6346)

L$_c$), L$_a$ = Me$_2$C-(Flu)(Cp-Me), L$_b$ = EtPhC-(Flu)(Cp), and L$_c$ = *rac*-{Me$_2$Si-(2-Me-4-Ph-Ind)$_2$. The ion pair reorganization processes for ion pairs **3a,b**-MeB(C$_6$F$_5$)$_3$, **3c**-MeB(C$_6$F$_5$)$_3$, **3c**-MeAl(C$_6$F$_5$)$_3$, and **6b,c** were studied by dynamic NMR. It was found that activation barriers are lower for the {SBI}-based systems as compared to those for the {Cp/Flu}-based congeners. Moreover, heterobimetallic complexes with the Zr(μ-Me)AlMe$_2$ core and various anions [MeB(C$_6$F$_5$)$_3$]$^-$, [B(C$_6$F$_5$)$_4$]$^-$, and [Me-MAO]$^-$ are more stable for {Cp/Flu}-based systems than for {SBI}-based counterparts. The less mobile ionic adducts of {Cp/Flu}-based systems can be responsible for the lower productivity of the corresponding polymerization catalysts.

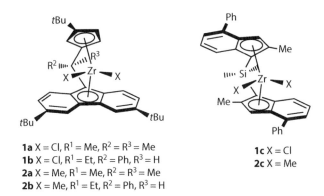

FIGURE 4.14 Metallocenes used for the study of activation with $E(C_6F_5)_3$ (E = B, Al), $[PhNMe_2H]^+[B(C_6F_5)_4]^-$, $[Ph_3C]^+[B(C_6F_5)_4]^-$, and MAO. (From Theurkauff, G. et al. 2016. *Organometallics* 35: 258–276)

4.1.8 Ion Pairs Formed upon Interaction of Cp_2TiCl_2 and *Rac*-$C_2H_4(Ind)_2TiCl_2$ with MAO

Detailed analysis of the ^1H-^{13}C NMR spectra of the catalyst systems Cp_2TiCl_2/MAO and Cp_2TiMe_2/MAO in toluene has shown that the following complexes can be identified: $Cp_2TiMeCl$, $[Cp_2MeTi(\mu\text{-}Cl)TiClCp_2]^+[Me\text{-}MAO]^-$ (II_1), $[Cp_2MeTi(\mu\text{-}Cl)TiMeCp_2]^+[Me\text{-}MAO]^-$ (II_2), $[Cp_2MeTi(\mu\text{-}Me)TiMeCp_2]^+[Me\text{-}MAO]^-$ (II_3), $[Cp_2Ti(\mu\text{-}Me)_2AlMe_2]^+[Me\text{-}MAO]^-$ (III), and a "contact ion pair" $[Cp_2TiMe^+\cdots Me\text{-}MAO]$ (IV) (Figures 4.15 and 4.16, Table 4.3).

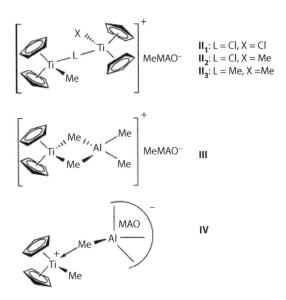

FIGURE 4.15 Structures proposed for intermediates II–IV. (Reproduced with permission from Bryliakov, K. P., Talsi, E. P., Bochmann, M. 2004. *Organometallics* 23: 149–152.)

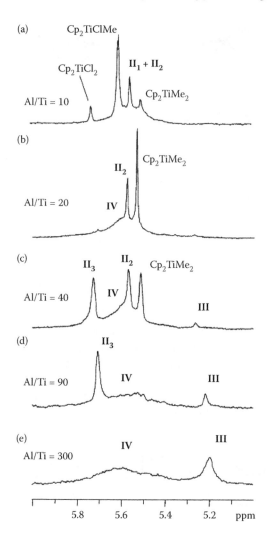

FIGURE 4.16 ¹H NMR spectra of Cp₂TiCl₂ + MAO in toluene at −15°C ([Al] = 0.75 M), in the range of Cp carbon signals, with Al:Ti ratios of (a) 10, (b) 20, (c) 40, (d) 90, and (e) 300. (Reproduced with permission from Bryliakov, K. P., Talsi, E. P., Bochmann, M. 2004. *Organometallics* 23: 149–152.)

At Al/Ti ratios modeling real polymerization conditions (100–300), complexes **III** and **IV** are the major species in the solution (Figure 4.15) [44]. As in the case of zirconium, heterobinuclear ion pairs **III** predominate in the catalyst systems *ansa*-titanocene/MAO at Al/Ti > 100. Titanium complexes [*rac*-C₂H₄(Ind)₂Ti(μ-Me)₂AlMe₂]⁺[Me-MAO]⁻ and [*rac*-C₂H₄(Ind)₂Ti(μ-Me)₂AlMe₂]⁺[B(C₆F₅)₄]⁻ display very similar ¹H NMR spectra [45].

The Ti species **III** and **IV** are far less stable than their Zr congeners: the latter persist for weeks at room temperature, while the former disappear at this temperature within several hours. In order to establish the origin of instability of these catalyst

NMR and EPR Spectroscopy of the Mechanisms of Metallocene

TABLE 4.3
^{13}C and ^{1}H NMR Chemical Shifts (ppm), Line Widths $\Delta v_{1/2}{}^a$ (Hz), and J_{CH} Coupling Constants (Hz) and for Complexes II–IV and Reference Complexes in Toluene

Number	Species	Al/Ti Ratio	T (°C)	^{13}C (Cp)	^{1}H (Cp)	^{13}C (Ti-Me)	^{1}H (Ti-Me)	^{13}C (μ-Me)	^{1}H (Al-Me)
1	Cp₂TiClMe	5	−20	116.03ᵇ	5.62ᵇ	53.06ᵇ	0.83ᵇ	—	—
	Cp₂TiClMe	10	−15	117.06ᵇ	5.60ᵇ	($\Delta v_{1/2}$ = 30) 59.56ᵇ ($\Delta v_{1/2}$ = 80)	0.85ᵇ	—	—
2	Cp₂TiMe₂	20	−15	112.07	5.50	46.25 ($\Delta v_{1/2}$ = 80) (J^1_{CH} = 123)	−0.21	—	—
3	[Cp₂TiMe(μ-Cl)Cp₂TiCl]⁺MeMAO⁻ **II₁**ᶜ	10	−15	117.9	5.55	64.38 (J^1_{CH} = 128)	0.79	—	—
	[Cp₂TiMe(μ-Cl)Cp₂TiMe]⁺MeMAO⁻ **II₂**ᶜ	20–40	−15	117.9	5.55	64.67 (J^1_{CH} = 128)	0.79	—	—
	[Cp₂TiMe(μ-Me)Cp₂TiMe]⁺MeMAO⁻ **II₃**ᶜ	90	−15	117.62	5.71	68.81 (J^1_{CH} = 128)	0.34	NFᵈ	—
4	[Cp₂Ti(μ-Me)₂AlMe₂]⁺MeMAO⁻ **III**ᶜ	90	−15	118.20	5.20	—	—	46.48ᵈ (J^1_{CH} = 119)	−0.74
5	Cp₂TiMe⁺←Me–Al≡MAO⁻ **IV**ᶜ	40–300	−15	117.9 ($\Delta v_{1/2}$ = 100)	5.6 ($\Delta v_{1/2}$ = 75)	66	NFᵈ	13ᵈ ($\Delta v_{1/2}$ = 150)	—
6	Cp₂TiCl₂ᶜ	—	25	119.50	5.86	—	—	—	—

(Continued)

TABLE 4.3 (Continued)

^{13}C and ^1H NMR Chemical Shifts (ppm), Line Widths $\Delta\nu_{1/2}{}^a$ (Hz), and J^1_{CH} Coupling Constants (Hz) and for Complexes II–IV and Reference Complexes in Toluene

Number	Species	Al/Ti Ratio	T (°C)	^{13}C (Cp)	^1H (Cp)	^{13}C (Ti-Me)	^1H (Ti-Me)	^{13}C (μ-Me)	^1H (Al-Me)
7	Cp$_2$TiClMe^{c,e}	–	–20	115.45	5.66	49.31 (J^1_{CH} = 129)	0.78	–	–
8	Cp$_2$TiMe$_2^{c,e}$	–	–20	113.10	5.60	46.20 (J^1_{CH} = 124)	0.01	–	–
9	[Cp$_2$TiMe(μ-Cl)Cp$_2$TiMe]$^+$[B(C$_6$F$_5$)$_4$]$^{-c}$	–	–15	117.7	5.50	63.6	0.69	–	–
10	[Cp$_2$Ti(μ-Me)$_2$AlMe$_2$]$^+$[B(C$_6$F$_5$)$_4$]$^{-c}$	50^f	–15	118.19	5.25	–	–	46.60^g	–0.72

Source: Bryliakov, K. P., Talsi, E. P., Bochmann, M. 2004. *Organometallics* 23: 149–152.

Note: NF, not found.

[a] For wide lines, either exchange-broadened (with AlMe$_3$ and MAO) or broadened due to nonuniformity of MAO sites.

[b] Chemical shift depends on the Al/Ti ratio, due to fast chemical exchange (cf. entries 1 and 7). Such a pronounced dependence is not observed for Cp$_2$TiMe$_2$; thus, Cp$_2$TiMeCl is expected to form Cl-bridged adducts with AlMe$_3$ and MAO.

[c] Narrow lines ($\Delta\nu_{1/2}$ ≤ 15–20 Hz).

[d] Respective ^1H peaks of μ-Me protons not found (either masked by solvent or MAO signals).

[e] Synthesized by reacting Cp$_2$TiCl$_2$ with LiMe in toluene at 0°C.

[f] Al present as AlMe$_3$.

[g] Respective ^1H peak of μ-Me protons at –0.30.

systems, we have monitored by EPR the interaction of rac-C$_2$H$_4$(Ind)$_2$TiCl$_2$ with AlMe$_3$ in toluene ([Ti] = 0.001 M, Al/Ti = 20) [45]. The EPR spectrum recorded one day after the onset of the reaction at room temperature displays a signal at g_0 = 1.978 with hyperfine structure from Ti (a_{Ti} = 12 G) and partially resolved hfs from Al (a_{Al} = 2.3 G) (Figure 4.17a). The intensity of the EPR signal corresponded to complete reduction of Ti(IV) into Ti(III). The hyperfine structure from Al clearly shows that the EPR spectrum at g_0 = 1.978 belongs to heterobinuclear complex incorporating both titanium(III) and aluminum atoms. The related complex [Cp$_2$Ti^{III}(µ-Me)$_2$AlMe$_2$] exhibits EPR spectrum at g_0 = 1.977 in toluene at −40°C with unresolved hfs from Al [46]. On the basis of these data, the Ti^{III} complex with

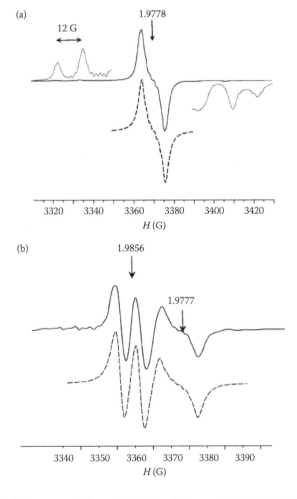

FIGURE 4.17 EPR spectra (toluene-d_8, 20°C) of (a) rac-C$_2$H$_4$(1-Ind)$_2$TiCl$_2$/AlMe$_3$, [Al]:[/Ti] = 20, 1 day after the onset of the reaction and (b) rac-C$_2$H$_4$(1-Ind)$_2$TiCl$_2$/Al(iBu)$_3$, [Al]:[Ti] = 10, 10 min after the onset of the reaction; dashed lines represent simulations of the experimental spectra. (Reproduced with permission from Bryliakov, K. P. et al. 2005. *Organometallics* 24: 894–904.)

EPR signal at $g_0 = 1.978$ was identified as the heterobinuclear titanium(III) species [rac-C$_2$H$_4$(Ind)$_2$Ti^{III}(μ-Me)$_2$AlMe$_2$].

The reducing agent in the rac-C$_2$H$_4$(Ind)$_2$TiCl$_2$/AlMe$_3$ system was proposed to originate from the admixture of Me$_2$AlEt, which is capable of generating metal hydride species (which in turn cause reduction of Ti^{IV}). To check this hypothesis, triisobutylaluminum (TIBA) was used as the reducing agent instead of AlMe$_3$. Commercial TIBA always contains some residual HAl(^{i}Bu)$_2$ that could reduce Ti^{IV}. Indeed, the addition of 10 equivalents of TIBA to the solution of rac-C$_2$H$_4$(Ind)$_2$TiCl$_2$ in toluene caused immediate and complete reduction of Ti^{IV} to Ti^{III}. The EPR spectrum observed exhibited the signal at $g_0 = 1.978$ with hfs from Al, $a_{Al} = 1.8$ G, and a doublet at $g_0 = 1.986$ from Ti(III) hydride with $a_H = 5$ G (Figure 4.17b). The reduction of cationic Ti^{IV} heterobinuclear intermediates into neutral Ti^{III} complexes seems to be an important channel of deactivation of $ansa$-titanocene/MAO catalysts.

4.1.9 Ion Pairs Formed upon Activation of (C$_5$Me$_5$)TiCl$_3$ and [(Me$_4$C$_5$)SiMe$_2$N^{t}Bu]TiCl$_2$ with MAO

The results presented above demonstrate that the activation of metallocenes with large excess of MAO usually leads to the formation of heterobinuclear ion pairs **III**. However, this is not the case for half-titanocenes and constrained-geometry catalysts, (C$_5$Me$_5$)TiCl$_3$ and [(Me$_4$C$_5$)SiMe$_2$N^{t}Bu]TiCl$_2$, respectively [44,47].

The starting complex (C$_5$Me$_5$)TiMe$_3$ displays three ^{13}C resonances due to C$_5$Me$_5$ and Ti-Me moieties (toluene-d_8, −20°C), δ 123.2 (Cp, 5C), 12.8 (Me-Cp, 5C, q, $J^1_{CH} = 127.5$ Hz), 62.1 (Ti-Me, 3C, q, $J^1_{CH} = 118.5$ Hz). The sample prepared by the addition of ^{13}C-labeled MAO to the solution of (C$_5$Me$_5$)TiMe$_3$ in toluene-d_8 displays Ti-Me resonances of complexes **Ia** and **IVa** (Figure 4.18) [47].

Species **Ia** is a complex between (C$_5$Me$_5$)TiMe$_3$ and Lewis acidic MAO (C$_5$Me$_5$) Me$_2$Ti-Me→Al≡MAO. A similar complex was observed previously for Cp$_2$ZrMe$_2$

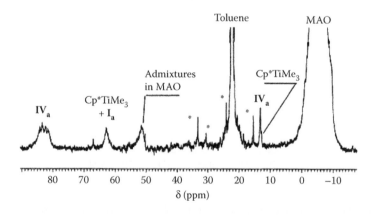

FIGURE 4.18 ^{13}C NMR spectrum (toluene at −10°C) of Cp*TiMe$_3$ recorded 1 h after the addition of ^{13}C-labeled sample of MAO, Al/Ti = 35, [Cp*TiMe$_3$] = 0.02 M. Asterisks denote admixtures in toluene. (Reproduced with permission from Bryliakov, K. P. et al. 2003. *J. Organomet. Chem.* 683: 23–28.)

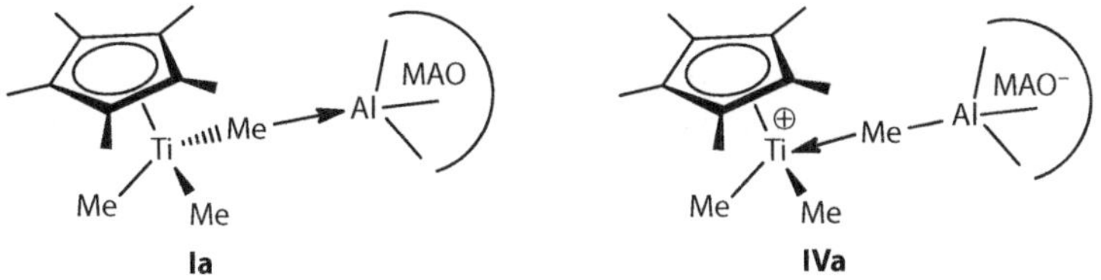

NMR and EPR Spectroscopy of the Mechanisms of Metallocene

FIGURE 4.19 Structures proposed for the intermediates **Ia** and **IVa**.

and MAO (complex **I**, Figure 4.6). "Contact ion pair" **IVa** displays a broad Ti-Me resonance centered at 83 ppm ($\Delta v_{1/2} = 400$ Hz), and that at 13.2 ppm from Cp-Me groups (Figure 4.18). The chemical shift of the Ti-Me resonance of **IVa** (δ 81–84) is close to that of the terminal Ti-Me resonance of the zwitterionic complex $[(C_5Me_5)TiMe_2^+\cdots MeB(C_6F_5)_3^-]$ (δ 80–81) [47]. Most probably, **IVa** is "contact ion pair" $[(C_5Me_5)Me_2Ti^+\cdots Me\text{-}MAO^-]$ (Figure 4.19). With the increase in the Al/Ti ratio up to 300, only one peak of Cp-Me carbons of **IVa** at 13.2 ppm is observed. The corresponding Ti-Me signal was not detected due to the low concentration of titanium. So, in conditions approaching to those of practical polymerization, the major part of titanium exists in the system $(C_5Me_5)TiMe_3/MAO$ as the "contact ion pair" **IVa**. This ion pair is far less stable than the related ion pair $[Cp_2ZrMe^+\cdots Me\text{-}MAO^-]$ (**IV**). The NMR resonances of **IVa** disappear within several hours at room temperature, whereas $[Cp_2ZrMe^+\cdots Me\text{-}MAO^-]$ is stable during weeks at this temperature. In the previously discussed catalyst systems, the heterobinuclear intermediates **III** were necessarily present in the reaction solution at high Al/metal ratios. On the contrary, the system $(C_5Me_5)TiMe_3/MAO$ predominantly displays broad resonances of the zwitterion-like species **IVa**.

The same behavior has been observed for the catalyst systems based on constrained-geometry titanium complexes [44]. Mixtures of ^{13}C-enriched MAO with $[(Me_4C_5)SiMe_2N^tBu]TiCl_2$ and $[(Me_4C_5)SiMe_2N^tBu]TiMe_2$ in toluene at 20°C (Al/Ti = 100) display inhomogeneously broadened ^{13}C NMR resonance at δ 68 ($\Delta v_{1/2} = 170$ Hz). On the basis of the data obtained for Cp_2TiCl_2/MAO system, this resonance could be reasonably assigned to the contact ion pair $[(Me_4C_5)SiMe_2N^tBuTiMe^{(+)}\cdots Me\text{-}MAO^-]$. Marks et al. observed NMR resonances in a similar region for two diastereomers of related species $[(Me_4C_5)SiMe_2N^tBu]TiMe^{(+)}\leftarrow PBA^{(-)}$ (PBA = tris(2,2′,2″-nonafluorobiphenyl)-fluoroaluminate), at δ 61.99 and 60.50 [48]. So, in contrast to the Cp_2TiCl_2/MAO and $rac\text{-}C_2H_4(Ind)_2TiCl_2/MAO$ systems, heterobinuclear pairs of the type $[LTi(\mu\text{-}Me)_2AlMe_2]^+[Me\text{-}MAO]^-$ are not observed for constrained-geometry catalysts. Apparently, in the more opened half-sandwich and constrained-geometry complexes, a closer approach of $[Me\text{-}MAO]^-$ counterions is favored, to afford ion pairs of the type **IV** upon the activation by MAO, while for the more restricted coordination-gap aperture of metallocenes, heterobinuclear ion pairs of type **III** are preferable.

4.1.10 Observation of Ion Pairs Formed in the Catalyst Systems Zirconocene/MMAO

The activation of zirconocene precatalysts for applications in olefin polymerization catalysis is most often achieved via reaction with a large excess of MAO. This activator can be partially replaced, without loss of activity, by less expensive aluminum alkyls, such as TIBA (AliBu$_3$) [49–53]. MMAO, a widely used polymerization cocatalyst, contains isobutyl groups originating from AliBu$_3$. Apparently, the understanding of the mechanism by which the addition of AliBu$_3$ may affect the cocatalytic activity of MAO is an important problem. We have applied EPR spin-probe technique and ^{1}H NMR spectroscopy for the comparison of Lewis acidic sites present in MAO and MMAO, and for the comparison of ion pairs formed upon the activation of metallocene with these cocatalysts.

As mentioned in Section 4.1.3, MAO contains two types of Lewis acidic sites, I_L and II_L, both being capable of forming adducts with TEMPO, with different EPR signals (a_N = 18.6 G, a_{Al} = 1 G for site I_L, and a_N = 19.6 G, a_{Al} = 1.9 G for site II_L), attributed to coordinatively unsaturated Al atoms in AlOMe$_2$ and AlO$_2$Me environment [28].

In MAO/AliBu$_3$ mixtures, the EPR spin-probe study likewise shows the presence of two types of Lewis acidic sites, denoted as I_{iBu} and II_{iBu}. The TEMPO adduct with site I_{iBu} displays a triplet (a_N = 19.5 G) without observed splitting from Al (Figure 4.20a). The adduct of TEMPO with site II_{iBu} exhibits a triplet of sextets, with a_N = 19.5 G and a_{Al} = 4 G (Figure 4.20b) [28].

So, the acidic sites II_{iBu} in the MAO/AliBu$_3$ mixtures appear to be more strongly Lewis acidic than sites II_L in MAO, which is reflected by the increased value of a_{Al} (4 G for MAO-AliBu$_3$ vs. 1.9 G for MAO). Crucially, the same acidic sites I_{iBu} and II_{iBu}

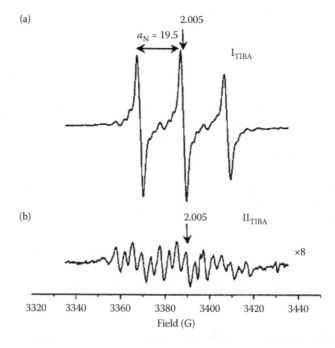

FIGURE 4.20 EPR spectra (toluene) of the MAO/AliBu$_3$/TEMPO system, with [Al]$_{TIBA}$/[Al]$_{MAO}$ = 1:10 and [TEMPO]/[Al] = 1:100, [MAO] = 0.2 M, recorded 2 min after melting in the resonator (a) and after 15 min at 20°C (b). (Reproduced with permission from Bryliakov, K. P. et al. 2006. *Macromol. Chem. Phys.* 207: 327–335.)

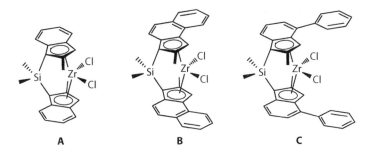

FIGURE 4.21 Zirconium precatalysts used for the study of activation with AliBu₃/MAO. (From Babushkin, D. E., Brintzinger, H. H. 2007. *Chem. Eur. J.* 13: 5294–5299.)

were observed for MMAO. By analogy with the assignment of the MAO sites II_L, sites II_{iBu} can be assigned to coordinatively unsaturated $AlO_2(iBu)$ centers. The increased Lewis acidity of sites II_{iBu}, as compared to that of sites II_L in MAO, is likely to contribute to the positive effect of the presence of AliBu₃ on the cocatalytic activity of MAO.

Babushkin and Brintzinger contributed a detailed investigation of the nature of the heterobinuclear ion pairs **III** formed in the catalyst systems zirconocene/MAO/AliBu₃ [54]. Complexes **A**–**C** (Figure 4.21) were used as precatalysts. The reaction system **A**/MAO in toluene ([Al]$_{MAO}$/[Zr]≈600) contains the cationic AlMe₃ adduct *rac*-[Me₂Si(Ind)₂Zr(μ-Me)₂AlMe₂]⁺ (denoted here as III_{Me2}). The addition of AliBu₃ to such a solution ([AliBu₃]/[Zr] = 14–108) results in the decrease on the intensity of the resonances of III_{Me2} and buildup of peaks of two new complexes *rac*-[Me₂Si(Ind)₂Zr(μ-Me)₂AlMeiBu]⁺ (III_{iBuMe}) and *rac*-[Me₂Si(Ind)₂Zr(μ-Me)₂AliBu₂]⁺ (III_{iBu2}). With the increase in the [AliBu₃]/[Zr] ratio, the proportion of III_{iBu2} increases. Complex III_{iBuMe} displays a set of two μ-Me, one terminal Al-Me, and four ligand–CH signals that have an integral ratio of 3:3:3:1:1:1:1 (Table 4.4). Complex III_{iBu2} exhibits a set of one μ-Me and two sharp ligand CH signals (3:1:1)

TABLE 4.4
¹H NMR Signals (δ) of *rac*-Me₂Si(Ind)₂Zr Derivatives[a]

Complex	H(3)	H(2)	Zr-Me	Al-Me$_{term}$
(SBI)ZrCl₂[b]	6.80	5.74	–	–
(SBI)ZrMe₂	6.69	5.68	–0.97	–
[(SBI)Zr(μ-Me)₂AlMe₂]⁺ (III_{Me2})	6.20	5.03	–1.34	–0.62
[(SBI)Zr(μ-Me)₂AliBuMe⁺] (III_{iBuMe})	6.18, 6.35	5.05, 5.15	–1.37, –1.23	–0.55
[(SBI)Zr(μ-Me)₂Al(iBu)₂]⁺ (III_{iBu2})	6.33	5.17	–1.28	–
(SBI)ZrMe⁺⋯MeMAO⁻ (**IV**)	5.7 to 6.0	5.1 to 5.2	–1.6 to –1.8	–

[a] C₆D₆, 23 °C, [Zr]$_{total}$ = 0.6 mM.
[b] SBI = *rac*-Me₂Si(Ind)₂.

152　Applications of EPR and NMR Spectroscopy in Homogeneous Catalysis

(Table 4.4). No signals assignable to the isobutyl groups of either III_{iBuMe} or III_{iBu2} were observed; these signals are probably masked by the other AliBu species present.

The studies of reaction systems **B**/MAO and **C**/MAO have shown that in this case, the monoisobutyl species III_{iBuMe} predominates even at the highest isobutyl loadings studied, and the formation of dibutylated species III_{iBu2} was not observed [54]. Apparently, this is due to the bulkiness of the corresponding substituted indenyls.

The so-called "oscillating" catalysts (bis(2-Ar-indenyl)zirconocenes), producing elastomeric polypropylenes, are an intriguing class of olefin polymerization catalysts. Apparently, the formation of stereoblock elastomeric plastics is a result of the dynamic character of the active species, with the general formula of the cationic part $[(2\text{-}Ar\text{-}indenyl)_2ZrP]^+$ (Ar = aryl; P = polymeryl) [7,55–57].

Busico and coworkers proposed that the generation of isotactic and atactic polypropylene stereosequences is a result of the interconversion of chiral *rac*-isomers at a rate competitive with the propylene insertion [57]. For one of the *rac*-isomers, the rate of indenyl ring rotation is faster than propylene insertion (atactic block is formed); for another *rac*-isomer, this rotation is slower than propylene insertion (isotactic block is formed). The key point of this hypothesis is the assumption that isomers differ in the nature of the counteranions. For some counteranions, their association with $[(2\text{-}Ar\text{-}indenyl)_2ZrP]^+$ leads to a locked conformation (the ligand rotation is restricted), leading to long isotactic sequences.

It was shown previously that $(2\text{-}PhInd)_2ZrCl_2$ (**1-Cl**), when activated with MMAO (containing isobutylaluminum groups), yielded polypropylenes with higher isotacticity than with MAO (containing only methylaluminum groups) [58]. We undertook a ^1H and ^{13}C NMR study of the ion pairs formed in the catalyst systems **1-Cl**/MAO, **1-Cl**/MMAO, and $\textbf{1-Cl}/AlMe_3/[CPh_3][B(C_6F_5)_4]$, in order to evaluate the possibility of the formation of ion pairs with a locked conformation of the 2-PhInd ligands [59].

The following ion pairs were characterized by ^1H and ^{13}C NMR: $[(2\text{-}PhInd)_2Zr(\mu\text{-}Me)_2AlMe_2]^+[Me\text{-}MAO]^-$ (**III**), $[(2\text{-}PhInd)_2Zr(\mu\text{-}Me)_2AlMe_2]^+[B(C_6F_5)_4]^-$ (**III'**), $[(2\text{-}PhInd)_2Zr(\mu\text{-}Me)_2AlMeiBu]^+[Me\text{-}MAO]^-$ (III_{MeiBu}), $[(2\text{-}PhInd)_2Zr(\mu\text{-}Me)_2AliBu_2]^+[Me\text{-}MAO]^-$ ($III_{iBu/iBu}$), and $[(2\text{-}PhInd)_2ZrMe^+\cdots MAO^-]$ (**IV**). In the temperature range of $-50°C$ to $20°C$, the rotation of indenyl ligands of complexes **III**, **III'**, and III_{MeiBu} is faster than the evaluated rate of propylene insertion. The situation with the ion pair $III_{iBu/iBu}$ is different. The latter ion pair exhibits two peaks from Cp protons at $-20°C$ (the distance between the peaks is about 10 Hz). These peaks coalesce at room temperature. The characteristic time of the indenyl ligands rotation, evaluated from these data, should be about 10^{-2} s at $20°C$. This time is comparable with the characteristic time of propylene insertion (10^{-2}–10^{-3} s) [56]. $III_{iBu/iBu}$ is a rare example of an ion pair with restricted rotation of indenyl ligands at room temperature, found for the "oscillating" catalyst, demonstrating the fundamental possibility of the existence of intermediates with "locked" conformation of 2-PhInd ligands.

4.1.11　ION PAIRS FORMED IN THE CATALYST SYSTEMS METALLOCENE/AliBu₃/ $[Ph_3C][B(C_6F_5)_4]$

Mixtures of $AliBu_3$ and cation-generating agents are widely used as cocatalysts in ternary systems for olefin polymerizations based on group IV metallocene complexes

[60–64]. Götz and coworkers reported an NMR investigation of the products of reaction of Ph$_2$C(Cp)(Flu)ZrCl$_2$ with AliBu$_3$ and [PhNMe$_2$H][B(C$_6$F$_5$)$_4$] at 60°C [65]. Surprisingly, they found the formation of hydrido-bridged Zr–Al binuclear cationic species [Ph$_2$C(Cp)(Flu)Zr(μ-H)(μ-C$_4$H$_7$)AliBu$_2$]$^+$, although the structures of some intermediates could only be tentatively proposed.

Using ^{13}C, ^{1}H, and ^{19}F NMR spectroscopy, we studied the formation of cationic species in ternary system (SBI)ZrX$_2$/AliBu$_3$/[Ph$_3$C][B(C$_6$F$_5$)$_4$], where X = Cl, Me [66]. The reaction of (SBI)ZrCl$_2$ with AliBu$_3$ at room temperature (Al/Zr = 5:1) in toluene leads to the formation of the monoalkylated complex (SBI)ZrCliBu. This complex is unstable in toluene solutions and decays within hours at room temperature (Figure 4.22a). The addition of [Ph$_3$C][B(C$_6$F$_5$)$_4$] to the *in situ*–formed (SBI)ZrCliBu at room temperature results in the formation of two zirconium species.

At low Al/Zr ratios (5–10), a new zirconium complex **V** was detected as the major product. In the ^{1}H NMR spectrum, **V** displays two characteristic doublets

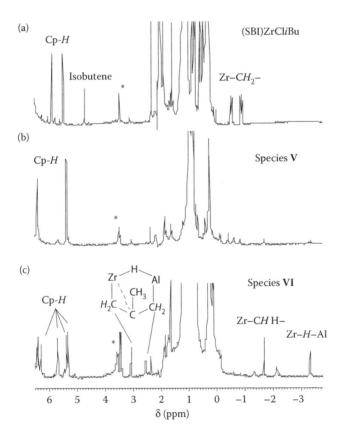

FIGURE 4.22 ^{1}H NMR spectra (20°C) of the systems (SBI)ZrCl$_2$/AliBu$_3$ = 1:5, toluene (a); (SBI)ZrCl$_2$/AliBu$_3$/[CPh$_3$][(C$_6$F$_5$)$_4$] = 1:5:1, toluene/1,2-dichlorobenzene (b); and sample (b) after 30 min stirring with extra 25-fold excess of AliBu$_3$ (c). Asterisks mark impurities in TIBA. (Reproduced with permission from Bryliakov, K. P. et al. 2007. *J. Organomet. Chem.* 692: 859–868.)

154 Applications of EPR and NMR Spectroscopy in Homogeneous Catalysis

from the cyclopentadienyl protons of the SBI ligand, at δ 5.39 and 6.42 (with the corresponding ^{13}C peaks at δ 121.6 and 122.6 [Figure 4.22b]), and a single resonance from the Si-Me groups at δ 0.84. Complex **V** is an ionic product with $[B(C_6F_5)_4]^-$ as a counterion, its cationic part exhibiting C_2-symmetry (^{19}F NMR chemical shifts of **V** correspond to $[B(C_6F_5)_4]^-$). **V** was successfully generated by the synthetic procedure described by Jordan et al. for similar zirconocene complexes [67]. X-ray diffraction witnessed that **V** is a binuclear ion pair $[(SBI)Zr(\mu-Cl)_2Zr(SBI)][B(C_6F_5)_4]_2$.

At high Al/Zr ratios (≥ 30), a new species **VI** is observed in the reaction mixture along with residual **V** and aluminum alkyls (Figure 4.22c). The NMR spectra of **VI** give evidence for the allyl-bridged heterobinuclear complex $[(SBI)Zr(\mu-H)(\mu-C_4H_7)AliBu_2][B(C_6F_5)_4]$, similar to the structure proposed by Götz for a $Me_2C(Cp)(Flu)$ complex [65]. In the presence of an excess $AliBu_3$, species **V** has been shown to polymerize propylene and is therefore the last detectable precursor of the active sites of propylene polymerization. Species **VI** is formed at high Al/Zr ratios in the absence of a monomer and is probably the thermodynamic sink of the catalyst (Scheme 4.1).

For titanium and zirconium complexes, there have been no examples of industrial applications of ternary systems of the type precatalyst/$AliBu_3$/$[Ph_3C][B(C_6F_5)_4]$. For hafnium catalysts, the situation is different. It has been shown that ultrahigh-molecular-weight polypropylene elastomers could be prepared using highly active C_1-symmetric hafnocene catalysts [68–70]. Pyridyl-amide hafnium complexes have recently been optimized for commercial high-temperature propylene polymerization [71,72]. In addition, the latter catalysts have been applied in the large-scale preparation of olefin block copolymers through chain shuttling polymerization [73]. Significantly, the aforementioned catalysts are effectively activated by mixtures of triisobutyl aluminum and cation-generating borates (e.g., $AliBu_3$/$[CPh_3][B(C_6F_5)_4]$), whereas MAO appeared to be an unexpectedly poor activator [68–73]. Apparently, it is very important to elucidate the origin of the difference between zirconium- and hafnium-based catalysts.

We have compared the structures of cationic species formed upon the activation of C_2-, C_s-, and C_1-symmetric hafnocenes **D–G** (Figure 4.23) with MAO and $AliBu_3$/$[CPh_3][B(C_6F_5)_4]$ by means of ^1H, ^{13}C, and ^{19}F NMR spectroscopy [74].

Thermally stable heterobinuclear intermediates of the type $[LHf(\mu-Me)_2AlMe_2]^+$ $[MeMAO]^-$ (**III$_{Hf}$**) and $[LHf(\mu-Me)_2AlMe_2]^+[B(C_6F_5)_4]^-$ (**III'$_{Hf}$**) have been

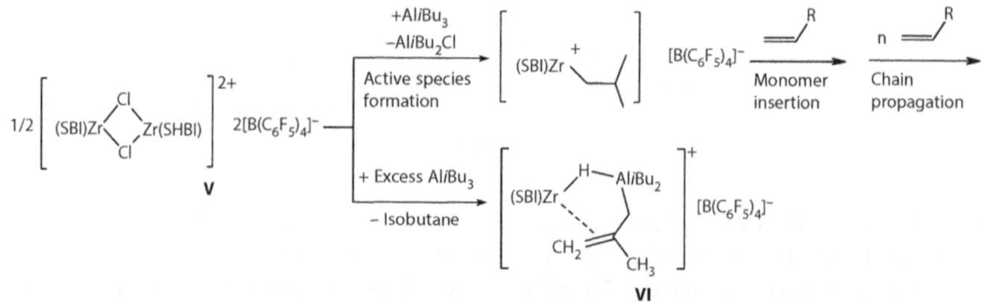

SCHEME 4.1 Possible reactions of complex **V**.

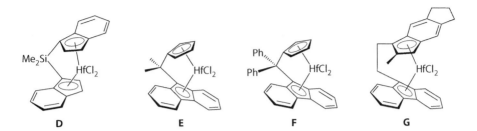

FIGURE 4.23 Hafnium precatalysts used for the study of activation with MAO, AlMe₃/[CPh₃][B(C₆F₆)₄], and AliBu₃/[CPh₃][B(C₆F₆)₄].

identified when using MAO and AlMe₃/[Ph₃C][B(C₆F₅)₄] as activators, respectively. These intermediates are thermally stable at room temperature and display low polymerization productivities. In the NMR tube experiments, **III**$_{Hf}$ and **III'**$_{Hf}$ polymerize propene only on warming to room temperature. The low reactivity of **III**$_{Hf}$ and **III'**$_{Hf}$ as compared to **III**$_{Zr}$ and **III'**$_{Zr}$ is thought to be responsible for the poor activity of hafnocenes as compared to zirconocenes when AlMe₃ is present (e.g., with AlMe₃-rich MAO). In full agreement with this assumption, Busico and coworkers showed that the activation of complex **E** (Figure 4.23) with AlMe₃-depleted MAO and AliBu₃/[CPh₃][B(C₆F₅)₄ affords catalysts with comparable propene polymerization activities, whereas unmodified MAO is a poor precatalyst for the activation of complex **E** [75].

The activation of hafnocene complexes with AliBu₃/[Ph₃C][B(C₆F₅)₄], on the other hand, generates heterobinuclear hydrido-bridged species with the proposed structures [LHf(μ-H)₂AliBu₂]⁺ or [LHf(μ-H)₂Al(H)iBu]⁺, which are much more reactive in olefin polymerizations. The hafnium hydride species were completely consumed in reactions with a few equivalents of 2,4,4-trimethyl-1-pentene and allylbenzene, thus confirming their role as the active-site precursors [75].

Brintzinger, Bercaw, and coworkers showed that the reaction of *ansa*-zirconocene complex **A** (Figure 4.21) with diisobutylaluminum hydride (HAliBu₂) and [Ph₃C][B(C₆F₅)₄] in hydrocarbon solution affords the cation [(SBI)Zr(μ-H)₃(AliBu₂)₂]⁺ (**VII**), the identity of which was derived from the NMR data and supported by crystallographic structure determination (Figure 4.24) [76].

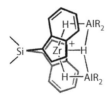

FIGURE 4.24 Structure of the alkylaluminum-complexed zirconocene trihydride cation [(SBI)Zr(μ-H)₃(AliBu)₂]⁺ (**VII**) formed via the reaction of (SBI)ZrCl₂ with [Ph₃C][B(C₆F₅)₄] and excess HAliBu₂ in toluene solution. (Reproduced with permission from Baldwin, S. M. et al. 2011. *J. Am. Chem. Soc.* 133: 1805–1813.)

In the ^{1}H NMR spectrum, **VII** displays two hydride resonances (benzene-d_6, 25°C, δ: –2.25 (2H, d, $^2J_{HH}$ = 8 Hz), 0.34 (1H, t, $^2J_{HH}$ = 8 Hz)). A rather similar pattern of hydride resonances was observed for the hafnium cation formed in the system **D**/Al*i*Bu$_3$/[Ph$_3$C][B(C$_6$F$_5$)$_4$] (toluene-d_8/1,2-difluorobenzene, 25°C, δ: –1.13 (2H, d, $^2J_{HH}$ = 5 Hz), 1.4 (1H, t, $^2J_{HH}$ = 5 Hz)) [74], and was tentatively assigned to the [LHf(μ-H)$_2$Al(H)*i*Bu]$^+$ species. Very likely, this cation in fact has the same structure as intermediate **VII**, [(SBI)Hf(μ-H)$_3$(Al*i*Bu$_2$)$_2$]$^+$.

4.1.12 Ion Pairs Operating in the Catalyst Systems Zirconocene/Activator/α-Olefin

Up to this moment, we discussed the ion pairs of the type [LM(Me)]$^+$ (or their adducts), thought to initiate the polymer chain growth. For mechanistic studies, it is important to detect and study the true chain-propagating species [LM(P)]$^+$, where P = polymeryl. However, *in situ* characterization of the propagating species is often complicated by rapid chain transfer and β-hydride elimination. The first spectroscopic detection of Zr-polymeryl species in the catalyst systems Cp$_2$Zr(^{13}CH$_3$)$_2$/B(C$_6$F$_5$)$_3$/^{13}C$_2$H$_4$ and Cp$_2$Zr(^{13}CH$_3$)$_2$/MAO/^{13}C$_2$H$_4$ was reported by Tritto and coworkers [77]. The ^{13}C NMR spectra of these systems (toluene-d_8, –20°C) displayed a doublet at about δ 65 (J = 29.65 Hz), which could be reasonably assigned to the methylene carbons of the polymer chain bonded to the zirconium complexes [Cp$_2$Zr-^{13}CH$_2$P]$^+$[^{13}CH$_3$B(C$_6$F$_5$)$_3$]$^-$ and [Cp$_2$Zr-^{13}CH$_2$P]$^+$[MeMAO]$^-$.

Landis and coworkers generated *in situ* the ion pair species *rac*-[C$_2$H$_4$(1-Ind)$_2$Zr(polymeryl)]$^+$[MeB(C$_6$F$_5$)$_4$]$^-$ (**VIII**) (polymeryl = polyhexyl chain) and studied by ^{1}H NMR the insertion of ethene and propene into the Zr-polymeryl moiety at –40°C, to form intermediates **VIII**$_{ethene}$ and **VIII**$_{propene}$ (Figure 4.25) [78]. Warming a

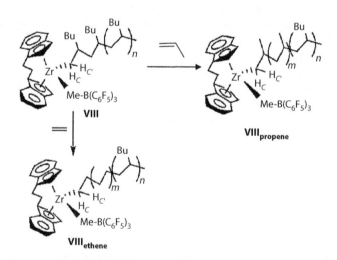

FIGURE 4.25 Zr-polymeryl species **VIII**, **VIII**$_{ethene}$, and **VIII**$_{propene}$. (Reproduced with permission from Landis, C. R., Rosaaen, K. A., Sillars, D. R. 2003. *J. Am. Chem. Soc.* 125: 1710–1711.)

NMR and EPR Spectroscopy of the Mechanisms of Metallocene   157

solution of **VIII** up to −20°C led to a chain release, to afford a vinylidene-terminated polymer. *In situ* kinetic measurements showed that at −40°C, the first insertion of 1-hexene into the Zr-Me bond is ca. 400 times slower than subsequent insertions. The chain propagation begins with reversible coordination of alkene and displacement of the $[MeB(C_6F_5)_4]^-$ anion from the inner coordination sphere. Irreversible insertion of the coordinated alkene is rate-determining. The data obtained by direct NMR observation of the $[C_2H_4(1\text{-Ind})_2Zr(\text{propenyl})]^+[MeB(C_6F_5)_4]^-$ intermediate were used to clarify the origin of stereoerrors in propene polymerization [79].

The same group studied 1-hexene polymerization catalyzed by $[(SBI)Zr(CH_2Si Me_3)]^+[B(C_6F_5)_4]^-$ and documented the formation of two types of cationic Zr-allyls that could not be observed with the relatively tightly coordinating $[MeB(C_6F_5)_3]^-$. Those species were characterized by chemical and NMR spectroscopic methods, and were considered as the "dormant" forms of the catalyst, due to their lower reactivity toward the α-olefin, as compared with cationic Zr-alkyls [80].

More recently, Vatanamu with coworkers detected $[Cp_2Zr^+\text{-allyl}]$ species in the reaction of either $[Cp_2ZrMe]^+[MeB(C_6F_5)_3]^-$ or $[Cp_2ZrMe]^+[B(C_6F_5)_4]^-$ with propylene by a combination of ^1H NMR spectroscopy and electrospray ionization mass spectrometry techniques. The proposed reaction pathways for the formation of $Cp_2Zr^+\text{-allyl}$ species are depicted in Scheme 4.2 [81].

Brintzinger, Babushkin, and coworkers, by using a combination of NMR and UV–Vis spectroscopic techniques, showed the formation of polymer-carrying Zr-allyl cationic species in the course of 1-hexene polymerization by *rac*-[Me_2Si(1-indenyl)_2ZrMe_2] activated with $[Ph_3C][B(C_6F_5)_4]$ and MAO. When using $[Ph_3C][B(C_6F_5)_4]$ as the activator, the resulting cationic Zr-allyl complexes account for about 90% of the total catalyst concentration. To clarify the nature of Zr-allyl species observed, the authors prepared model Zr-allyl complexes, replacing 1-hexene as substrate with 2-butyl-1-hexene. Three types of isomeric Zr-allyl complexes were found and thoroughly identified by means of 2D NMR spectroscopy. The UV–Vis spectra of the model complexes were similar to those for the Zr-allyl complexes formed in the course of 1-hexene polymerization. The samples containing Zr-alkyl

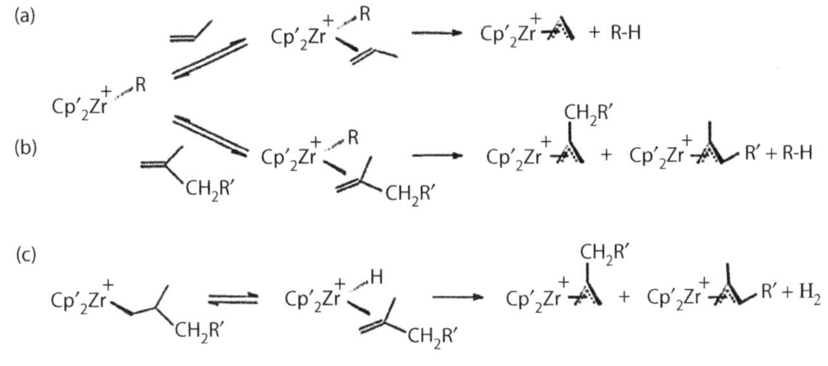

SCHEME 4.2 Possible reaction pathways for the formation of $Cp_2Zr^+\text{-allyl}$ species during zirconocene-catalyzed propylene polymerization. (Reproduced with permission from Vatanamu, M. 2015. *J. Catal.* 323: 112–120.)

158   Applications of EPR and NMR Spectroscopy in Homogeneous Catalysis

and Zr-allyl cationic species displayed similar polymerization activities. On the basis of these experimental data, it was assumed that Zr-alkyl and Zr-polymeryl species, participating in polymerization, can be readily generated from the allyl pool [82,83].

The investigation of species actually arising as resting states in polymerization catalysts under conditions approaching the real polymerization is a very difficult task because a large excess of substrate and activator (most often MAO) complicates the application of NMR spectroscopic techniques. Nevertheless, Babushkin and Brintzinger were able to identify by ^{1}H and ^{13}NMR spectroscopy the bridging Zr-polymeryl species [(SBI)Zr(μ-Me)(μ-Pol)AlMe$_2$]$^+$ (**IX**) formed during the 1-hexene polymerization by the system (SBI)ZrMe$_2$/MAO [84].

The ^{1}H NMR spectra of (SBI)ZrMe$_2$/MAO mixtures in C$_6$D$_6$ exhibit the signals of the cationic AlMe$_3$ adduct [(SBI)Zr(μ-Me)$_2$AlMe$_2$]$^+$. The addition of 1-hexene to this solution ([hexene]/[Zr] = 1000) results in a decline of the [(SBI)Zr(μ-Me)$_2$AlMe$_2$]$^+$ signals, accompanied by the appearance of a new set of signals assigned to **IX** (Table 4.5). As judged by the integrals of its C$_5$-H ligand and Zr(μ-Me) signals, **IX** contains only a single μ-Me moiety. The ^{1}H NMR spectrum of the model species [(SBI)Zr(μ-CH$_2$SiMe$_3$)(μ-Me)AlMe$_2$]$^+$ was similar to that of **IX** (Table 4.5).

Convincing evidence for the presence of the Zr–CH$_2$ bond in **IX** was obtained using 1-^{13}C-1-hexene as a substrate. Two small ^{13}C NMR signals were observed in the course of 1-^{13}C-1-hexene polymerization with the catalyst system (SBI)ZrMe$_2$/MAO. The signal at δ 84.6 (J_{CH} = 149 Hz) was assigned to the Zr-allyl species on the basis of characteristic J_{CH} coupling constant. The signal at δ 80.6 (J_{CH} = 107 Hz) associated with intermediate **IX** was assigned to the Zr–CH$_2$ group of its μ-polymeryl moiety.

Integration of the corresponding ^{1}H NMR resonances shows that the overall concentration of the NMR-detectable species in the catalyst system (SBI)ZrMe$_2$/

TABLE 4.5

^{1}H NMR Signals of Cationic (SBI)Zr Complexes[a]

Complex	C$_6$-H	C$_5$-H(3)	C$_5$-H(2)	Me$_{term}$	μ-Me
[(SBI)Zr(μ-Me)$_2$AlMe$_2$]$^{+b}$	6.71[c]	6.13[d]	4.98[d]	−0.65	−1.39[e]
[(SBI)Zr(μ-R)(μ-Me)AlMe$_2$]$^+$	6.55[c]	6.39[d]	4.91[d,f]	N/O	−1.53[g]
(**IX**)[b]		6.38[d]	5.26[d]		
[(SBI)Zr(μ-CH$_2$SiMe$_3$)(μ-Me)	6.52[c]	6.08[d]	5.16[d]	−0.63	−1.37[g]
AlMe$_2$]$^{+b,h}$	6.73[c]	6.58[d]	5.17[d]	−0.80	

Note: N/O: not observed

[a] C$_6$D$_6$, 27°C, 400 MHz, δ in ppm, [Zr$_{total}$] = 0.6 mM, [hexene]$_{init}$/[Zr] = 1000, [Al]$_{MAO}$/[Zr] = 960.

[b] MeMAO$^-$ as counterion.

[c] Pseudotriplets with J_{HH} 7.6–8 Hz.

[d] Doublets with J_{HH} 3.2–3.5 Hz.

[e] Two μ-Me groups per zirconocene unit.

[f] Observed during polymerization of 1-^{13}C-1-hexene.

[g] One μ-Me groups per zirconocene unit.

[h] At 20°C, further signals: Zr–CH$_2$ −0.91 (d, 5.7 Hz), −3.54 (d, 5.7 Hz), Me$_2$Si 0.81 (3H), 0.72 (3H), Me$_3$Si −0.11 (9H).

NMR and EPR Spectroscopy of the Mechanisms of Metallocene

MAO/1-hexene corresponds to less than one-half of the initial zirconocene concentration [84]. One can assume that these "NMR-silent" species are Zr-allyl complexes with very broad resonances (due to the presence of a range of different allyl-bonded Zr-polymeryl species).

As it was mentioned (Section 4.1.8), one of the main channels of *ansa*-titanocene catalysts deactivation is reduction of Ti^{IV} to Ti^{III}. On the contrary, reduction of Zr^{IV} to Zr^{III} has been rather rarely considered as a feasible route of deactivation of *ansa*-zirconocene catalysts. Nevertheless, sufficiently strong reducing agents (such as $HAliBu_2$, inevitably present in MMAO) are capable of reducing zirconocenes.

Bercaw, Brintzinger, and coworkers investigated the formation of trivalent zirconocene complexes from *ansa*-zirconocene-based olefin polymerization precatalysts [85]. EPR "fingerprints" of various Zr^{III} metallocene complexes that could form in working catalyst systems were obtained. The chloride-bridged heterobinuclear cation $[(SBI)Zr^{IV}(\mu\text{-Cl})_2AlMe_2]^+$ was reduced with sodium amalgam (NaHg) to the complex $(SBI)Zr^{III}(\mu\text{-Cl})_2AlMe_2$, the identity of which was derived from EPR data ($g_0 = 1.958$, $a_{Al} = 3.4$ G), supported by X-ray structure determination. The reduction of $[(SBI)Zr^{IV}(\mu\text{-Me})_2AlMe_2]^+$ and $[(SBI)Zr^{IV}(\mu\text{-H})_3(AliBu_2)_2]^+$ results in the formation of $(SBI)Zr^{III}\text{-Me}$ ($g_0 = 1.971$) and $(SBI)Zr^{III}(\mu\text{-H})_2AliBu_2$ ($g_0 = 1.970$, $a_{Al} \sim 8$ G), respectively. These products can also be accessed when $(SBI)Zr^{IV}Me_2$ is allowed to react with $HAliBu_2$. Furthermore, complexes $(SBI)ZriBu$ ($g_0 = 1.97$, $a_H = 4$ G) and $[(SBI)Zr^{III}]^+AlR_4^-$ ($g_0 = 1.983$) are formed. The role Zr^{III} species play in zirconocene-catalyzed polymerizations is thus far unclear and requires further studies.

4.2 POST-METALLOCENE CATALYSTS

4.2.1 BIS(IMINO)PYRIDINE IRON ETHYLENE POLYMERIZATION CATALYSTS

The emergence of metallocene catalysts has been one of the two major milestones in the progress of coordination polymerization of olefins after the Ziegler–Natta catalysts. The other milestone is the discovery of *post-metallocenes*, the name that encloses a plethora of transition metal complexes, capable of performing as single-site olefin polymerization catalysts.

One of the first (and one of the most important) classes of post-metallocene polymerization catalysts is based on bis(imino)pyridine complexes of first-row transition metals. More than 15 years after their discovery [86–90], bis(imino) pyridine complexes of iron and cobalt continue attracting great attention from academic and industrial laboratories [91–93]. This interest is due to the high activity of bis(imino)pyridine catalysts (especially iron complexes) in the polymerization and oligomerization of ethylene, and the possibility of using $AliBu_3$ as the activator instead of the much more expensive MAO. For the iron compounds, precatalysts bearing two large 2,6-substituents at the aryl rings ($L^{2iPr}FeCl_2$, Figure 4.26) are known to produce linear polyethylene (PE), whereas those with only one *ortho*-aryl substituent ($L^{Me}FeCl_2$, Figure 4.26) are selective for the production of a mixture of α-olefins with the Schultz–Flory distributions [94].

In contrast to group 4 metallocene catalysts, much less mechanistic details have been reported for bis(imino)pyridine iron-catalyzed polymerizations. The catalytic

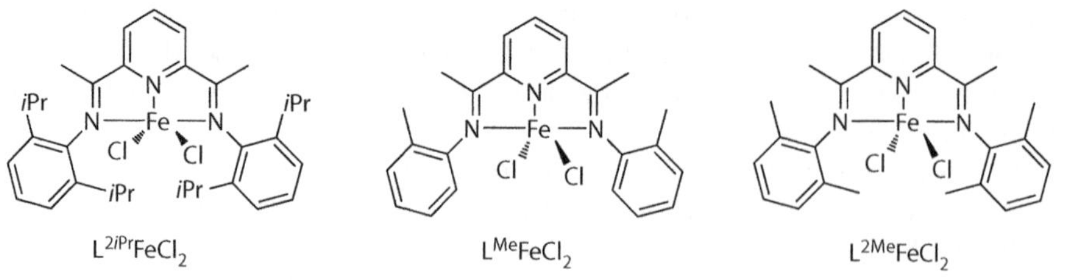

160 Applications of EPR and NMR Spectroscopy in Homogeneous Catalysis

FIGURE 4.26 Bis(imino)pyridine iron(II) precatalysts.

behavior of bis(imino)pyridine iron-based systems is rather complicated. From the kinetic studies of Barabanov [95] and Kissin [96], it follows that at least two types of active sites may be present in the systems $L^{2iPr}FeCl_2/MAO$ and $L^{2iPr}FeCl_2/Al(iBu)_3$. The more active (and unstable) sites, operating during the early stages of polymerization and affording low-molecular weight polyethylene (low-MW PE), in the course of polymerization are transformed into the less active sites, responsible for the formation of the higher-MW fraction of PE.

4.2.1.1 Activation of $L^{2iPr}FeCl_2$ with MAO

Talsi, Bryliakov, and coworkers conducted a series of ^1H and ^2H NMR spectroscopic studies, aimed at identifying the catalytic intermediates following the addition of MAO, $AlMe_3$, $AlMe_3/B(C_6F_5)_3$, $AlMe_3/[Ph_3C][B(C_6F_5)_4]$, or other trialkylaluminum compounds to $L^{2iPr}FeCl_2$ [97–99]. In spite of the paramagnetism ($S = 2$), the $L^{2iPr}FeCl_2$ complex displays a characteristic ^1H NMR spectrum, which allows unambiguous assignment of all resonances that can be used to identify bis(imino)pyridine iron compounds (Figure 4.27a). $L^{2iPr}FeCl_2$, when reacted with DMAO in toluene-d_8, is converted into new iron(II) complexes X and XI, both having the ^1H NMR spectra with the same range of chemical shifts as $L^{2iPr}FeCl_2$ (−150 to +150 ppm, Table 4.6). Complex X could be detected at an Al/Fe ratio of 20 (Figure 4.27b), complex XI predominates at higher ratios (Figure 4.27c). The characteristic resonance from the $AlMe_2$ moiety at δ 45 (Figure 4.27b, c) witnesses that X and XI are most likely heterobinuclear species of the type $[L^{2iPr}Fe(\mu-Me)(\mu-Cl)AlMe_2]^+[Me-MAO]^-$ and $[L^{2iPr}Fe(\mu-Me)_2AlMe_2]^+[Me-MAO]^-$, respectively. To confirm the ionic nature of the intermediates X and XI, $L^{2iPr}FeCl_2$ was activated with the $AlMe_3/B(C_6F_5)_3$ combination.

The ^1H NMR spectrum of the resulting complex $[L^{2iPr}Fe(\mu-Me)_2AlMe_2]^+$ $[MeB(C_6F_5)_3]^-$ (XII) was found to be nearly identical to that of XI (Figure 4.27c, d and Table 4.6). The proposed routes for the formation of species X–XII are presented in Scheme 4.3. Species X–XII persist in the solution for hours at room temperature. However, their concentration decreased with time (this decay was accompanied by a color change from initial light brown to dark brown). In line with the decay of the concentration of X–XII, the appearance of a new EPR-active iron species ($g_0 = 2.08$, $\Delta v_{1/2} = 330$ G) was detected (Figure 4.28). $AlMe_3/MAO$ combination is a reducing agent. Therefore, it is reasonable to assign the EPR signal at $g = 2.08$ to an iron(I) species (XIII). The EPR resonance of XIII is within the range characteristic of iron(I) species [100]. Tentatively, XIII can be represented as low-spin iron(I) complex $L'Fe^I(\mu-Me)_2AlMe_2$, where L' is intact or modified bis(imino)pyridine ligand.

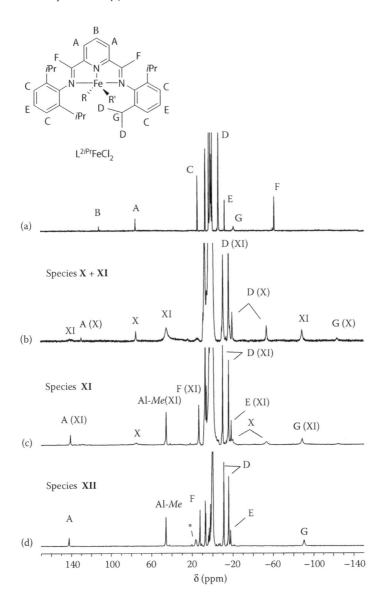

FIGURE 4.27 ¹H NMR spectra (toluene-d_8, 296 K) of complex L^{2iPr}FeCl$_2$ (a) and L^{2iPr}FeCl$_2$ after the interaction with different activators: DMAO, Al/Fe = 20 (b); DMAO, Al/Fe = 50 (c); and AlMe$_3$/B(C$_6$F$_5$)$_3$, Fe:Al:B = 1:15:1.1 (d).

4.2.1.2 Activation of L^{2iPr}FeCl$_2$ with AlMe$_3$

¹H NMR monitoring of the interaction of L^{2Me}FeCl$_2$ (Figure 4.6) with AlMe$_3$ revealed the formation of intermediate **XIV** with the proposed structure L^{2Me}(R)Fe^{II}(μ-Me)$_2$AlMe$_2$ (R = Cl or Me). This intermediate displayed unexpectedly large paramagnetic shifts of some bis(imino)pyridine ligand resonances (up to δ 358

162 Applications of EPR and NMR Spectroscopy in Homogeneous Catalysis

TABLE 4.6
^{1}H NMR Data for the Complexes Considered (Toluene-d_8, δ)[a]

Complex	T (K)	A Py-H_m	B Py-H_p	C Ar-H_m	D ArCHMe_2	E Ar-H_p	F N=C(Me)	G ArCH Me_2	X
L^{2iPr}FeCl$_2$	296	75.6	111.8	14.6	−5.7, −5.9	−11.8	−61.0	−20.8	−
X[b]	296	128.8	NA	NA	−19.7, −53.8	NA	NA	−124	NA
XI[c]	263	171.0	NA	NA	−13.8, −20.5	−23.7	12.8	−119.0	60.0[d]
XI[c]	296	140.6	NF	NF	−9.9, −15.6	−18.2	13.4	−88.6	45.6[d]
XII[e]	263	170.9	NF	NF	−15.3, −20.3	−22.6	11.0	−119.5	59.0[d]
XII[e]	296	142.2	NF	NF	−11.2, −15.8	−17.7	12.4	−90.5	45.7[d]
XIII[f]	263	116.2	422	NF	−15.9, −20.5	−17.1	−287.7	−105.0	44.6[d]
XIII[f]	296	99.3	370.6	NF	−13.1, −17.3	−13.9	−249.1	−86.8	37.4[d]
XIV[g]	253	70.3	432.9	9.5	−1.6, −7.8	−10.9	−267.4	NA	25.6[d]
XIV	263	66.0	409.8	9.2	−1.5, −7.2	−9.9	−252.1	NA	23.9[d]
L^{iPr}FeCl	263	77.4	442.0	−8.9	−25.8, −40.2	−16.6	−243.0	−133.0	−
L^{iPr}FeCl	296	67.3	381.7	−6.0	−22.8, −33.2	−13.1	−211.0	−108.2	−
L^{iPr}FeMe	296	59.6	211.2	−2.3	−11.5, −20.9	NA	−164.8	−71.6	−

[a] NF, not found; NA, not reliably assigned.
[b] In the system 1/DMAO = 20.
[c] In the system 1/DMAO = 50.
[d] X stands for Al(CH$_3$)$_2$.
[e] In the system 1/AlMe$_3$/B(C$_6$F$_5$)$_3$ = 1:15:1.1, additional broad peak assigned to [CH$_3$B(C$_6$F$_5$)$_3$]$^-$ found at δ 16.5 (296 K) and 20.7 (263 K).
[f] In the system 1/AlMe$_3$ = 10.
[g] In the system L^{iPr}FeMe/AlMe$_3$ = 10.

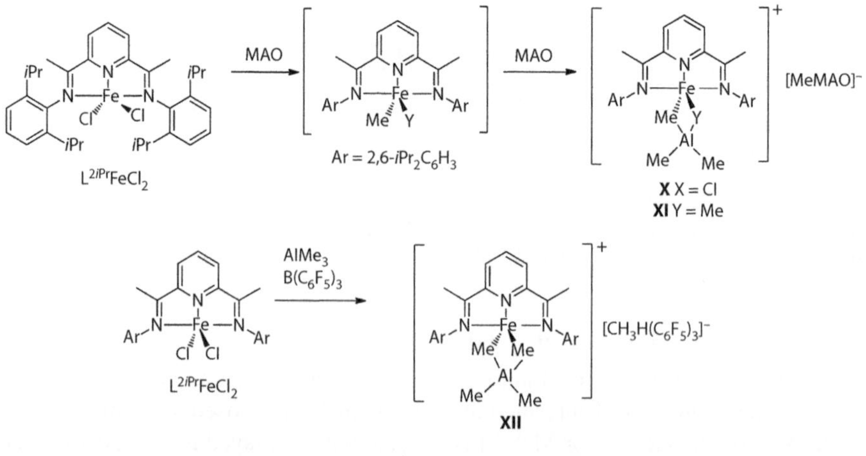

SCHEME 4.3 Interaction of L^{2iPr}FeCl$_2$ with MAO and AlMe$_3$/B(C$_6$F$_5$)$_3$.

NMR and EPR Spectroscopy of the Mechanisms of Metallocene 163

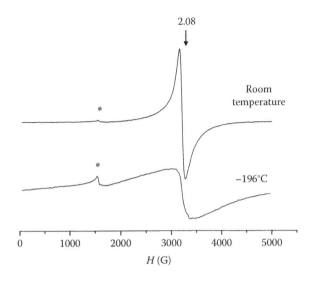

FIGURE 4.28 EPR spectra (toluene) measured after the interaction (10 min) of $L^{2iPr}FeCl_2$ with DMAO (Al/Fe = 20). Asterisks mark Fe^{3+} impurities in the NMR tube glass.

and −276) [98]. The origin of such unexpectedly large paramagnetic shifts became clear after the seminal studies of Chirik and coworkers [101–105]. They have shown that a one-electron reduction of $L^{2iPr}FeCl_2$ resulted in the formation of complex $L^{2iPr}FeCl$. In the latter, iron retains the 2+ oxidation state, one electron being distributed over the tridentate ligand to yield a complex of the type $(L^{2iPr})^{(•-)}Fe^{(+)}Cl$. It has μ_{eff} = 3.5 μB, indicative of the high-spin Fe(II) ion (S_{Fe} = 2) antiferromagnetically coupled to the ligand anion radical ($S = ½$). The unpaired electron spin density on the pyridine part of the ligand leads to the increased paramagnetic shifts in the ^{1}H NMR spectra [102,103]. Very similar large paramagnetic shifts for protons B and F, indicative of similar structure, were established for complex $L^{2iPr}FeMe$ [103].

This background allowed us to define more exactly the structure of intermediate **XIV** formed in the reaction of $L^{2iPr}FeCl_2$ with large excess of $AlMe_3$ (Al/Fe = 30, Scheme 4.4). The ^{1}H NMR spectrum of **XIV** displayed resonances at δ

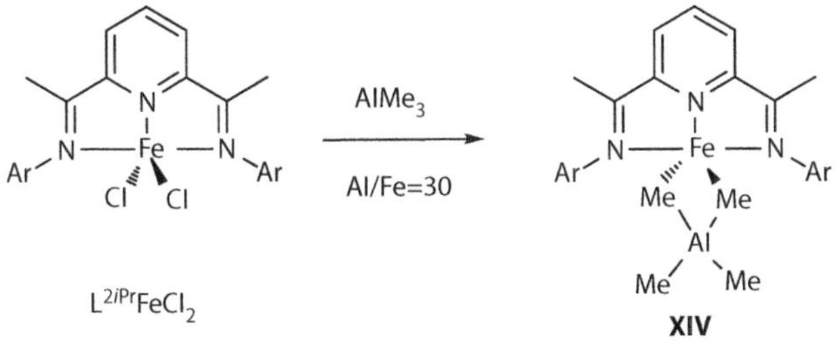

SCHEME 4.4 Interaction of bis(imino)pyridine iron complexes with $AlMe_3$.

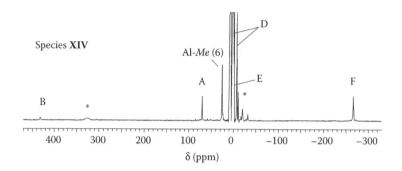

FIGURE 4.29 ^{1}H NMR spectrum (toluene-d_8, 253 K) of $L^{2iPr}FeCl_2/AlMe_3$ (Al/Fe = 30). Asterisks mark peaks that were not reliably assigned.

422 (B, Py-H_p) and −288 (F, N=C(Me)) at 263 K resembling those for the $L^{2iPr}FeCl$; besides, the new peak at δ 45.5 with an integral intensity of 6H was observed, which is typical for the Al(CH$_3$)$_2$ moiety (Figure 4.29, Table 4.6). So, **XIV** can be assigned to a heterobinuclear complex $(L^{2iPr})^{(\bullet-)}Fe^{(+)}(\mu\text{-Me})_2AlMe_2$ [99]. For a related complex $(L^{2Me})^{(\bullet-)}Fe^{(+)}(\mu\text{-CD}_3)_2Al(CD_3)_2$, we succeeded in the detection of both deuterium resonances by ^{2}H NMR spectroscopy. The ^{2}H NMR resonance of the μ-CD$_3$ bridges (with integral intensity of 6D) was found at δ 622 and that of Al(CD$_3$)$_2$ at δ 45.5 [97,98].

Intermediate **XIV** is far less stable than intermediates **X** and **XI**. To disclose the ways of the intermediate **XIV** degradation, the reaction of $L^{2iPr}FeCl_2$, $L^{2iPr}FeCl$, and $L^{2iPr}FeMe$ with AlMe$_3$ was studied by EPR spectroscopy. In all cases, a similar picture was observed. At the first stage of the reaction at −40°C, a narrow signal at $g = 2.003$ was observed (Figure 4.30a). Then, after warming the sample to room temperature, a new broad resonance at $g = 2.08$ appears and grew up (Figure 4.30b, c). The signal at g 2.003 belongs to the complex $(L^{2iPr})^{(\bullet-)}Al^{(+)}Me_2$, reported by Gambarrotta and coworkers [106]. The same spectrum at g 2.003 can be observed upon reacting L^{2iPr} ligand with AlMe$_3$ (Figure 4.30a, insert). Complex with the EPR resonance at g 2.08 was observed in the system $L^{2iPr}FeCl_2/MAO$ and was tentatively assigned to L′Fe^I(μ-Me)$_2$AlMe$_2$, where L′ is intact or modified L^{2iPr} ligand. Ethylene polymerization studies have shown that the systems $L^{2iPr}FeCl_2/MAO$ and $L^{2iPr}FeCl_2/AlMe_3$ display very similar catalytic activities and afforded linear PEs with similar properties.

As it was shown, at high Al/Fe ratios, neutral complex $(L^{2iPr})^{(\bullet-)}Fe(II)^{(+)}(\mu\text{-Me})_2AlMe_2$ is present in the catalyst system $L^{2iPr}FeCl_2/AlMe_3$, and ion pair $[L^{2iPr}Fe(II)(\mu\text{-Me})_2AlMe_2]^+[Me\text{-MAO}]^-$ prevails in the system $L^{2iPr}FeCl_2/MAO$. Both intermediates convert with time to the same iron(I) species with the proposed structure L′Fe^I(μ-Me)$_2$AlMe$_2$ (Scheme 4.5). One can assume that in the case of $(L^{2iPr})^{(\bullet-)}Fe(II)^{(+)}(\mu\text{-Me})_2AlMe_2$ intermediate, reduced bis(imino)pyridine ligand plays the role of the counteranion instead of [Me-MAO]$^-$. It is generally accepted for group 4 complexes that they should be converted to cationic species to form the active polymerization catalyst. However, it is not necessarily the case for late transition metal complexes. For example, neutral nickel(II) alkyl complexes can conduct polymerization of ethylene without any activator (see Section 4.2.5).

NMR and EPR Spectroscopy of the Mechanisms of Metallocene

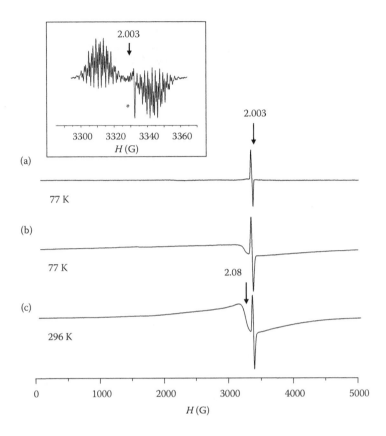

FIGURE 4.30 EPR spectra measured after the interaction of $L^{2iPr}FeCl$ with $AlMe_3$ (Al/Fe = 3): after mixing the reagents at low temperature (–40°C) (a), after warming the previous sample to room temperature (b), and the same sample stored at room temperature for 1 h (c). Inset for Figure 4.1a: EPR spectrum (21°C) of the mixture of $AlMe_3/L^{2iPr}$ (asterisk marks an impurity in the sample tube).

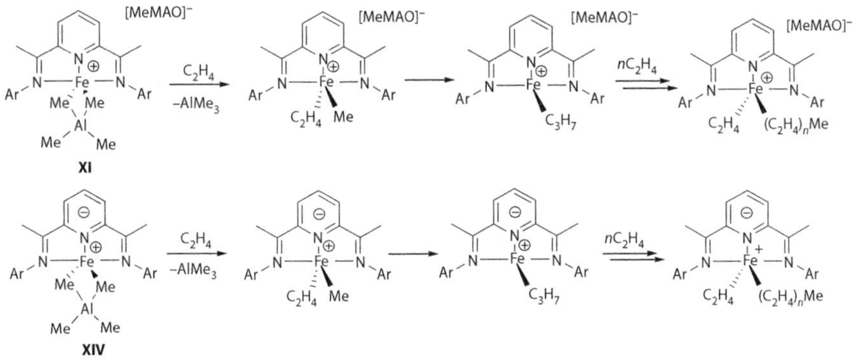

SCHEME 4.5 Polymerization initiation and propagation in the systems $L^{2iPr}FeCl_2/MAO$ and $L^{2iPr}FeCl_2/AlMe_3$.

166   Applications of EPR and NMR Spectroscopy in Homogeneous Catalysis

As it was mentioned, two types of active sites are present in the systems $L^{2iPr}FeCl_2$/MAO and $L^{2iPr}FeCl_2$/AlMe₃; the more active (and unstable) sites operating during the early stage of polymerization and giving low-MW PE are transformed into the less active but more stable sites responsible for the higher-MW fraction of PE [95]. Available experimental data give evidence that the more active sites are intermediates stemming from the heterobinuclear species $(L^{2iPr})^{(\cdot-)}Fe(II)^{(+)}(\mu-Me)_2AlMe_2$ and $[L^{2iPr}Fe(II)(\mu-Me)_2AlMe_2]^+[Me-MAO]^-$. The less active sites can be EPR-active iron species displaying the $g = 2.08$ resonance, formed upon degradation of the initially formed intermediates. The coexistence of (at least) two types of intermediates accounts for the observed multisite nature of the bis(imino)pyridine iron-based catalyst systems.

4.2.2 BIS(IMINO)PYRIDINE COBALT ETHYLENE POLYMERIZATION CATALYSTS

According to the results of Gibson [107,108] and Gal [109], in the $L^{2iPr}Co^{II}Cl_2$/MAO system, the reduction of the initial cobalt(II) precatalyst to cobalt(I) halide is followed by the conversion to a cobalt(I) methyl and ultimately to a cobalt(I) cationic species that contains no cobalt-alkyl bond. Addition of ethylene affords ethylene adduct $[L^{2iPr}Co^I(\eta-C_2H_4)]^+[Me-MAO]^-$, which is considered to be the immediate precursor to the active species. Contrariwise, our ^1H and ^2H NMR studies of the $L^{2Me}Co^{II}Cl_2$/MAO system showed that Co(II) species strongly predominates in the toluene-d_8 solution at 20°C for at least several hours after mixing the reagents [110]. To resolve this contradiction, we have undertaken a ^1H, ^2H, and ^{19}F NMR study of the catalyst systems $LCo^{II}Cl_2$/MAO, $LCo^{II}Cl_2$/AlMe₃, and $LCo^{II}Cl_2$/AlMe₃/[Ph₃C][B(C₆F₅)₄] (L = L^{2iPr}, L^{3Me}, L^{tBu}, Figure 4.31) [111].

4.2.2.1 Activation of $L^{2iPr}Co^{II}Cl_2$ with MAO

The ^1H NMR spectra of the catalyst systems $L^{2iPr}Co^{II}Cl_2$/MAO and $L^{2iPr}Co^{II}Cl_2$/AlMe₃/[CPh₃][B(C₆F₅)₄], recorded at various times after mixing the reagents, show that the starting cobalt complex $L^{2iPr}Co^{II}Cl_2$ quantitatively converted to cobalt(II) complexes $[L^{2iPr}Co^{II}Me(S)]^+[Me-MAO]^-$ (**XV**) and $[L^{2iPr}Co^{II}Me(S)]^+[B(C₆F₅)₄]^-$ (**XV′**), respectively (S = solvent or vacancy). The ^1H NMR spectra of **XV** and **XV′** are very similar (Figures 4.32a, b and Table 4.7). The broad peak at ca. δ 150 in the ^1H NMR spectra of **XV** and **XV′** belongs to a Co–CH₃ group. A similar resonance was earlier observed for the related complex $[L^{2Me}Co^{II}Me(S)]^+[MeMAO]^-$ (S = solvent or vacancy); its assignment to the Co-CD₃ group was confirmed by ^2H NMR spectroscopy, using deuterated MAO [110]. The ^{19}F spectrum of **XV′** was typical for an outer-sphere perfluoroaryl borate anion $[B(C₆F₅)₄]^-$, thus confirming the ionic structure of **XV** and **XV′** [111].

Complexes **XV** and **XV′** are rather stable and remain the major cobalt species in the reaction solution even one day after mixing the reagents at 20°C. Prolonged storing of the samples at room temperature (ca. 1 week) gives rise to a gradual decrease in the concentration of **XV** and **XV′**, accompanied by the increase of the concentration of diamagnetic cobalt(I) complexes **XVI** and **XVI′**. The ^1H NMR spectra of **XVI** and **XVI′** display sharp resonances, typical for diamagnetic species (Table 4.7). The ^1H NMR spectra of this type were reported for cobalt(I) species with coordinated ethylene

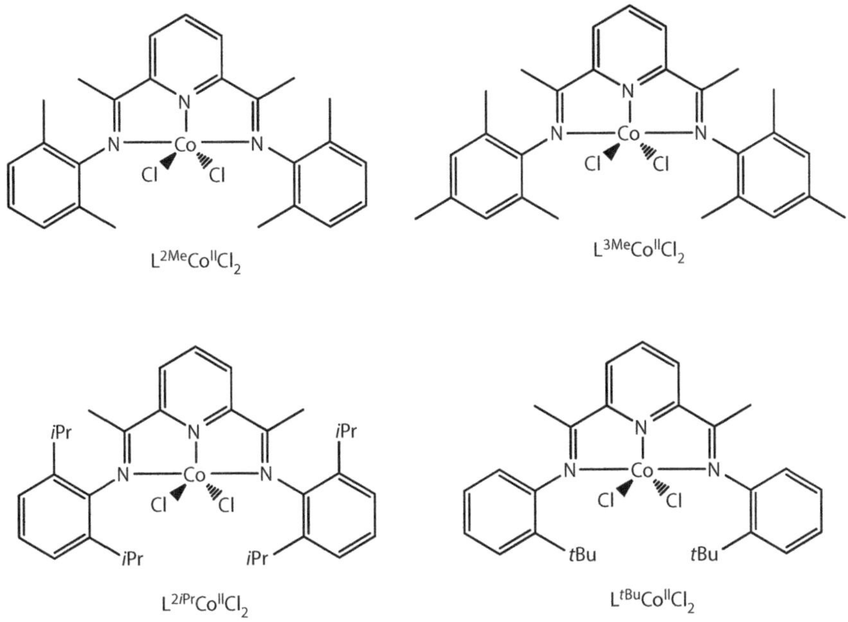

NMR and EPR Spectroscopy of the Mechanisms of Metallocene 167

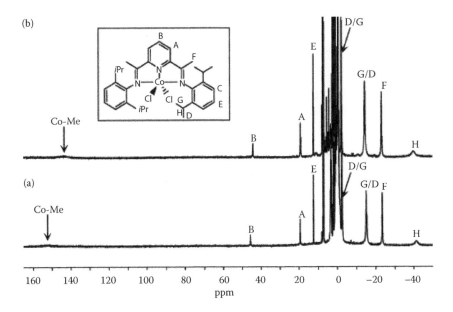

FIGURE 4.31 Structures of cobalt bis(imino)pyridine precatalysts studied.

FIGURE 4.32 ^{1}H NMR spectra (toluene-d_8, 20°C) of the samples L^{2iPr}Co^{II}Cl$_2$/MAO ([Al]:[Co] = 100:1, [Co] = 5 × 10^{-3} M) (a) and L^{2iPr}Co^{II}Cl$_2$/AlMe$_3$/[CPh$_3$][B(C$_6$F$_5$)$_4$] ([Al]:[Co]:[B] = 10:1:2, [Co] = 10^{-2} M) (b), recorded 16 h after mixing the reagents. The inset shows the structure of the precatalyst, with the atom-numbering scheme. (Reproduced with permission from Soshnikov, I. E. et al. 2009. *Organometallics* 28: 6003–6013.)

168 Applications of EPR and NMR Spectroscopy in Homogeneous Catalysis

TABLE 4.7

^{1}H NMR Data for the Complexes Observed upon Activation of $L^{2iPr}Co^{II}Cl_2$ with MAO, $AlMe_3/[CPh_3][B(C_6F_5)_4]$, and $AlMe_3$ (Toluene-d_8, 20°C)

Complex		A Py-H_m	B Py-H_p	C Ar-H_m	D Me$_2$CH	E Ar-H_p	F N=C(Me)	G Me$_2$CH	H Me$_2$CH	Co-Me	μ-Me
$L^{2iPr}Co^{II}Cl_2$ [a]		116.5	49.4	10.0	-17.8	-8.8	4.4	-19.0	-84.6		
$[L^{2iPr}Co^{II}Me(S)]^+[Me\text{-}MAO]^-$	XV	19.4	45.9	NA	-2.4	12.4	-23.6	-15.2	-41.5	~152	
$[L^{2iPr}Co^{II}Me(S)]^+[B(C_6F_5)_4]^-$	XV'	19.1	44.3	NA	-2.2	12.3	-23.3	-14.6	-40.0	~143	
$[L^{2iPr}Co^I(S)]^+[Me\text{-}MAO]^-$	XVI	7.71	8.35	NA	1.20	NA	-0.80	1.03	2.44		
$[L^{2iPr}Co^I(S)]^+[B(C_6F_5)_4]^-$	XVI'	7.48	7.92	NA	1.15	NA	-0.81	0.91	2.44		
$L^{2iPr}Co^I(\mu\text{-}Me)AlMe_2$ [b]	XVII	7.5	9.7	NA	1.1	NA	N/O	0.6	3.1		-2.8
$[L^{2iPr}Co^I(\eta^2\text{-}C_2H_4)]^+[AlMe_3Cl]^-$	XVIII	NA	8.1	NA	1.1	NA	0.6	0.8	3.1		
$[L^{2iPr}Co^I(\eta^1\text{-}N_2)]^+[MeB(C_6F_5)_3]^-$ [c]		6.69	7.55	6.87	1.05	7.00	1.11	0.98	2.85		N/O
$[L^{2iPr}Co^I(C_2H_4)]^+[MeB(C_6F_5)_3]^-$ [c]		~7.0	8.00	6.77	1.09	~7.0	0.75	0.76	3.02		N/O
$L^{2iPr}CoMe$ [c]		7.86	10.19	7.37	1.19	7.49	-1.14	0.62	3.13	N/O	
$L^{2iPr}CoCl$ [c]		6.91	9.54	7.27	1.18	7.41	0.05	1.06	3.33		

Note: N/O, not observed; NA, not reliably assigned.

[a] Solution in CH_2Cl_2.

[b] Spectra recorded at −10°C, only broad singlets in the each case.

[c] Data from Reference [108].

NMR and EPR Spectroscopy of the Mechanisms of Metallocene

molecule $[L^{2iPr}Co^I(\eta^2-C_2H_4)]^+[MeB(C_6F_5)_4]^-$ [108]. Based on these data, **XVI** and **XVI′** were assigned to the ion pairs $[L^{2iPr}Co^I(S)]^+[MeMAO]^-$ and $[L^{2iPr}Co^I(S)]^+[B(C_6F_5)_4]^-$ (S = solvent or vacancy), respectively. Interestingly, the addition of ethylene sharply accelerated the conversion of Co(II) species **XV** and **XV′** to Co(I) species **XVI** and **XVI′**, leading to the presence of only Co(I) species of the type $[L^{2iPr}Co^I(\eta^2-C_2H_4)]^+[A]^-$ ($[A]^- = [MeMAO]^-$ or $[B(C_6F_5)]^-$) in the catalyst systems $L^{2iPr}Co^{II}Cl_2/MAO/C_2H_4$ and $L^{2iPr}Co^{II}Cl_2/AlMe_3/[CPh_3][B(C_6F_5)_4]/C_2H_4$. So, in agreement with the assumption of Gibson [108] and Gal [109], it is cobalt(I) species that are the most likely closest precursors of the active polymerizing species in the catalyst systems considered.

4.2.2.2 Activation of $L^{2iPr}Co^{II}Cl_2$ with $AlMe_3$

In contrast to the catalyst systems $L^{2iPr}Co^{II}Cl_2/MAO$ and $L^{2iPr}Co^{II}Cl_2/AlMe_3/[CPh_3]$ $[B(C_6F_5)_4]$, where Co(II) species predominate in the reaction solution for at least several hours after the reaction onset, the activation of $L^{2iPr}Co^{II}Cl_2$ with $AlMe_3$ ([Al/ [Co] = 50:1, toluene-d_8) leads to a very rapid and quantitative reduction of the starting complex to diamagnetic Co(I) species of the type $L^{2iPr}Co^I(\mu-Me)_2AlMe_2$ (**XVII**). In the presence of C_2H_4 (according to the ^1H NMR data), this species converts to the $[L^{2iPr}Co^I(\eta^2-C_2H_4)]^+[AlMe_3Cl]^-$ (**XVIII**) species.

The catalyst systems based on $LCo^{II}Cl_2$ (L = L^{2iPr}, L^{3Me}, L^{tBu}) produce highly linear PE containing one $-CH_3$ and one $-CH=CH_2$ group per one PE molecule. The polymerization activities of $L^{2iPr}Co^{II}Cl_2/MAO$ and $L^{2iPr}Co^{II}Cl_2/AlMe_3$ systems are similar, and the obtained PEs have almost identical characteristics (Table 4.8, runs

TABLE 4.8
Ethylene Polymerization over the $LCo^{II}Cl_2$ and $LFe^{II}Cl_2$ Complexes[a]

Run	Complex	Cocatalyst	PE Yield[b]	M_w (g/mol)	M_w/M_n	Content per PE Molecule[c]	
						CH_3	$CH_2=CH_2$
1	$L^{2iPr}Co^{II}Cl_2$	MAO	1250	18,000	2.0	0.8	1.0
2		$AlMe_3$	750	18,000	2.0	0.8	1.0
3[d]		$AlMe_3/$"B"	980	18,000	2.0	0.9	1.0
4	$L^{3Me}Co^{II}Cl_2$	MAO	5880	1700	1.8	1.0	1.0
5		DMAO	2400	1700	1.9	1.0	1.0
6		$AlMe_3$	1920	1400	2.6	1.0	1.0
7	$L^{tBu}Co^{II}Cl_2$	MAO	4500	330,000	2.5	1.0	1.0
8		$AlMe_3$	2100	420,000	3.9	1.0	1.0
9[e]	$L^{2Me}Fe^{II}Cl_2$	MAO	3100	1500	1.8	1.0	1.0
10[e]		$AlMe_3$	3600	1600	1.8	1.0	1.0

[a] Polymerization in toluene at 40°C, ethylene pressure 5 bar (runs 5,9,10, ethylene pressure 2 bar), for 30 min, [M] = 1.4×10^{-5} M, [Al]/[M] = 500.

[b] kg PE/(mol(M)·bar).

[c] IRS data.

[d] "B" = $[CPh_3][B(C_6F_5)_4]$, molar ratio [Co]:[Al]:[B] = 1:100:1.1.

[e] Data of Reference [95].

1 and 2). These data correspond well to the observed close similarity of the active sites present in L^{2iPr}Co^{II}Cl$_2$/MAO/C$_2$H$_4$ and L^{2iPr}Co^{II}Cl$_2$/AlMe$_3$/C$_2$H$_4$. These sites [L^{2iPr}Co^I(η^2-C$_2$H$_4$)]$^+$[MeMAO]$^-$ and [L^{2iPr}Co^I(η^2-C$_2$H$_4$)]$^+$[AlMe$_3$Cl]$^-$ differ in only the nature of the outer-sphere counteranions. The existence of only one type of active sites in the considered catalyst systems is in good agreement with the observed M_w/M_n value of 2, which is characteristic of single-site polymerization catalysts (Table 4.8).

Like most homogeneous catalysts, the systems studied demonstrate high initial activity, which rapidly decreases in the course of polymerization (Figure 4.33). The EPR data show that the concentration of cobalt complexes in the L^{2iPr}Co^{II}Cl$_2$/AlMe$_3$ system rapidly decays via transfer of the L^{2iPr} ligand to AlMe$_3$, to afford the paramagnetic complex (L^{2iPr})$^{(\bullet-)}$Al$^{(+)}$Me$_2$, exhibiting characteristic EPR signal at $g = 2.003$ (cf. Figure 4.30). Another possible route of deactivation is the reduction of Co^I into Co^0, which is accompanied by the appearance of a very broad and intense ferromagnetic resonance signal at $g \sim 2.2$ ($\Delta H_{1/2} = 1700$ G) in the EPR spectrum, typical for Co^0. The black cobalt(0) residue was observed in the NMR tube upon storing the sample L^{2iPr}Co^{II}Cl$_2$/AlMe$_3$ for 1 h at 20°C.

It is noteworthy that the structure of the ion pair [L^{2iPr}Co^{II}Me(S)]$^+$[Me-MAO]$^-$, formed in the L^{2iPr}Co^{II}Cl$_2$/MAO system, differs from the structure of the ion pair [L^{2iPr}Fe^{II}(μ-Me)$_2$AlMe$_2$]$^+$[Me-MAO]$^-$ formed in the L^{2iPr}Fe^{II}Cl$_2$/MAO system. Interestingly, for L^{3Me}Co^{II}Cl$_2$, the formation of both ion pairs [L^{3Me}Co^{II}Me(S)]$^+$[Me-MAO]$^-$ and [L^{3Me}Co^{II}(μ-Me)$_2$AlMe$_2$]$^+$[Me-MAO]$^-$ has been documented. The latter species can be observed just after mixing L^{3Me}Co^{II}Cl$_2$ with MAO at low temperature (−20°C), and further rapidly converts to the former at room temperature.

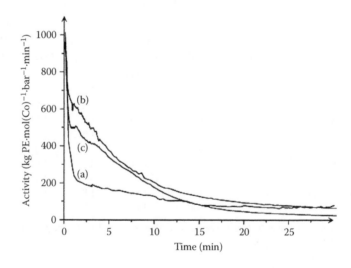

FIGURE 4.33 Ethylene polymerization rate versus time for the catalysts L^{2iPr}Co^{II}Cl$_2$ (a), L^{3Me}Co^{II}Cl$_2$ (b), and L^{tBu}Co^{II}Cl$_2$ (c), activated with MAO (runs 1, 4, and 8 in Table 4.8). (Reproduced with permission from Soshnikov, I. E. et al. 2009. *Organometallics* 28: 6003–6013.)

NMR and EPR Spectroscopy of the Mechanisms of Metallocene

In all other respects, the behavior of the catalyst systems $L^{2iPr}Co^{II}Cl_2/MAO/C_2H_4$ and $L^{3Me}Co^{II}Cl_2/MAO/C_2H_4$ is very similar. The closest precursors of the active polymerizing species in both cases are ion pairs of the type $[LCo^I(\eta^2-C_2H_4)]^+[MeMAO]^-$. At the same time, the detailed mechanism of ethylene polymerization in the presence of species of this type (containing no metal-alkyl bonds) remains unclear.

4.2.3 α-Diimine Vanadium(III) Ethylene Polymerization Catalysts

In spite of the stable interest in vanadium-based polymerization catalysts of ethylene homo- and copolymerization, little is known about the polymerization mechanism [10,92,112–117]. The major technical drawback is the paramagnetism and instability of vanadium alkyl compounds, formed upon the interaction of vanadium precatalyst with alkylaluminum activators.

For catalyst systems based on V^{III} precatalysts, complexes of trivalent vanadium are usually implicated as active sites of polymerization [112,113,118]. For the catalyst systems based on vanadium(V) precatalysts, trivalent, tetravalent, and pentavalent alkyl vanadium complexes have been considered as potential active species of polymerization [113,119].

Previously, the treatment of bis(imino)pyridine vanadium(III) complex $L^{2iPr}VCl_3$ (Figure 4.34) with MAO or $AlEt_2Cl$ was shown to afford the active catalyst for ethylene polymerization [120–122]. Gambarotta and coworkers found that the reaction of $L^{2iPr}VCl_3$ with 1 equivalent of MeLi or 2 equivalents of MAO in toluene resulted in the formation of V(III) complex $(mL^{2iPr})VCl_2$ (which was X-ray-characterized), where mL^{2iPr} is modified L^{2iPr} ligand (Figure 4.34). The activity of the system $(mL^{2iPr})VCl_2/MAO$ was the same as that of the system $L^{2iPr}VCl_3/MAO$, and identical polymers were produced with both systems. Therefore, it was concluded that $L^{2iPr}VCl_3$ is the precursor of $(mL^{2iPr})VCl_2$, which is in turn a precursor of the catalytically active species [120]. However, the true active species of the catalyst system $L^{2iPr}VCl_3/$

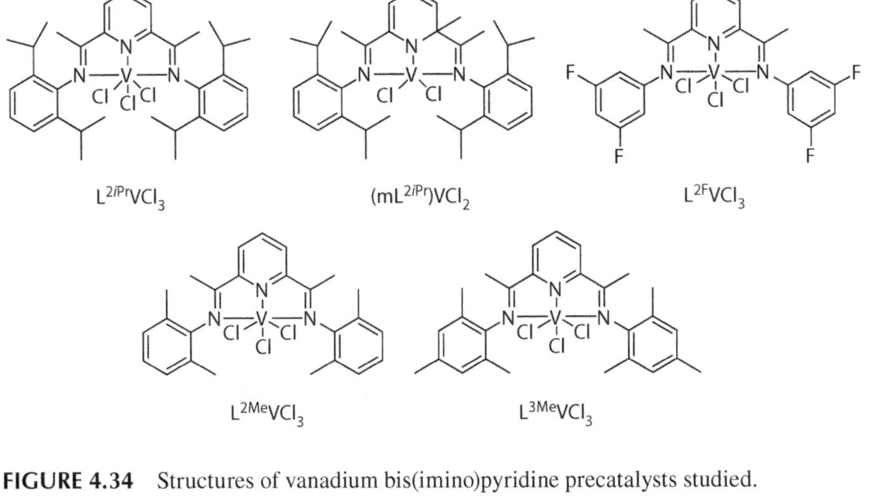

$L^{2iPr}VCl_3$ $(mL^{2iPr})VCl_2$ $L^{2F}VCl_3$

$L^{2Me}VCl_3$ $L^{3Me}VCl_3$

FIGURE 4.34 Structures of vanadium bis(imino)pyridine precatalysts studied.

172   Applications of EPR and NMR Spectroscopy in Homogeneous Catalysis

MAO were not identified. We undertook a ^{1}H and ^{2}H NMR spectroscopic study of vanadium(III) species formed upon the activation of $LVCl_3$ precatalysts ($L = L^{2iPr}$, L^{2Me}, L^{3Me}, L^{2F}) (Figure 4.34) with $AlMe_3/[Ph_3C][B(C_6F_5)]_4$ and with MAO [123].

The activation of complexes $LVCl_3$ ($L = L^{2iPr}$, L^{2Me}, L^{3Me}, L^{2F}) with MAO led to ethylene polymerization catalysts (Table 4.9). Complexes $L^{2Me}VCl_3$ and $L^{3Me}VCl_3$ (runs 2 and 3, Table 4.9) afforded much more active catalysts than complexes $L^{2iPr}VCl_3$ and $L^{2F}VCl_3$ (runs 1 and 4). Low-molecular-weight polymers were obtained with vanadium complexes, except $L^{2F}VCl_3$ (which gave liquid oligomeric products). The activity of $L^{2Me}VCl_3$ was substantially higher with MAO as activator as compared with $AlMe_3/[Ph_3C][B(C_6F_5)_4]$ as activator (runs 2 and 5, Table 4.9). For both activators, polymers with similar M_w, M_n, and MWD were obtained [123]. Previously, it was shown that ion pairs formed upon the activation of metallocene and post-metallocene catalysts with MAO and $AlMe_3/[Ph_3C][B(C_6F_5)_4]$ feature the same cationic parts [6,11,13]. For the present NMR study, we mostly focused on the vanadium precatalysts, activated with $AlMe_3/[Ph_3C][B(C_6F_5)_4]$, because of the more informative NMR spectra.

The starting precatalysts $LVCl_3$ ($L = L^{2iPr}$, L^{2Me}, L^{3Me}, L^{2F}) are paramagnetic (d^3 configuration, $S = 3/2$), displaying ^{1}H NMR resonances in the range of $+110$ to -32 ppm. As an example, Figure 4.35a shows the ^{1}H NMR spectrum of the most active complex $L^{2Me}VCl_3$ in $CDCl_3$ at $-20°C$ (Table 4.10). The ^{1}H resonances of complexes $LVCl_3$ ($L = L^{2iPr}$, L^{2Me}, L^{3Me}) were previously assigned on the basis of integration and comparison of the spectra of different complexes, containing various substituents in the benzene rings [121–122]. Complex $(mL^{2iPr})VCl_2$, prepared as described in Reference [120], displayed no observable ^{1}H NMR resonances.

4.2.3.1 System $L^{2Me}VCl_3/AlMe_3/[Ph_3C][B(C_6F_5)_4]$

Mixing the components of the sample $L^{2Me}VCl_3/AlMe_3/[Ph_3C][B(C_6F_5)_4]$ ($[V]:[Al]:[B] = 1:10:1.2$) in toluene-d_8 for several minutes at $-20°C$ resulted in the conversion of $L^{2Me}VCl_3$ into a new complex **XIX** (Figure 4.35b), assigned to

---

TABLE 4.9
Ethylene Polymerization in the Presence of $LVCl_3$ ($L = L^{2iPr}$, L^{2Me}, L^{3Me}, L^{2F})[a]

Run	Precatalyst	Time (min)	$P(C_2H_4)$ (bar)	m(PE) (g)	Activity ($g_{PE}/$ (mmol$_V$·bar·h))
1	$L^{2iPr}VCl_3$	60	2	2.2	550
2	$L^{2Me}VCl_3$	30	1	15.5	15,500
3	$L^{3Me}VCl_3$	30	2	13.5	6750
4	$L^{2F}VCl_3$	30	2	1.0^b	500
5	$L^{2Me}VCl_3{}^c$	30	2	3.3	1650

[a] Conditions: toluene (50 mL), 60°C, 2 μmol of V, activator: MAO ([Al]:[Ti] = 500:1) for runs 1–4 and $AlMe_3/[Ph_3C][B(C_6F_5)_4]$ ([Al]:[B]:[Ti] = 100:1.2:1) for run 5.

[b] Liquid oligomeric products were obtained (ca. 70 mol.% of 1-butene and 30 mol.% of 1-hexene).

[c] Activator: $AlMe_3/[Ph_3C][B(C_6F_5)_4]$.

NMR and EPR Spectroscopy of the Mechanisms of Metallocene

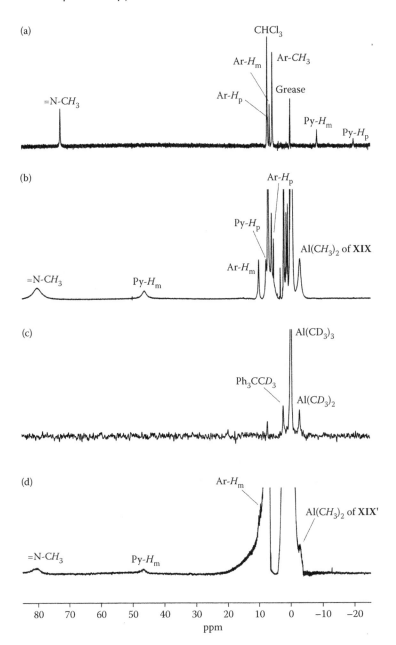

FIGURE 4.35 ¹H NMR spectrum (chloroform-*d*, −20°C) of a saturated solution of L²ᴹᵉVCl₃ (a). ¹H NMR spectrum (toluene-*d*₈, −20°C) of L²ᴹᵉVCl₃/AlMe₃/[Ph₃C]⁺[B(C₆F₅)₄]⁻ ([V]:[Al]:[B] = 1:10:1.2, [V] = 0.01 M) (b). ²H NMR spectrum (toluene, −20°C) of L²ᴹᵉVCl₃/Al(CD₃)₃/[Ph₃C]⁺[B(C₆F₅)₄]⁻ ([V]:[Al]:[B] = 1:10:1.2, [V] = 0.01 M), asterisks mark peak of CH₃C₆H₄*D* (c). ¹H NMR spectrum (toluene-*d*₈, −20°C) of 2/MAO ([V]:[Al] = 1:20, [V] = 0.01 M) (d). Spectra (b)–(d) were measured 10 min after mixing the reagents. (Reproduced with permission from Soshnikov, I. E. et al. 2014. *Organometallics* 33: 2583–2587.)

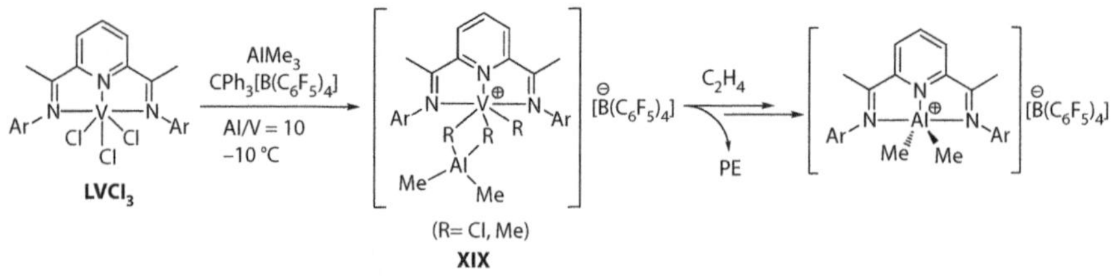

174 Applications of EPR and NMR Spectroscopy in Homogeneous Catalysis

SCHEME 4.6 Reactions in the systems $LVCl_3/AlMe_3/[Ph_3C][B(C_6F_5)_4]$.

a heterobinuclear ion pair $[L^{2Me}(R)V^{III}(\mu\text{-}R)_2AlMe_2]^+[B(C_6F_5)_4]^-$ (R = Me or Cl) (Scheme 4.6 and Table 4.10). The integral intensities of the observed resonances corroborated this assignment. Apparently, **XIX** is an ion pair rather than a neutral complex of V(III): it was formed only in the presence of the cationizing reagent $[Ph_3C][B(C_6F_5)_4]$ and was not found in the reaction of $L^{2Me}VCl_3$ with $AlMe_3$.

For the reliable assignment of the ^1H NMR peaks of paramagnetic complexes, selective introduction of deuterium labels can provide valuable information. The ^1H NMR spectrum of the sample $L^{2Me}VCl_3/Al(CD_3)_3/[Ph_3C][B(C_6F_5)_4]$ ([V]:[Al]:[B] = 1:10:1.2) displayed the same peaks as **XIX**, besides that at δ –2.9. At the same time, the resonance at δ –2.9 ($\Delta v_{1/2} = 25$ Hz) from the $Al(CD_3)_2$ moiety was found in the ^2H NMR spectrum of the same sample (Figure 4.35c), thus confirming its assignment to the $AlMe_2$ moiety of the heterobinuclear ion pair **XIX**. We have also partially replaced the $=N\text{-}CH_3$ protons of $L^{2Me}VCl_3$ with the $=N\text{-}CD_3$ congeners. Activation of the thus obtained complex $L^{2Me}VCl_3$ with $AlMe_3/[Ph_3C][B(C_6F_5)_4]$ affords complex **XIX**, exhibiting the ^2H NMR resonance at δ 78 ($\Delta v_{1/2} = 180$ Hz) (toluene, –20°C). This result confirms the assignment of the ^1H resonance of **XIX** at δ 80.2 (Figure 4.35b) to the $=N\text{-}CH_3$ group. The evaluated concentration of **XIX** was ca. 2 times lower than that expected on the basis of known initial amount of precatalyst $L^{2Me}VCl_3$. Apparently, this discrepancy is due to the low solubility of **XIX**, which precipitates at the bottom of the NMR tube.

The same approach was applied to the systems $LVCl_3$ (L = L^{iPr}, L^{3Me}, L^{2F}), activated with $AlMe_3/[Ph_3C][B(C_6F_5)_4]$, where ion pairs analogous to **XIX** were found. Comparison of the ^1H NMR spectra of ion pairs of type **XIX** for L = L^{iPr}, L^{2Me}, L^{3Me}, L^{2F} allowed the assignment of most part of their ^1H NMR resonances [123]. A comparison of the ^1H NMR spectra of the samples formed upon the activation of

TABLE 4.10
^1H NMR Data, δ ($\Delta v_{1/2}$, Hz) for the Complexes $L^{2Me}VCl_3$ and XIX at –20°C

	$=N\text{-}CH_3$	Py-H_m	Py-H_p	Ar-H_m	Ar-H_p	Ar-CH_3	$AlMe_2$
$L^{2Me}VCl_3$,[a]	72.6 (100)	–8.3 (90)	–19.8 (100)	6.5 (60)	7.1 (70)	5.7 (80)	–
XIX[b]	80.2 (1100)	47.0 (580)	7.7 (140)	10.1 (100)	6.0 (65)	~6 (~1000)	–2.9 (290)

[a] In $CDCl_3$.
[b] Sample $L^{2Me}VCl_3/AlMe_3/[Ph_3C][B(C_6F_5)_4]$ ([V]:[Al]:[B] = 1:10:1.2) in toluene-d_8 at –20°C.

NMR and EPR Spectroscopy of the Mechanisms of Metallocene

$L^{2iPr}VCl_3$ and $(mL^{2iPr})VCl_2$ with $AlMe_3/[Ph_3C][B(C_6F_5)_4]$ ([V]:[Al]:[B] = 1:10:1.2) in toluene-d_8 at $-20°C$ witnesses that the same complex **XIX** is formed in both samples.

4.2.3.2 System $L^{2Me}VCl_3$/MAO

Some of the ^1H resonances of the ion pair **XIX'** formed upon the activation of $L^{2Me}VCl_3$ with MAO are obscured by the intense resonances of MAO and toluene. Nevertheless, the key resonances of this ion pair can be detected. Those resonances coincide with the corresponding resonances of **XIX** (Figure 4.35d). Hence, **XIX'** can be reasonably assigned to the ion pair $[L^{2Me}(R)V^{III}(\mu\text{-}R)_2AlMe_2]^+[MeMAO]^-$ (R = Me or Cl). Similar ion pairs were observed upon the activation of LVCl$_3$ (L = L^{2iPr}, L^{3Me}, L^{2F}) with MAO. The ion pairs of the type **XIX** and **XIX'** provides the first reported example of spectroscopic characterization of intermediates formed in the catalyst systems $LV^{III}Cl_3$/activator.

When ethylene was injected into the NMR tubes containing freshly generated ion pairs of the type **XIX** and **XIX'** at $-10°C$, monomer consumption was observed, accompanied by precipitation of PE onto the inner surface of the tubes. It is logical to conclude that it is ion pairs $[L(R)V^{III}(\mu\text{-}R)_2AlMe_2]^+[A]^-$ (R = Me, Cl; $[A]^- = [B(C_6F_5)_4]^-$ or $[MeMAO]^-$) that are the direct precursors of the true active species of polymerization.

This picture is analogous to that for bis(imino)pyridine complexes of Fe^{II}, where heterobinuclear ion pairs $[LFe^{II}(\mu\text{-}Me)_2AlMe_2]^+[A]^-$ ($[A]^- = [B(C_6F_5)_4]^-$ or $[MeMAO]^-$) were proposed to be the precursors of the active species of polymerization of the catalyst systems $LFeCl_2/AlMe_3/[Ph_3C][B(C_6F_5)_4]$ and $LFeCl_2$/MAO.

Ion pairs **XIX** and **XIX'** are unstable and convert with time into other species. Three possible routes of this transformation can be proposed: (1) modification of the starting bis(imino)pyridine ligand of the ion pairs **XIX** and **XIX'** upon the reaction with $AlMe_3$, (2) reduction of vanadium(III) to a lower valence state, and (3) bis(imino)pyridine ligand transfer (from V to Al).

The analysis of the ^1H NMR spectra of the sample $L^{2Me}VCl_3/AlMe_3/[Ph_3C]$ $[B(C_6F_5)_4]/C_2H_4$ ([V]:[Al]:[B] = 1:10:1.2; $[C_2H_4]$:[V] = 50:1) showed that in the presence of ethylene, the rate of decay of **XIX** markedly increased ($\tau_{1/2}$ = 15 min at $-10°C$ in the presence of ethylene, and $\tau_{1/2}$ > 80 min at $-10°C$ in the absence of ethylene). In the course of the decay of complex $L^{2Me}VCl_3$, resonances of complex $[L^{2Me}AlMe_2]^+$ $[B(C_6F_5)_4]^-$ grew up. The identity of the latter complex was established by independent synthesis. The ^1H NMR spectrum of $[L^{2Me}AlMe_2]^+[B(C_6F_5)_4]^-$ witnesses that the structure of L^{2Me} does not undergo modifications in the course of the formation of this ion pair. The evaluated concentration of $[LAlMe_2]^+[B(C_6F_5)_4]^-$ accounts for at least 30% of the concentration of the starting complex $L^{2Me}VCl_3$; so, the ligand scrambling (to form $[L^{2Me}AlMe_2]^+[B(C_6F_5)_4]^-$) may be regarded as one of the major deactivation pathways of **XIX**.

For catalyst systems based on V^{III} precatalysts, complexes of trivalent vanadium with the alkylaluminum activator were often considered as the active sites of polymerization [112,113,118]. These results represent the first experimental proof of this hypothesis. However, this does not mean that the formation of heterobinuclear ion pairs of the type **XIX** is a necessary step for all vanadium(III) precatalyst activation. Very recently, we have found that the activation of α-diimine complex **XX**

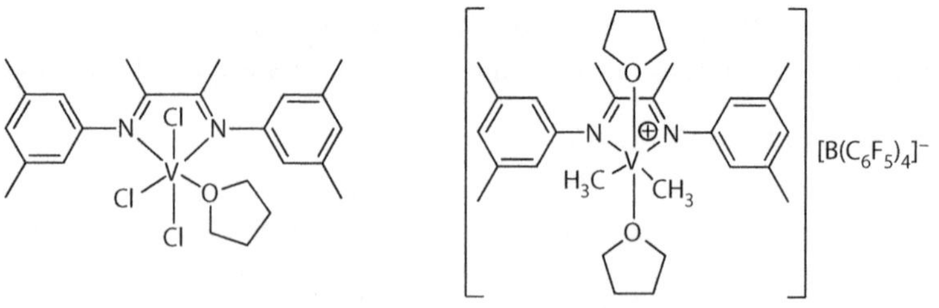

176 Applications of EPR and NMR Spectroscopy in Homogeneous Catalysis

FIGURE 4.36 Example of α-diimine vanadium complex [ArN = C(Me)-C(Me) = NAr] V^{III}(THF)Cl$_3$ (**XX**) and of the ion pair [{ArN = C(Me)-C(Me) = NAr}V^{III}MeCl]$^+$[B(C$_6$F$_5$)$_4$]$^-$ (**XXI**) generated therefrom.

(Figure 4.36) with AlMe$_3$/[Ph$_3$C][B(C$_6$F$_5$)$_4$] or MAO leads to the formation of ion pairs of type [LV^{III}Me$_2$(THF)]$^+$[A]$^-$ (**XXI**), rather than of heterobinuclear ion pairs of type **XIX**. A similar situation was previously observed for bis(imino)pyridine complexes L^{2iPr}CoCl$_2$ (Section 4.2.2).

To conclude this section, the presented data show that the activation of V^{III} precatalysts LVCl$_3$ with AlMe$_3$/[Ph$_3$C][B(C$_6$F$_5$)$_4$] or MAO results in the formation of ion pairs of type **XIX** or **XXI**, depending on the structure of L. The situation is more complicated for V^V precatalysts. In this case, V^V, V^{IV}, and V^{III} species have been previously considered as possible active sites of polymerization [113,119].

4.2.4 ETHYLENE POLYMERIZATION PRECATALYST BASED ON CALIX[4]ARENE VANADIUM(V) COMPLEX

We undertook several attempts to clarify the nature of V^{IV} species formed upon the activation of V(V) precatalysts with AlEt$_2$Cl and AlMe$_2$Cl [125–127]. The most clear results were obtained for calix[4]arene complex **1^V** (Figure 4.37). The stable calix[4]arene ligand of **1^V** was not released or transformed in the course of activation with AlEt$_2$Cl or AlMe$_2$Cl. The EPR spectra of V(IV) species formed were intense and well resolved, thus providing reliable information on their structure [127]. The polymerization data showed that only in the presence of the reactivator (ethyl trichloroacetate, ETA), the catalyst systems **1^V**/AlEt$_2$Cl/ETA and **1^V**/AlMe$_2$Cl/ETA display high initial activity in ethylene polymerization (about 500 kgPE·[mol V bar min]$^{-1}$, calculated according to the kinetic curve for PE yield within 2 min). The catalyst system **1^V**/AlEt$_3$/ETA was inactive in ethylene polymerization.

4.2.4.1 Reaction of Calix[4]arene Vanadium(V) Complex with AlEt$_2$Cl

The EPR spectrum recorded 10 min after mixing the oxo-calix[4]arene complex **1^V** with AlEt$_2$Cl at −70°C ([AlEt$_2$Cl]:[1] = 10:1, [1] = 10^{-2} M, toluene) displays intense resonances of complex **1^{V}a**, as well as relatively weak resonances of complex **1^{V}b** (Figure 4.38a). Storing the sample in Figure 4.38a for 20 min at −40°C led to the EPR spectrum, which showed resonances of complex **1^{V}b** and complex

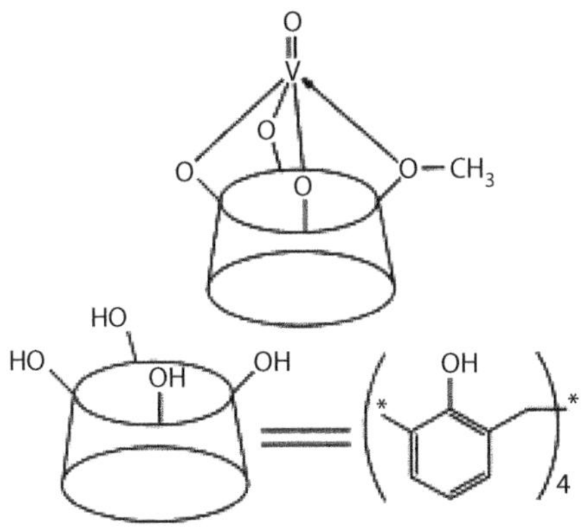

NMR and EPR Spectroscopy of the Mechanisms of Metallocene

FIGURE 4.37 Structure of complex 1^V. (Reproduced with permission from Soshnikov, I. E. et al. 2009. *Organometallics* 28: 6714–6720.)

1^Vc (Figure 4.38b). Further warming this sample to room temperature gave rise to the EPR spectrum, exhibiting only resonances of 1^Vc (Figure 4.38c) [127]. The concentration of 1^Vc decayed with a half-life time of 30 min at 20°C via the reduction of vanadium(IV). The EPR parameters of $1^Va–c$ are collected in Table 4.11. The EPR parameters were derived by theoretical simulation of the experimental spectra (Table 4.11) [127]. One example of such simulation is presented in Figure 4.38d (dotted line). Quantitative EPR measurements show that up to $(50 \pm 15)\%$ of the starting vanadium(V) complex was converted into vanadium(IV) species upon the interaction with $AlEt_2Cl$.

At a high $[AlEt_2Cl]:[1^V]$ ratio (100:1), that is, under conditions closer to those of practical polymerizations, the EPR spectrum of the sample $1^V/AlEt_2Cl$, recorded 10 min after storing at −50°C, displayed resonances of 1^Vb (Figure 4.39a). Further storing this sample at −10°C led to the EPR spectrum exhibiting resonances of 1^Vb and 1^Vc (Figure 4.39b), and only 1^Vc was observed at higher temperatures (Figure 4.39c). The maximum concentration of vanadium(IV) was $70 \pm 20\%$ of the initial vanadium concentration. However, the concentration of vanadium(IV) species sharply decreased upon warming the sample from −50°C to 20°C (Figure 4.39a–d). Storing this sample for 1 day at 20°C led to complete disappearance of the EPR spectra due to the reduction of vanadium(IV) into EPR-silent vanadium species.

The addition of reactivator (ETA) retards the reduction of vanadium(IV). The EPR spectrum of the sample $([AlEt_2Cl]:[ETA]:[1^V] = 100:100:1$, $[1^V] = 10^{-2}$ M, toluene, −50°C) displays resonances of complexes 1^Vb and 1^Vc (Figure 4.40a). Warming this sample to 5°C gave rise to the disappearance of 1^Vb, such that only 1^Vc was observed (Figure 4.40b). In contrast to the sample containing no reactivator, the decay of 1^Vc in the sample with ETA at 20°C occurred much slower (cf. Figures 4.39a–d and 4.40a–c). Moreover, storing the sample containing ETA for 1 day at 20°C resulted in the conversion of 1^Vc back to 1^Vb (Figure 4.40d), rather than in the reduction of vanadium(IV). This result shows that the calix[4]arene ligand is not released or modified during the transformation $1^V \rightarrow 1^Va \rightarrow 1^Vb \rightarrow 1^Vc$. Otherwise, it is difficult

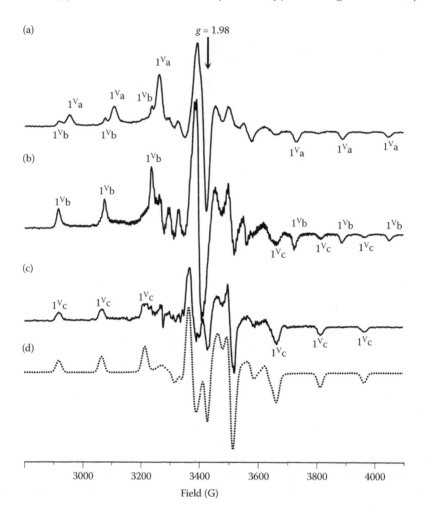

FIGURE 4.38 EPR spectra (toluene, −196°C) of the sample 1^V/AlEt$_2$Cl ([AlEt$_2$Cl]:[1^V] = 10:1, [1^V] = 10^{-2} M) after various treatments: 10 min after mixing of reagents at −70°C (a), 20 min after storing sample in "a" at −40°C (b), and 10 min after storing sample in "b" at 20°C (c). Dotted line shows simulated spectrum of 1^Vc with parameters presented in Table 4.11 (d). (Reproduced with permission from Soshnikov, I. E. et al. 2009. *Organometallics* 28: 6714–6720.)

to explain the mechanism of transformation of 1^Vb to 1^Vc, and then, after prolonged storing, of conversion of 1^Vc back to 1^Vb (Figure 4.40a–d). The proposed transformation of 1^V to 1^Vc is presented in Scheme 4.7. To get additional data in favor of the structures presented in Scheme 4.7, we have compared the EPR spectra of the intermediates formed in the systems 1^V/AlEt$_2$Cl, 1^V/AlMe$_2$Cl, and 1^V/AlEt$_3$.

4.2.4.2 Reaction of Calix[4]arene Vanadium(V) Complex with AlMe$_2$Cl

The EPR spectrum recorded 5 min after mixing 1^V with AlMe$_2$Cl at −70°C ([AlMe$_2$Cl]:[1] = 10:1, [1] = 10^{-2} M) displays resonances from 2 species, namely,

TABLE 4.11
Selected EPR Data (−196°C) for Vanadium(IV) Species Formed in the Systems 1^V/AlR$_3$ and 1^V/AlR$_2$Cl in Toluene (R = Me, Et)

Complex		$g_1 \pm 0.002$	$A_1 \pm 1$, G	$g_2 \pm 0.002$	$A_2 \pm 1$, G	$g_3 \pm 0.002$	$A_3 \pm 1$, G
L$_1$V^{IV}(OAlEtCl)	1^{V}a	1.980	36	1.955	50	1.936	153
L$_1$V^{IV}Cl(AlEt$_2$Cl)	1^{V}b	1.984	41	1.960	49	1.947	160
L$_1$V^{IV}Et(AlEt$_2$Cl)	1^{V}c	1.978	34	1.949	47	1.969	148
L$_1$V^{IV}Cl(AlMe$_2$Cl)	1^{V}b$_{Me}$					1.947	160
L$_1$V^{IV}Me(AlMe$_2$Cl)	1^{V}c$_{Me}$	1.982	33	1.955	45	1.967	149
L$_1$V^{IV}(OAlEt$_2$)	1^{V}a′	1.979	34	1.964	46	1.932	151
L$_1$V^{IV}Et(AlEt$_3$)	1^{V}c′	1.974	25	1.949	45	1.969	145

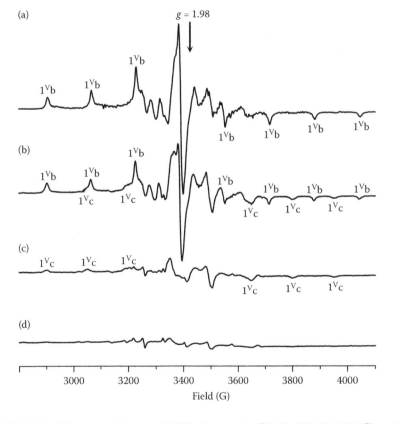

FIGURE 4.39 EPR spectra (toluene, −196°C) of the sample 1^V/AlEt$_2$Cl ([AlEt$_2$Cl]:[1^V] = 100:1, [1^V] = 10^{-2} M) after various treatments: 10 min after mixing of reagents at −50°C (a), 10 min after storing sample in "a" at −10°C (b), 10 min after storing sample in "b" at 5°C (c), and 5 min after storing sample in "c" at 20°C (d). (Reproduced with permission from Soshnikov, I. E. et al. 2009. *Organometallics* 28: 6714–6720.)

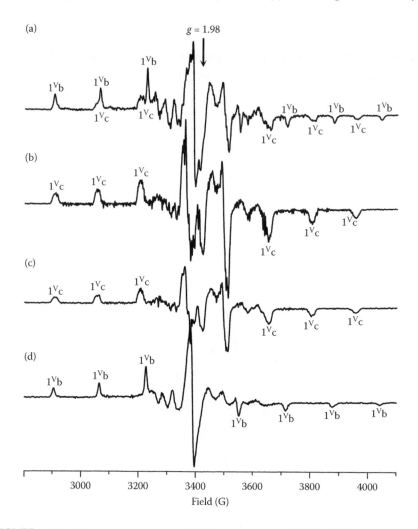

FIGURE 4.40 EPR spectra (toluene, −196°C) of the sample 1^V/AlEt$_2$Cl/ETA ([AlEt$_2$Cl]:[ETA]:[1] = 100:100:1, [1^V] = 10^{-2} M) after various treatments: 10 min after mixing the reagents at −50°C (a), 10 min after storing sample in "a" at 5°C (b), 10 min after storing sample in "b" at 20°C (c), and 24 h after storing sample in "c" at 20°C (d). (Reproduced with permission from Soshnikov, I. E. et al. 2009. *Organometallics* 28: 6714–6720.)

1^{V}b$_{Me}$ and 1^{V}c$_{Me}$, analogous to complexes 1^{V}b and 1^{V}c observed in the 1^V/AlEt$_2$Cl system (Table 4.11). The observed resonances of 1^{V}b$_{Me}$ and 1^{V}b coincide. At the same time, the EPR parameters of 1^{V}c$_{Me}$ and 1^{V}c are very similar but not identical (Table 4.11). This difference supports the incorporation of various alkyls into 1^{V}c$_{Me}$ and 1^{V}c {LV^{IV}Me(S) and LV^{IV}Et(S$_1$)}, respectively (L is calix[4]arene ligand, and S and S$_1$ are donor molecules occupying the sixth coordination site of vanadium). Possible nature of S and S$_1$ will be discussed below. The identical parameters of the observed resonances of 1^{V}b$_{Me}$ and 1^{V}b well reflect their close structures

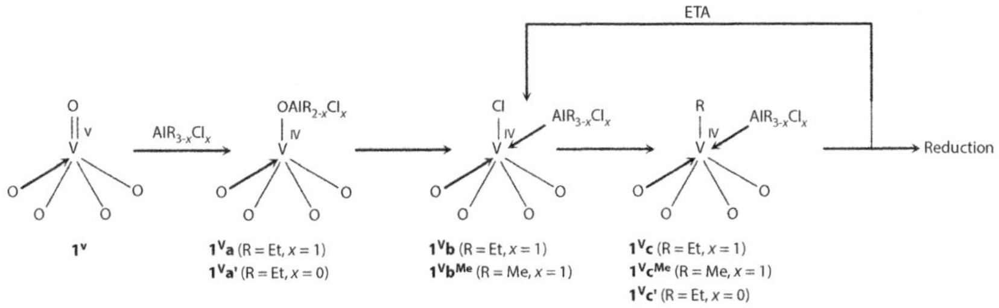

NMR and EPR Spectroscopy of the Mechanisms of Metallocene   181

SCHEME 4.7 Proposed transformations of 1^V after the interaction with $AlR_{3-x}Cl_x$ ($x = 0, 1$; R = Me, Et). (Reproduced with permission from Soshnikov, I. E. et al. 2009. *Organometallics* 28: 6714–6720.)

(presumably $LV^{IV}Cl(S)$ and $LV^{IV}Cl(S_1)$). The proposed structures of $1^V b_{Me}$ and $1^V b$ contain chloride ion as a ligand. If this is correct, complexes of the type $1^V b$ should be absent in the catalyst system $1^V/AlEt_3$. The EPR spectra of the system $1^V/AlEt_3$ have confirmed this prediction (see below).

4.2.4.3 Reaction of Calix[4]arene Vanadium(V) Complex with AlEt₃

The EPR spectrum recorded 10 min after mixing 1^V with $AlEt_3$ at $-70°C$ ([$AlEt_3$]: [1^V] = 10:1, [1] = 10^{-2} M) is a superposition of the EPR spectra of two vanadium(IV) complexes, $1^V a'$ and $1^V c'$ (Table 4.11). Warming up the sample to $-50°C$ results in the disappearance of complex $1^V a'$, while species $1^V c'$ remains in the solution. The EPR parameters of $1^V a'$ are very close yet not identical to those of $1^V a$, whereas the EPR parameters of $1^V c'$ are close yet not identical to those of $1^V c$ (Table 4.11). As predicted, the system $1^V/AlEt_3$ contains no complexes resembling $1^V b$, confirming the assignment of $1^V b$ to species of the type $LV^{IV}Cl(S)$. The structures of $1^V a$ and $1^V a'$ can be tentatively represented as $LV^{IV}(OAlEtCl)$ and $LV^{IV}(OAlEt_2)$, respectively. These species exist in the reaction solution only during the early stage of activation and do not participate in polymerization. It is interesting to compare the structures of complexes $1^V c'$ and $1^V c$. At $20°C$, $1^V c'$ is present in the inert system $1^V/AlEt_3/ETA$, while $1^V c$ is found in the active system $1^V/AlEt_2Cl/ETA$. The small difference between the EPR parameters of $1^V c'$ and $1^V c$ can be explained by the different nature of the ligand S at the sixth coordination site of vanadium complexes $LV^{IV}Et(AlEt_2Cl)$ ($1^V c$) and $LV^{IV}Et(AlEt_3)$ ($1^V c'$). The drastic difference between the reactivities of $1^V c'$ and $1^V c$ may be caused by various degrees of polarization of the V-Et bond upon the interaction with $AlEt_2Cl$ and $AlEt_3$ (the former has higher Lewis acidity [28]). However, further studies are needed to verify this assumption.

Ethylene polymerization activity of the catalyst system $1^V/AlEt_2Cl/ETA$ is 20 times higher than that of the system $1^V/AlEt_2Cl$. This correlates with the slower decay and higher concentration of $1^V c$ in the former system (cf. Figures 4.39 and 4.40). On the basis of these data, one can assume that $1^V c$ {$LV^{IV}Et(AlEt_2Cl)$} is the active species of the catalyst systems $1^V/AlEt_2Cl/ETA$ and $1^V/AlEt_2Cl$. However, at this stage, we cannot exclude that V^{III} intermediates formed upon reduction of $1^V c$ may act as the active species of polymerization.

182   Applications of EPR and NMR Spectroscopy in Homogeneous Catalysis

4.2.5 NEUTRAL $Ni^{II}\kappa^2$-(N,O)-SALICYLALDIMINATO OLEFIN POLYMERIZATION CATALYSTS

The majority of transition metal–based olefin polymerization catalysts are deactivated by polar compounds due to their high oxophilicity. This problem can be solved by using late transition metal catalysts that are less oxophilic [128,129]. By varying the catalyst structures and polymerization conditions, the microstructures of the resulting PE or oligoethylene can be varied from strictly linear to highly branched (as a result of a chain walking mechanism) [128,130]. Neutral κ^2-(N,O)-salicylaldiminato Ni^{II} complexes are an example of catalysts of this type: they tolerate the presence of polar, oxygen-containing compounds and can conduct polymerizations even in aqueous media [131–133].

Mechanistic studies of neutral Ni^{II} polymerization catalysts have so far been rather rare in the literature. Jenkins and Brookhart reported a ^1H NMR spectroscopic investigation of a Ni^{II} anilinotropone system [134]. In their study, the kinetic parameters of the first insertion of ethylene into the Ni-aryl bond of [(N,O)Ni(Ph)(PPh$_3$)] and of the second insertion of ethylene into the Ni-alkyl bond of [(N,O)Ni(CH$_2$CH$_2$Ph)(PPh$_3$)] were evaluated on the basis of the NMR data. Grubbs with coworkers studied the interaction of a [(N,O)Ni(Ph)(PPh$_3$)]-type catalyst with methyl acrylate [135]. Mecking's group reported the dimerization of ethylene to butenes over Ni^{II} salycilaldiminato complex of the type [(N,O)Ni(CH$_3$)(DMSO)] in dimethylsulfoxide and documented the formation of [(N,O)Ni(CH$_2$CH$_3$)(DMSO)] intermediate upon the interaction of the starting catalyst with ethylene, accompanied by the formation of propene [136]. Subsequently, the insertion of polar vinyl monomers into the [(N,O)Ni(CH$_2$CH$_3$)(DMSO)] and [(N,O)Ni(H)(PMe$_3$)] precatalysts was investigated [137]. Delferro and Marks with coworkers reported a bimetallic Ni^{II} catalyst and discussed its deactivation pathways [138]. More recently, Mecking and coworkers documented the involvement of Ni(I) intermediates in salicylaldiminato Ni(II) catalyst deactivation [139]. In none of those mechanistic studies, however, the true catalytically active sites (Ni-polymeryl chain-propagating species) were detected spectroscopically, and their evolution in the course of polymerization was not traced.

We have attempted to get an insight into the processes occurring in the course of chain propagation and termination (and also catalyst deactivation). To do this, we undertook a direct NMR spectroscopic study of the nickel species formed upon the interaction of three Ni^{II} precatalysts of the type [(N,O)Ni(CH$_3$)(Py)] (complexes 1^{Ni}–3^{Ni}, Figure 4.41) with ethylene [140].

4.2.5.1 Chain-Propagating Species Formed upon Ethylene Polymerization with Neutral Salicylaldiminato Nickel(II) Catalysts

The nickel(II) precatalysts 1^{Ni}–3^{Ni} exhibit the ^1H and ^{13}C NMR spectra typical for diamagnetic Ni^{II} complexes. The ^1H NMR spectrum of 1^{Ni} is presented in Figure 4.42a, b. Without added monomer, all considered catalysts are stable at room temperature in toluene-d_8.

The following experimental technique was used to monitor the Ni^{II}–polymeryl species in the reaction solution. Ethylene was injected into the NMR tube

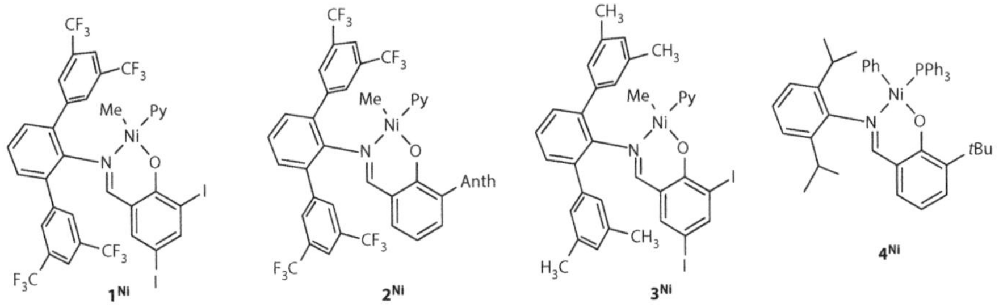

NMR and EPR Spectroscopy of the Mechanisms of Metallocene 183

FIGURE 4.41 Catalysts $\mathbf{1^{Ni}}$–$\mathbf{4^{Ni}}$. Anth = 9-anthranyl.

containing the catalyst solution at low temperature (in an acetone/liquid N_2 mixture), and the sample was placed in a thermostatted (+60°C) bath for (controlled) short periods of time (2 or 3 min) to trigger the formation of [(N,O)Ni(polymeryl)] species. The sample was then quickly cooled with liquid N_2 and placed into the NMR probe (maintained at –20°C). Further measurements were conducted at this temperature.

After storing for 2 min at +60°C, the emergence of new resonances in the range of +0.8 to –1.0 ppm (Figure 4.42b, c) indicated the formation of new nickel-alkyl species $\mathbf{1^{Ni}a}$. The most upfield resonance of $\mathbf{1^{Ni}a}$ (δ –0.19) is close to the Ni–CH_3 peak in $\mathbf{1^{Ni}}$ and can be assigned to the protons of the Ni–CH_2–CH_2– methylene group of the Ni–polymeryl species formed through ethylene insertion into the nickel-methyl bond of catalyst $\mathbf{1^{Ni}}$. ^{1}H COSY spectrum reveals direct spin–spin coupling between the Ni–CH_2–CH_2– protons at δ –0.19 and those at δ 0.73 ppm, which could be assigned to the adjacent Ni–CH_2–CH_2– methylene group of intermediate $\mathbf{1^{Ni}a}$. At the same time, ^{1}H TOCSY spectrum shows that protons of the nickel-bound methylene group Ni–CH_2–CH_2– correlate with protons of at least three remote methylene groups of the same molecule, thus confirming that $\mathbf{1^{Ni}a}$ is indeed a Ni–polymeryl species rather than Ni–propyl or Ni–ethyl products resulting from either a single ethylene insertion or from a single insertion followed by propene elimination and ethylene reinsertion into the resulting nickel hydride species [140]. In the ^{13}C NMR spectrum, resonances of the corresponding Ni–CH_2–CH_2– and Ni–CH_2–CH_2– carbon atoms were found at δ 15.34 and 31.94, respectively (Figure 4.43, Table 4.12; to facilitate the ^{13}C NMR measurements, ethylene-$^{13}C_2$ was used as the monomer).

During the early stages of interaction (2–5 min at +60°C), only very small NMR signals assignable to unsaturated products of ethylene oligomerization were observed in the ^{1}H and ^{13}C NMR spectra, the spectrum in Figure 4.43b mostly representing the alkyl part of the Ni–($^{13}CH_2$$^{13}CH_2$)$_n$–$^{12}CH_3$ chain-carrying species $\mathbf{1^{Ni}a}$. The assignment of ^{13}C and ^{1}H NMR peaks was confirmed by the 2D heteronuclear ^{13}C,^{1}H-correlation spectra [140]. We note that even during this early stage of polymerization, the existence of chain walking is corroborated by the presence of Me branches (marked as B1 in Figure 4.43; the corresponding tertiary CH carbons are marked as CH_B). Further storing the chain-propagating species $\mathbf{1^{Ni}a}$ in the presence of ethylene at +60°C resulted in the appearance of intense PE signals at δ 1.3 (–CH_2–) and 0.9 (–CH_3) (Figure 4.42d, e). At the same time, peaks of $\mathbf{1^{Ni}a}$ declined,

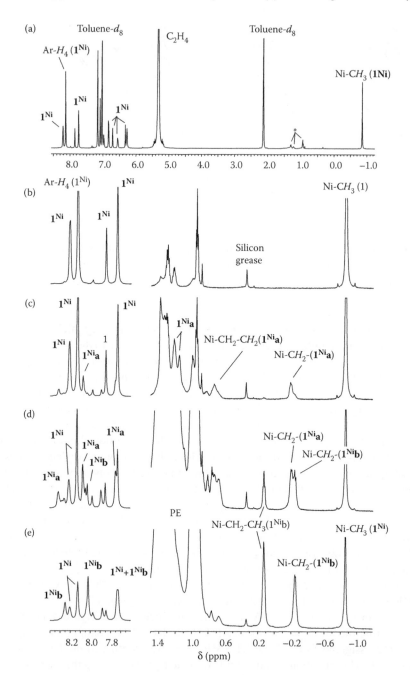

FIGURE 4.42 ¹H NMR spectra (toluene-d_8, −20°C) of a 0.012 M solution of catalyst **1**[Ni] after the injection of 30 equivalents of C_2H_4: full spectrum (a) and regions of δ 8.4 to 7.6 and 1.5 to −1.2 ppm (b); the same sample after storing at +60°C for 2 min (c), 22 min (d), and 51 min (e). The impurities (grease, pentane) in toluene-d_8 are marked with an asterisk. PE—polyethylene. (Reproduced with permission from Soshnikov, I. E. et al. 2013. *Chem. Eur. J.* 19: 11409–11417.)

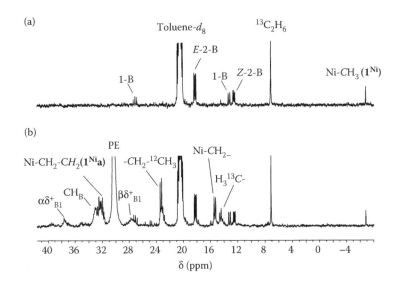

FIGURE 4.43 $^{13}C\{1H\}$ NMR spectra (toluene-d_8, −20°C) of a 0.012 M solution of catalyst 1^{Ni} after various treatments: after the injection of 30 equivalents of $^{13}C_2H_4$ (a) and the same sample after storing for 3 min at +60°C (b). Minor impurities originally present in $^{13}C_2H_4$, that is, ^{13}C-enriched ethane, 1-butene, cis-2-butene, and trans-2-butene, are marked as $^{13}C_2H_6$, 1-B, Z-2-B, and E-2-B, respectively. PE—polyethylene. (Reproduced with permission from Soshnikov, I. E. et al. 2013. Chem. Eur. J. 19: 11409–11417.)

whereas those of a new complex $1^{Ni}b$ grew up (Figure 4.42d, e). ^1H and ^{13}C NMR parameters of $1^{Ni}b$ (Table 4.12) suggest that $1^{Ni}b$ is a Ni-Et complex.

Compound $1^{Ni}b$ is relatively stable and partially survives in the solution even after storing the sample for several days at room temperature or for hours at +60°C. One possible reason for this high stability is the presence of equatorially coordinated pyridine in the coordination sphere of Ni, which hampers β-hydride elimination, so that $1^{Ni}b$ could be reasonably formulated as [(N,O)Ni(Et)(Py)]. The latter gradually declines upon prolonged storing for hours at +60°C or for days at room temperature, yielding new paramagnetic nickel species (see below). Complex 2^{Ni} reacted with ethylene in a similar way, yielding species [(N,O)Ni(polymeryl)] ($2^{Ni}a$) and [(N,O)Ni(Et)(Py)] ($2^{Ni}b$) (Table 4.12). On the contrary, the [(N,O)Ni(polymeryl)] intermediate was not detected for the catalyst 3^{Ni}. Apparently, 3^{Ni} is more prone to β-hydride elimination so that after storing the sample $3^{Ni}/C_2H_4$ for 3 min at 60°C, the majority of 3^{Ni} undergoes ethylene insertion, with subsequent fast formation of the free polymer and [(N,O)Ni(Et)(Py)] ($3^{Ni}b$); some residual 3^{Ni} can also be detected. For the free polymer, intense NMR signals of olefinic groups (cis- and trans-vinylene groups) were observed. This result is in a qualitative agreement with the report that catalyst 3^{Ni} yields a much shorter (and more densely branched) PE than with 1^{Ni} and 2^{Ni} due to its higher tendency to β-H elimination [132]. NMR spectroscopic analysis of PEs formed indicated the presence of one olefinic group (predominantly cis- and trans-vinylene groups rather than terminal vinyl groups) per one PE molecule [140].

TABLE 4.12
NMR Data (−20°C, Toluene-d_8) of Ni(II)-Alkyl Species Considered in Reference [140]

Species	Ni–CH_2–		Ni–CH_2–CH_2–		Ni–CH_2–CH_3		Ar-H_4	Other	
	δ ^{1}H ($^1J_{CH}$, Hz)	δ ^{13}C ($^1J_{CC}$, Hz)	δ ^{1}H	δ ^{13}C ($^1J_{CC}$, Hz)	δ ^{1}H ($^1J_{CH}$, Hz)	δ ^{13}C ($^1J_{CC}$, Hz)	δ ^{1}H	δ ^{1}H	δ ^{13}C
1^{Ni}	—	—	—	—	—	—	8.13	−0.86 (Ni–CH$_3$)	−6.99 (Ni–CH$_3$)
1^{Ni}a	−0.19 (126 Hz)	15.34 (d, 32.3 Hz)	0.73	31.94 (t, 32 Hz)	—	—	8.07	1.21, 1.47 (–CH$_2$–)	30.6 (–CH$_2$–)
1^{Ni}b	−0.25 (126 Hz)	7.53 (d, 33.5 Hz)	—	—	0.13 (124.5 Hz)	15.85 (d, 33.5 Hz)	8.02	—	—
2^{Ni}	—	—	—	—	—	—	8.28	−1.08 (Ni–CH$_3$)	−7.47 (Ni–CH$_3$)
2^{Ni}a	−0.44	14.40 (d, 32.8 Hz)	0.48	31.83 (t, 33 Hz)	—	—	8.23	1.00, 1.22 (–CH$_2$–)	30.5 (–CH$_2$–)
2^{Ni}b	−0.48	6.69 (d, 33.3 Hz)	—	—	—	15.85 (d, 33.3 Hz)	8.22	—	—
3^{Ni}	—	—	—	—	0.01	—	7.37	−0.59 (Ni–CH$_3$), 2.26 (Ar–(CH$_3$)$_4$)	−7.18 (Ni–CH$_3$)
3^{Ni}b	0.06	7.08 (d, 33.2 Hz)	—	—	0.33	16.67 (d, 33.2 Hz)	7.31	2.26 (Ar–(CH$_3$)$_4$)	—

NMR and EPR Spectroscopy of the Mechanisms of Metallocene 187

4.2.5.2 Evaluation of the Size of Ni–Polymeryl Species by PFG NMR Spectroscopy

In recent years, ^{1}H pulsed field-gradient spin echo (PFGSE) NMR spectroscopy [141–142] has been successfully applied to the evaluation of coefficients of translational diffusion and hydrodynamic radii of ion pair precursors of catalytically active sites in metallocene-based olefin polymerization catalysts [40,143–145]. We have used ^{1}H PFGSE NMR for the determination of the size of chain-propagating species $\mathbf{1^{Ni}a}$ and $\mathbf{2^{Ni}a}$ formed upon the reaction of catalysts $\mathbf{1^{Ni}}$ and $\mathbf{2^{Ni}}$ with ethylene in toluene-d_8. Using the dstegp3s1d pulse sequence (which takes advantage of convection compensation), the diffusion-average radii of $\mathbf{1^{Ni}}$, $\mathbf{2^{Ni}}$, $\mathbf{1^{Ni}b}$, and $\mathbf{2^{Ni}b}$ and of $\mathbf{1^{Ni}a}$ and $\mathbf{2^{Ni}a}$ at various ethylene consumptions were estimated (Table 4.13). It appears that even at this rather approximate level (the technique does not take into account the presence of branches in the polymeryl chain, or conformation of the latter), PFGSE NMR provides a reasonably plausible direct size estimate of the chain-propagating intermediates. The estimated average hydrodynamic radii of the chain-propagating intermediates demonstrated a nonmonotonic behavior; they increased in intensity during the early stages of the interaction (2–5 min for catalyst $\mathbf{1^{Ni}}$ and 2–20 min for catalyst $\mathbf{2^{Ni}}$), and decreased at high ethylene consumption, thus reflecting the nonstationary

TABLE 4.13
Estimated Radii, Volumes, and Chain Lengths of Various Nickel Species (Toluene-d_8, −20°C)[a]

Entry	Species	t (min[b])	r_H (Å[c])	Polymeryl, n-(C_2H_4)–[d]
1	$\mathbf{1^{Ni}}$ [(N,O)Ni(CH$_3$)(Py)][e]	–	5.96	–
2	$\mathbf{1^{Ni}a}$ [(N,O)Ni(Polymeryl)][f]	2	6.80	11
3	$\mathbf{1^{Ni}a}$ [(N,O)Ni(Polymeryl)][f]	5	7.07	14–15
4	$\mathbf{1^{Ni}a}$ [(N,O)Ni(Polymeryl)][f]	10	7.00	14
5	$\mathbf{1^{Ni}a}$ [(N,O)Ni(Polymeryl)][f]	20	6.73	10–11
6	$\mathbf{1^{Ni}a}$ [(N,O)Ni(Polymeryl)][f]	32	6.37	6–7
7	$\mathbf{1^{Ni}b}$ [(N,O)Ni(Et)(Py)][e]	60	5.96	–
8	$\mathbf{2^{Ni}}$ [(N,O)Ni(CH$_3$)(Py)][e]	–	6.34	–
9	$\mathbf{2^{Ni}a}$ [(N,O)Ni(Polymeryl)][g]	2	7.40	16
10	$\mathbf{2^{Ni}a}$ [(N,O)Ni(Polymeryl)][g]	5	7.48	17
11	$\mathbf{2^{Ni}a}$ [(N,O)Ni(Polymeryl)][g]	20	7.56	18
12	$\mathbf{2^{Ni}a}$ [(N,O)Ni(Polymeryl)][g]	40	7.33	15
13	$\mathbf{2^{Ni}b}$ [(N,O)Ni(Et)(Py)][e]	60	6.68	–

[a] For measurements details, see Reference [140].
[b] Time of interaction at +60°C.
[c] Hydrodynamic radius.
[d] For the evaluation of the polymeryl chain lengths, see Reference [140].
[e] At 303 K.
[f] $C_2H_4/\mathbf{1^{Ni}} = 30$.
[g] $C_2H_4/\mathbf{2}^{\ Ni} = 30$.

character of the polymerization process in the NMR tube in which no continuous ethylene flow was present. The highest observed radii corresponded to polymeryl chain lengths of 15–18 -C_2H_4- units. Upon prolonged storage at +60°C (40–60 min), the concentration of the [(N,O)Ni(polymeryl)] intermediate decreased, whereas that of the relatively stable (N,O)Ni–Et intermediates increased and became sufficient for the PFGSE measurements. Significantly, the estimated radii of the Ni–Et intermediates **1^{Ni}b** and **2^{Ni}b** were very close to those of the parent catalysts **1^{Ni}** and **2^{Ni}**, respectively (Table 4.13, entries 1, 7 and 8, 13), in line with our suggestion that a pyridine molecule is likely to persist in the structures of **1^{Ni}b** and **2^{Ni}b** [(N,O)Ni(C_2H_5)(Py)].

4.2.5.3 Catalyst Deactivation

In the absence of ethylene, complexes **1^{Ni}–3^{Ni}** are stable in the toluene solution at +60°C for hours. In the course of ethylene polymerization at +60°C, however, these catalysts gradually decomposed. In the system **1^{Ni}/C_2H_4**, the ^{1}H NMR peaks of the diamagnetic Ni-Me, Ni-polymeryl, and Ni-Et complexes decreased, and those of a new paramagnetic Ni(II) complex **1^{Ni}c** grew up. The ^{1}H NMR spectrum of **1^{Ni}c** (Figure 4.44a) exhibits a set of relatively sharp paramagnetically shifted peaks, most

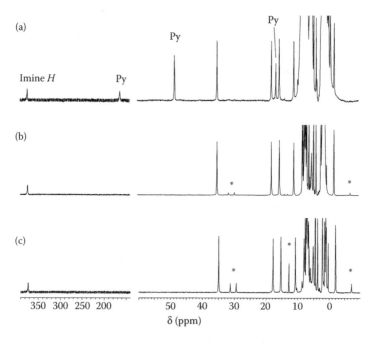

FIGURE 4.44 ^{1}H NMR spectra (toluene-d_8, 0°C) of a 0.012 M solution of catalyst **1^{Ni}** after the injection of 30 equivalents of $^{13}C_2H_4$ and storing for 82 min at +60°C (a). ^{1}H NMR spectrum of crystals of **1^{Ni}c** + pyridine-d_5 (Py/Ni ca. 1.5:1) (b). ^{1}H NMR spectrum of independently prepared bis-ligated complex **1^{Ni}c′** ([(N,O)$_2$Ni]) + pyridine-d_5 (Py/Ni of 5:1) (c). Peaks of coordinated pyridine are marked as Py. Small peaks of another paramagnetic complex (**1^{Ni}c·2Py**) are marked with asterisks. (Reproduced with permission from Soshnikov, I. E. et al. 2013. *Chem. Eur. J.* 19: 11409–11417.)

of them having integral intensities of 2H, except the resonance at δ 16.4, which has an integral intensity of 1H. The shifts and line widths are characteristic of a paramagnetic (high-spin) nickel(II) complex having nonplanar geometry [146]. The spectrum shown in Figure 4.44a did not change when C_2D_4 was used instead of C_2H_4, thus indicating that ethylene was not incorporated into the structure of **1^{Ni}c**. Layering the toluene solution of **1^{Ni}c** with pentane allowed its isolation in single-crystalline form. X-ray analysis indicated that the crystals had a mononuclear bis-ligated structure [(N,O)$_2$Ni]. The quality of crystals of **1^{Ni}c** obtained from the reaction mixture was insufficient. A related complex **1^{Ni}c′** was prepared independently by the reaction of deprotonated ligand with anhydrous NiCl$_2$ and was characterized by X-ray crystallography (Figure 4.45). NMR spectroscopic analysis indicated that **1^{Ni}c** and **1^{Ni}c′** had identical ^{1}H NMR spectra under similar conditions; furthermore, all their resonances were located within the diamagnetic region. The origin of this behavior became clear when pyridine-d_5 was added to samples of **1^{Ni}c** (Figure 4.44b) or **1^{Ni}c′** (Figure 4.44c). Intriguingly, the ^{1}H spectra (Figure 4.44b and c) are paramagnetic and identical to that of Figure 4.44a, besides the lack of three pyridine peaks. This indicates that one pyridine molecule is incorporated into the structure of the major (paramagnetic) product of decay of **1^{Ni}**, and it can be presented as **1^{Ni}c·Py**. Additional pyridine-d_5 caused the formation of another paramagnetic species (presumably **1^{Ni}c·2Py**); see peaks marked with asterisks in Figure 4.44c. At high pyridine/Ni ratios (ca. 100:1, spectrum not shown), **1^{Ni}c·2Py** became the major species in the solution. On the basis of these data, the likely deactivation pathway of the catalytically active sites can be schematically represented by Scheme 4.8. Apparently, under our experimental conditions, the deactivation is kinetically limited by the formation of the elusive [(N,O)Ni(H)] hydride species from the relatively stable [(N,O)Ni(Et)(Py)], followed by reductive elimination of the chelating ligand (N,O)H. The resulting (N,O)

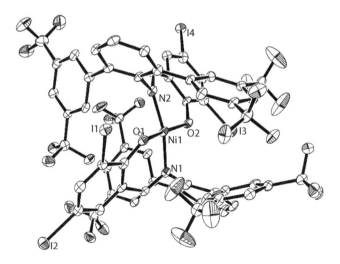

FIGURE 4.45 Molecular structure of complex **1^{Ni}c′**, with thermal ellipsoids drawn at 25% probability level. (Reproduced with permission from Soshnikov, I. E. et al. 2013. *Chem. Eur. J.* 19: 11409–11417.)

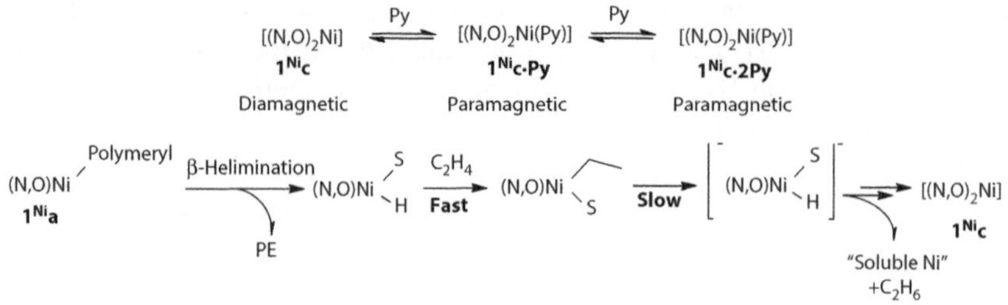

190 Applications of EPR and NMR Spectroscopy in Homogeneous Catalysis

SCHEME 4.8 Equilibria between $1^{Ni}c$, $1^{Ni}c\cdot Py$, and $1^{Ni}c\cdot 2Py$, and the proposed catalyst deactivation pathway in the system $1^{Ni}/C_2H_4$. S = pyridine, solvent molecule, or vacancy.

H ligand can further react with the abundant [(N,O)Ni(Et)(Py)], releasing ethane and bis-ligated [(N,O)₂Ni(Py)] ($1^{Ni}c\cdot Py$). Similar deactivation pathways involving the reductive elimination of the chelating ligand were considered previously [134,135]. More recently, Mecking and coworkers studied catalyst systems 4^{Ni}/ethylene and 4^{Ni}/methyl methacrylate (Figure 4.41) by means of NMR and EPR spectroscopy and showed that Ni complexes [Ni(II)-R] and [Ni(II)-R′] formed in the course of ethylene polymerization can react to form Ni(I) species and release the coupled product R-R′ [139]. Ni(I) complex {[(N,O)Ni(I)(PPh₃)₂]} was directly observed by EPR in a reaction mixture $4^{Ni}/C_2H_4$ and characterized by various techniques (including X-ray structure determination). Complex [(N,O)Ni(I)(PPh₃)₂] underwent further disproportionation, ultimately leading to the observed formation of (N,O)₂Ni(II) and [Ni(0) (PPh₃)₄] [139]. This deactivation pathway can be significant at relatively high [Ni] concentrations (as present in NMR experiments).

4.2.6 FORMATION OF CATIONIC INTERMEDIATES UPON THE ACTIVATION OF BIS(IMINO)PYRIDINE NICKEL CATALYSTS

The nature of the precursors of catalytically active sites in bis(imino)pyridine iron [99] and cobalt [111] systems has been reliably established (see Sections 4.2.1 and 4.2.2). On the contrary, bis(imino)pyridine nickel catalysts have remained relatively underexplored. Norbornene polymerization over bis(imino)pyridine nickel complexes in the presence of MAO was studied in a series of papers and was reported to proceed with low to moderate activities [147,148]. We have demonstrated that the introduction of one or more electron acceptors into the aryl rings drastically enhanced the polymerization activity {up to 1.16×10^7 g of PNB/[(mol cat.) h] in chlorobenzene}, affording high-molecular-weight polynorbornene (Table 4.14) [149]. The ^1H, ^{13}C, and ^{19}F NMR spectroscopic study of the activation of various bis(imino) pyridine nickel catalysts (Figure 4.46) with Lewis acidic cocatalysts (MAO and AlMe₃/[CPh₃][B(C₆F₅)₄]) was undertaken to establish the nature of active species conducting norbornene polymerization [150].

The activation of bis(imino)pyridine nickel catalysts 5^{Ni}–9^{Ni} (Figure 4.46) with MAO and AlMe₃/[CPh₃][B(C₆F₅)₄] has been monitored by NMR spectroscopy. The starting complexes are paramagnetic (high-spin d⁸ configuration), exhibiting the ^1H NMR spectra in the range of δ −10 to +100 [149]. On the contrary, the interaction

NMR and EPR Spectroscopy of the Mechanisms of Metallocene

TABLE 4.14
Norbornene Polymerization on Bis(Imino)Pyridine Ni Complexes[a]

Ni Complex	NB/Ni	T (°C)	Reaction Time (min)	PNB Yield (g)	Activity ($\times 10^6$ g/ [mol Ni h])
5[Ni]	50,000	60	15	2.91	11.6
6[Ni]	50,000	60	15	0.17	0.68
7[Ni]	50,000	60	15	0.95	3.81
8[Ni]	50,000	60	15	1.82	7.30
9[Ni]	50,000	60	15	2.35	9.43

Source: Data from Antonov, A. A. et al. 2012. *Organometallics* 31: 1143–1149.

[a] Reaction conditions: chlorobenzene, 50 mL; Ni complex, 1 µmol; MAO (Al/Ni = 500/1).

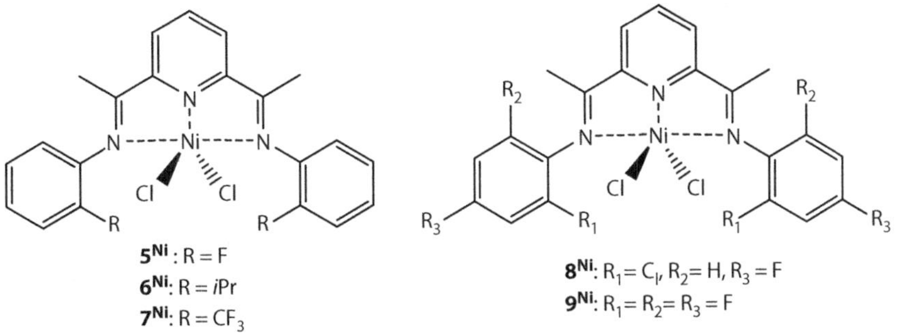

5[Ni]: R = F
6[Ni]: R = *i*Pr
7[Ni]: R = CF$_3$

8[Ni]: R$_1$= C$_l$, R$_2$= H, R$_3$ = F
9[Ni]: R$_1$= R$_2$= R$_3$= F

FIGURE 4.46 Structures of nickel(II) complexes 5[Ni]–9[Ni].

of complexes 5[Ni]–9[Ni] with MAO in toluene-d_8/1,2-difluorobenzene mixture or in chlorobenzene at relatively high Al/Ni ratios (Al/Ni ≥ 30) quantitatively converted the paramagnetic catalysts 5[Ni]–9[Ni] into the new diamagnetic nickel(II) species 5[Ni]a–9[Ni]a. The ^{1}H and ^{13}C NMR spectra of complexes 5[Ni]a and 6[Ni]a are presented in Figure 4.47. The ^{1}H spectra of 5[Ni]a and 6[Ni]a both exhibit sharp peaks, suggesting that bulky oligomeric MAO does not enter into the first coordination sphere of nickel. Interestingly, while the spectrum of the 6[Ni]/MAO system witnesses the formation of a major and minor 6[Ni]a species in a ca. 87/13 ratio, the activation of 5[Ni] with MAO results in the formation of a 54/46 mixture of isomeric nickel complexes.

Both isomers exhibit ^{1}H and ^{13}C NMR peaks of the Ni–CH$_3$ groups (Table 4.15). Their positions and coupling constants ($^1J_{CH}$ = 129–131 Hz) are close to those reported for cationic α-diimine nickel(II)-methyl intermediates [151]. The cationic nature of species 5[Ni]a–9[Ni]a was confirmed by independent experiments: AlMe$_3$/[CPh$_3$][B(C$_6$F$_5$)$_4$] was used as the activator instead of MAO. The resulting intermediates exhibited NMR patterns of the cationic parts, very close to those in the systems with MAO, along with the NMR signals of outer-sphere tetrakis-(perfluorophenylborate) anion.

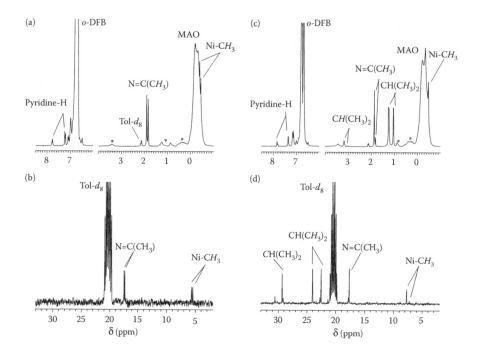

FIGURE 4.47 ^{1}H and ^{13}C NMR spectra (toluene-d_8/1,2-diflourobenzene, 10°C [$6^{Ni}a$] or 2°C [$5^{Ni}a$]) of complexes $5^{Ni}a$—(a, b) and $6^{Ni}a$—(c, d) obtained by reaction of 5^{Ni} or 6^{Ni} with MAO ([Ni] = 1.9×10^{-3} M, Al/Ni = 40). Asterisks mark minor impurities in MAO: oxygenates and vacuum grease. (Reproduced with permission from Antonov, A. A. et al. 2012. *Organometallics* 31: 1143–1149.)

One can conclude that bis(imino)pyridine nickel dichloride species react with MAO or with AlMe$_3$/[CPh$_3$][B(C$_6$F$_5$)$_4$] according to Scheme 4.9. In contrast to the conversion of analogous iron catalysts by those activators into species of the type [LFe(μ-Me)$_2$AlMe$_2$]$^+$[X]$^-$ (where X = MeMAO$^-$, [B(C$_6$F$_5$)$_4$]$^-$) [97–99], no evidence for the formation of heterobinuclear species was gained in the case of Ni catalysts, apparently due to the smaller ionic radius of Ni(II) [152] and hence to greater steric hindrance at the Ni center. The origin of major and minor isomers was studied using complexes 5^{Ni}–9^{Ni} bearing various substituents on the aryl rings. It was found (Table 4.15) that for complex $6^{Ni}a$ with bulky o-CF$_3$ substituents, the ratio of major/minor isomers was 94/6, while for complex $7^{Ni}a$ with a smaller o-Cl substituent, this ratio was 59/41. This led us to a suggestion that isomers of these types could differ in the relative orientations of the ortho-substituents (syn and anti, Scheme 4.9). For the cationic intermediate $9^{Ni}a$ (featuring 2,4,6-F$_3$-substituted aryl rings), no *syn* and *anti* conformers could be expected; indeed, the formation of a single species was observed by NMR (Table 4.15). In the absence of monomer, ion pair species $5^{Ni}a$–$9^{Ni}a$ are stable for at least several hours at room temperature. The reactivity of these intermediates toward norbornene was probed as follows: 4 equivalents of norbornene (dissolved in 0.1 mL of chlorobenzene) was injected at room temperature into an NMR sample containing species $5^{Ni}a$ or $6^{Ni}a$ in chlorobenzene, and the

NMR and EPR Spectroscopy of the Mechanisms of Metallocene

TABLE 4.15

Selected ^{13}C and ^1H NMR Chemical Shifts (δ, ppm), Multiplicities, and Coupling Constants ($^1J_{CH}$, Hz) for Complexes $5^{Ni}a$–$9^{Ni}a$[a]

	^1H				^{13}C			
Ni	Py-H_p	Ar-H	=C(Me)	Ni$-CH_3$	N=C(Me)	Ni$-CH_3$	^{19}F	Major/Minor
$5^{Ni}a$	7.77	7.09, 6.47	1.89	−0.39	17.56	5.53 (J = 131)	−123.17	54/46
			1.83	−0.44	17.40	5.76 (J = 131)	−122.79	
$5^{Ni}a'$[b]	7.57	6.41	1.67	−0.50	17.03	5.21 (J = 131)		53/47
	7.52		1.69	−0.47	17.04	5.18 (J = 131)		
$6^{Ni}a$	7.80	−[c]	1.84	−0.50	17.67	7.68 (J = 129)		87/13
			1.79	−[c]	17.78	7.21		
$6^{Ni}a'$[b]	7.66	6.48	1.74	−0.56	17.32	7.17 (J = 129)		90/10
	7.69		1.71	−0.48	17.46	−[b]		
$7^{Ni}a$[d]	7.66	7.41	1.82	−0.54	18.08	5.11 (J = 130.5)		94/6
			1.74	−[b]	−[b]	5.07		
$8^{Ni}a$	7.80	6.4	1.82	−0.46	17.57	5.04 (J = 128.2)	−110.01	59/41
	7.84		1.75	−0.52	17.50	4.95 (J = 129.2)	−109.99	
$9^{Ni}a$	7.63	6.33	1.80	−0.43	17.68	2.43 (J = 128.6)	−104.69	−[e]
							−115.75	

[a] Unless otherwise stated: in toluene-d_8/1,2-difluorobenzene, using solid MAO as the activator, Al/Ni = 40; at 2°C ($5^{Ni}a$), 21.5°C ($5^{Ni}a'$), 5°C ($8^{Ni}a$), or at 10°C (all other cases). Chemical shifts of minor isomers are given below those for major isomers.

[b] Not found.

[c] With AlMe$_3$/[CPh$_3$][B(C$_6$F$_5$)$_4$] as activator (Al:B:Ni = 30:1.2:1.0).

[d] In chlorobenzene.

[e] Not applicable.

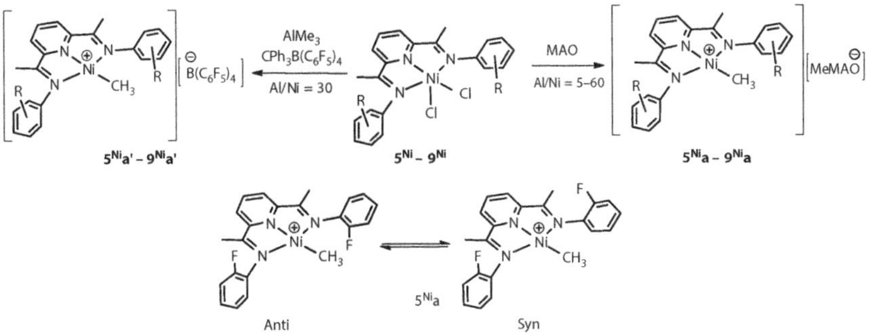

SCHEME 4.9 Interaction of complexes of the type 5^{Ni} with MAO and AlMe$_3$/[CPh$_3$][B(C$_6$F$_5$)$_4$] (top). *Syn* and *anti* conformers of complex $5^{Ni}a$ (bottom).

194   Applications of EPR and NMR Spectroscopy in Homogeneous Catalysis

sample was placed in an NMR probe thermostatted at 20°C. The temperature was increased (in 10°C increments), and the ^{1}H NMR spectra were measured at each temperature. Intermediate $6^{Ni}a$ exhibited rather low reactivity; the formation of a significant amount of polynorbornene could be detected only after storing the sample for several hours at +60°C. On the contrary, intermediate $5^{Ni}a$ polymerized norbornene within several minutes already at 50°C, in accordance with a much higher reactivity of catalyst system 5^{Ni}/MAO as compared to 6^{Ni}/MAO (Table 4.14). It is reasonable to conclude that intermediates $5^{Ni}a$–$9^{Ni}a$ are the closest precursors of the chain-propagating species in the catalyst systems based on bis(imino)pyridine nickel complexes. Further storing the samples $5^{Ni}a$/NB and $6^{Ni}a$/NB led to a drastic broadening of all ^{1}H NMR peaks, apparently due to accumulation of some new paramagnetic species. The EPR spectra of those samples documented the presence of very broad resonances of nickel(0) black; deposition of the latter in the tube was observed, accompanied by the formation of CH_4 (δ 0.16) and C_2H_6 (δ 0.78).

So, the closest precursors of the active species of norbornene polymerization with bis(imino)pyridine nickel complexes in the presence of MAO or AlMe$_3$/[CPh$_3$] [B(C$_6$F$_5$)$_4$], are ion pairs of the types [LNiMe]$^+$[MeMAO]$^-$ and [LNiMe]$^+$[B(C$_6$F$_5$)$_4$]$^-$. These ion pairs demonstrate the capability of polymerizing norbornene under the conditions of NMR-tube experiment; upon prolonged storage in the presence of norbornene, they decompose via reduction of Ni(II) to metallic nickel.

4.2.7   CATIONIC INTERMEDIATES FORMED UPON THE ACTIVATION OF Ni(II) CATALYSTS WITH AlMe$_2$Cl AND AlEt$_2$Cl

Highly branched ethylene oligomers have attracted significant interest as components of lubricants and surface modifiers [130,153,154]. Sun and coworkers have shown that 8-arylimino-5,6,7-trihydroquinolinylnickel chlorides (complexes 10^{Ni}–17^{Ni}, Figure 4.48) activated with MAO or AlEt$_2$Cl are potentially useful catalysts for the production of branched PE waxes [155]. Recently, cationic intermediates formed upon the activation of 10^{Ni} with AlEt$_2$Cl and AlMe$_2$Cl were characterized spectroscopically [156]. Before that study, there were no reported examples of spectroscopic

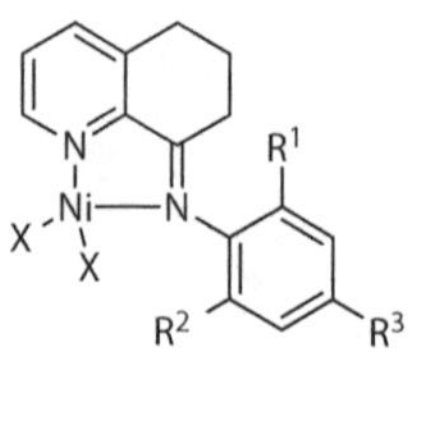

10^{Ni} R$_1$ = R$_2$ = R$_3$ = Me, X = Cl
11^{Ni} R$_1$ = R$_2$ = Et, R$_3$ = H, X = Cl
12^{Ni} R$_1$ = R$_2$ = ^{i}Pr, R$_3$ = H, X = Cl
13^{Ni} R$_1$ = R$_2$ = Me, R$_3$ = H, X = Cl
14^{Ni} R$_1$ = R$_2$ = Et, R$_3$ = Me, X = Cl
15^{Ni} R$_1$ = R$_2$ = C$_6$; R$_3$ = H, X = Br
16^{Ni} R$_1$ = R$_2$ = C$_5$; R$_3$ = H, X = Br
17^{Ni} R$_1$ = C$_6$; R$_2$ = R$_3$ = Me, X = Br
(C$_5$ = cyclopentyl, C$_6$ = cyclohexyl)

FIGURE 4.48   Structures of complexes 10^{Ni}–17^{Ni}.

NMR and EPR Spectroscopy of the Mechanisms of Metallocene   195

TABLE 4.16
Ethylene Polymerization Catalyzed by $10^{Ni}/AlR_2Cl^a$

Number	n(Ni) (µmol)	Cocatalyst	m(PE) (g)	Activity (g_{PE}/[mmol$_{Ni}$· bar·h])	M_n (g/mol)	M_w (g/mol)	M_w/M_n	Branches/ 1000 C^b
1	1.8	$AlMe_2Cl$	6.8	756	310	620	2.0	44 (B_{2+}) 17 (B_1)
2	2.0	$AlEt_2Cl$	5.7	570	310	580	1.9	45 (B_{2+}) 17 (B_1)

^{a} Conditions: heptane (50 mL), 50°C, P_{C2H4} = 5 bar, [Al]/[Ni] = 200.
^{b} ^{13}C NMR data.

characterization of intermediates formed upon the activation of post-metallocene precatalysts with AlR_2Cl (R = Me, Et).

It was found that the activation of 10^{Ni} with $AlMe_2Cl$ and $AlEt_2Cl$ leads to active ethylene polymerization catalysts (Table 4.16). The rate of polymerization for both catalyst systems decreased 3 times within a 30-min period. The polymers obtained had low M_w values (~600 g/mol, M_w/M_n ~ 2) and high content of branches (~45 branches/1000 C).

Precatalyst 10^{Ni} displayed broad paramagnetically shifted resonances in the range of δ +250 to −45 (Figure 4.49a). Most of them can be readily assigned on the basis of integral values and analysis of line width. The reaction of 10^{Ni} with $AlMe_2Cl$ ([Al]/ [Ni] = 120) in toluene-d_8 for 1 min at 0°C results in a conversion of 10^{Ni} to the new paramagnetic complex $10^{Ni}a$ (Figure 4.49b, Scheme 4.10). $10^{Ni}a$ exhibits ^1H resonances from the protons of the starting bidentate ligand and a new broad resonance from the $AlMeCl$ moiety (3H, $\Delta v_{1/2}$ = 540 Hz), which supports its assignment to the heterobinuclear ion pair of the type $[LNi^{II}(\mu\text{-}R)_2AlMeCl]^+[AlMe_3Cl]^-$ (R = Cl or Me).

Further warming the sample of Figure 4.49b to +25°C leads to the disappearance of $10^{Ni}a$ and the formation of the EPR active complex $10^{Ni}b$. The EPR spectrum of $10^{Ni}b$ is characteristic of Ni(I) species (Figure 4.50a, Scheme 4.10). The g-tensor of $10^{Ni}b$ is close to an axially symmetric one (g_1 = 2.046, g_2 = 2.077, g_3 = 2.223). Therefore, $10^{Ni}b$ should display nearly axial symmetry of the ligand environment of the Ni(I) center. The g_1 and g_2 components of the EPR spectrum display partially resolved hfs from nitrogen atoms. The simulation of the EPR spectrum of complex $10^{Ni}b$ (Figure 4.50b) shows that the observed hfs is well described by the interaction of unpaired electron with two equivalent nitrogen atoms (a_1^N = 9.28 G, a_2^N = 10.85 G, a_3^N = 8.85 G). Such picture can be the case for the $[L_2Ni^I]^+[AlMe_3Cl]^-$ ion pair, assuming that only one of the nitrogen atoms of each ligand L causes hfs splitting.

The reaction of 10^{Ni} with $AlEt_2Cl$ at 0°C ([Al]/[Ni] > 50) afforded a diamagnetic ion pair $[LNi^{II}Et]^+[AlEt_3Cl]^-$ ($10^{Ni}c$). The diamagnetic nature of $10^{Ni}c$ is evidence of its square-planar structure. Well-resolved ^1H NMR spectrum of $10^{Ni}c$ is in full agreement with the proposed structure (Figure 4.51, Scheme 4.10). Prolonged storing of

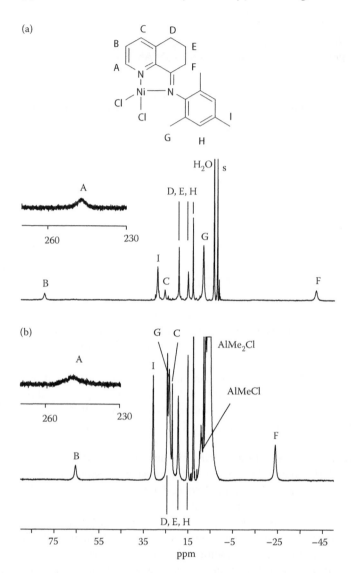

FIGURE 4.49 ¹H NMR spectrum (0°C, acetone-d_6) of **10^{Ni}** (a) and ¹H NMR spectrum (0°C, toluene-d_8) of the catalyst system **10^{Ni}**/AlMe$_2$Cl ([Al]/[Ni] = 120) (b). "s" marks residual solvent peak. (Reproduced with permission from Soshnikov, I. E. et al. 2015. *Organometallics* 34: 3222–3227.)

10^{Ni}c at −20°C results in the decrease in its concentration, accompanied by the formation of EPR active complex **10^{Ni}b′** with the proposed structure [L$_2$Ni^I]$^+$[AlEt$_3$Cl]$^−$. **10^{Ni}b** and **10^{Ni}b′** have identical EPR parameters.

So, different intermediates have been observed during the early stages of reaction of **10^{Ni}** with AlMe$_2$Cl and AlEt$_2$Cl at high Al/Ni ratios ([Al]/[Ni] > 50): paramagnetic heterobinuclear ion pair [LNi^{II}(μ-R)$_2$AlMeCl]$^+$[AlMe$_3$Cl]$^−$ (**10^{Ni}a**) and diamagnetic

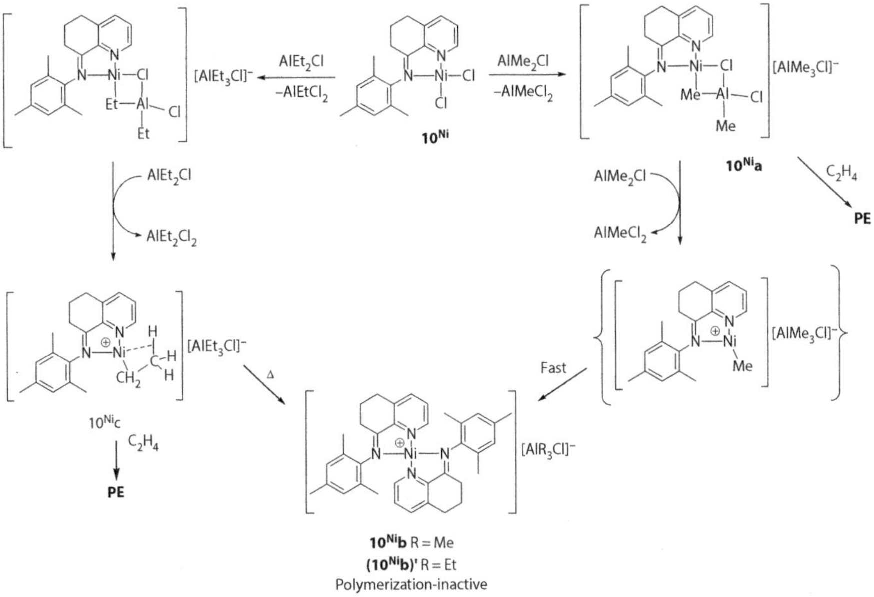

NMR and EPR Spectroscopy of the Mechanisms of Metallocene

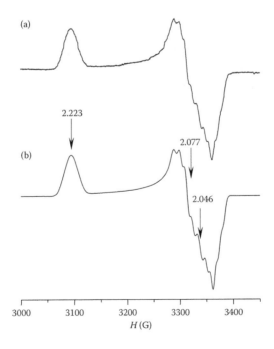

SCHEME 4.10 Transformations of complex **10^Ni** in the catalyst systems **10^Ni/AlMe_2Cl** and **10^Ni/AlEt_2Cl**. (Reproduced with permission from Soshnikov, I. E. et al. 2015. *Organometallics* 34: 3222–3227.)

FIGURE 4.50 Experimental (a) and simulated (b) EPR spectra of the complex **10^Nib**. (Reproduced with permission from Soshnikov, I. E. et al. 2015. *Organometallics* 34: 3222–3227.)

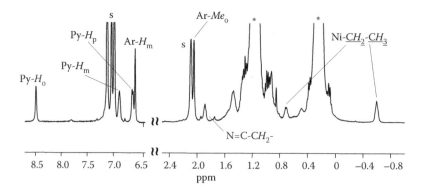

FIGURE 4.51 ^{1}H NMR spectrum (−20°C, toluene-d_8) of the sample **10^{Ni}**/AlEt$_2$Cl recorded after 10 min mixing of the reagents ([Al]/[Ni] = 50, [**10^{Ni}**] = 10^{-2} M). Asterisks mark signals of AlEt$_2$Cl. "s" marks residual solvent peaks. (Reproduced with permission from Soshnikov, I. E. et al. 2015. *Organometallics* 34: 3222–3227.)

ion pair [LNi^{II}Et]$^+$[AlEt$_3$Cl]$^-$ (**10^{Ni}c**), respectively. The catalytic properties of the systems **10^{Ni}**/AlMe$_2$Cl and **10^{Ni}**/AlEt$_2$Cl are very close (Table 4.16). Probably, the reaction of **10^{Ni}a** and **10^{Ni}c** with ethylene results in the formation of similar active species [LNi^{II}(polymeryl)]$^+$[A]$^-$, where [A]$^-$ = [AlMe$_3$Cl]$^-$ or [AlEt$_3$Cl]$^-$.

4.3 ON THE ORIGIN OF LIVING POLYMERIZATION OVER o-FLUORINATED POST-TITANOCENE CATALYSTS

One of the challenging goals of synthetic polymer chemistry is the development of chain-growth polymerization methods that enable consecutive enchainment of monomer units without chain termination ("living polymerization"). Such methods provide precise molecular weight control and allow the synthesis of special polymer architectures (e.g., ultra-high-molecular-weight polyolefins, polyolefins with very narrow MWD, linear block copolymers, stereoregular polypropylene). The conditions for living chain growth are never completely fulfilled in practice. However, for some catalyst systems, the termination processes can be efficiently hampered so that polyolefins with high molecular weights and very narrow molecular weight distributions (M_w/M_n = 1.01–1.10) can be obtained in the course of a quasi-living polymerization [157–159].

Coates [160] and Fujita [161] independently reported that fluorinated titanium catalysts **1^{Ti}** and **2^{Ti}** (Figure 4.52) are capable of conducting living syndiospecific polymerization of propylene. Fujita and coworkers showed that complex **1^{Ti}** catalyzes living polymerization of ethylene at 25°C [162]. The presence of at least one ortho-fluorine substituent was compulsory for the living character of ethylene polymerization, while complexes with *m*- and *p*-F substituents at the aniline moiety demonstrated similarly high activities, but did not exhibit living nature [163]. Coates and coworkers reported that *N*-fluoroaryl bis(phenoxyketimine) Ti complex **3^{Ti}** (Figure 4.52) conducts moderately isoselective living polymerization of propene (73% "mmmm" pentad content) [164]. Again, *o*-F substituents were necessary for

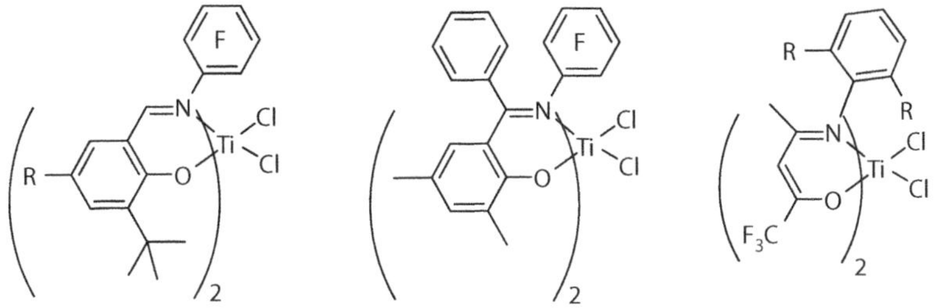

NMR and EPR Spectroscopy of the Mechanisms of Metallocene   199

1^{Ti} R=H
2^{Ti} R=tBu

3^{Ti}

4^{Ti}a R=F
4^{Ti}b R=H

FIGURE 4.52   Structure of complexes 1^{Ti}–4^{Ti}.

living polymerization [165]. Recently, o-F-substituted bis(enolatoimine) titanium complex 4^{Ti}a (R = F) was reported to induce living ethylene polymerization in the presence of MAO at hitherto unattainably high temperatures and afforded PEs with unprecedented narrow molecular weight distributions and high molecular weights at the same time ($M_w/M_n = 1.01$, $M_n = 3 \times 10^5$ g mol^{-1}) [166], while its unsubstituted counterpart 4^{Ti}b (R = H) displayed nonliving behavior (Figure 4.52) [167].

The mechanisms responsible for living polymerization in the presence of o-F substituents have been debated controversially. Based on DFT calculations, Fujita and coworkers hypothesized that the formation of a weak hydrogen bond between an o-F atom and a β-hydrogen atom of the growing polymeryl chain could suppress β-hydrogen elimination and polymeryl transfer to the cocatalyst [163]. Since then, other types of weak ligand–polymer interactions have been invoked to explain the origin of living polymerizations. Talarico and coworkers suggested that weak repulsive interactions between the o-F atoms and the growing polymer chain might be responsible for living propene polymerization with o-fluorinated bis-(phenoxyimine) titanium catalysts [168]. More recently, in a theoretical study of Villani and Giammarino, it was proposed that living polymerization could originate from a three-center Ti$\cdots$F$\cdots\beta$-H interaction, which inhibited β-H transfer to the metal or to the monomer [169]. Chan and coworkers used CF$_3$-substituted group 4 model complexes 5^M (M = Ti, Zr) (Figure 4.53) supported by phenolate-pyridine-(σ-aryl) [O,N,C] ligands to get evidence for the existence of intramolecular C–H$\cdots$F–C interactions. Multinuclear NMR, X-ray, and neutron diffraction studies were interpreted in favor of feasibility of such "contacts" [170,171]. More recently, on the basis of multinuclear NMR and X-ray studies of complexes 6^M (M = Zr, Ti, Hf) (Figure 4.53), Chan and coworkers suggested that, depending on the nature of the metal, the CF$_3$ group of 6^M is involved in various types of interactions: C–H$\cdots$F–C for M = Ti, and o-F$\cdots$M for M = Zr and Hf (**I** and **II**, Figure 4.53) [172]. One could expect that the less rigid ligand environment of complexes 1^{Ti}–4^{Ti} may facilitate the o-F$\cdots$Ti contacts, making these contacts more likely than in the case of complexes 5^M and 6^M.

We undertook direct NMR spectroscopic studies of the active sites of ethylene polymerization formed in the catalyst systems 4^{Ti}a,b/MAO/C$_2$H$_4$ and 4^{Ti}a,b/AlMe$_3$/ [Ph$_3$C][B(C$_6$F$_5$)$_4$]/C$_2$H$_4$ in order to elucidate the interactions that can lead to the suppression of the chain termination processes [173,174]. At high Al/Ti ratios (i.e., at

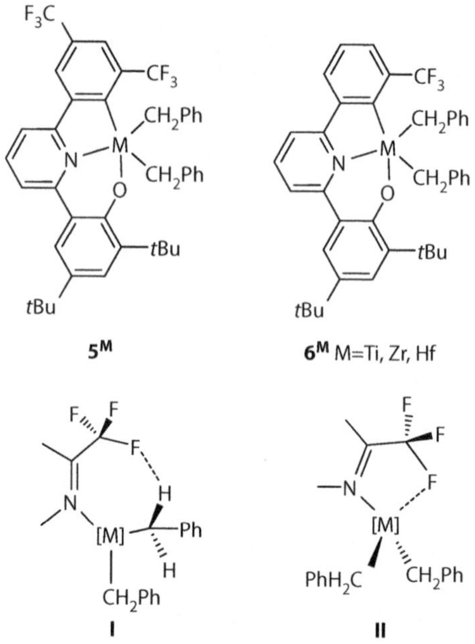

200 Applications of EPR and NMR Spectroscopy in Homogeneous Catalysis

FIGURE 4.53 Structure of complexes **5^M** and **6^M**, and various types of interactions discussed [172] for precatalysts of the type **6^M**.

Al/Ti > 25), complexes of the type **4^Ti a,b** reacted with ^{13}C-MAO to yield ion pairs of the type [LTi(^{13}CH$_3$)]$^+$[MeMAO]$^-$ (**7^Ti a,b**, where L stands for the fluorinated and nonfluorinated ligands, Scheme 4.11). NMR study showed that the Ti-^{13}CH$_3$ group of **7^Ti a** appears at δ 126.6 as a triplet due to coupling with two ^{19}F nuclei, with a coupling constant of 7 Hz [173]. It was assumed that this coupling is due to the weak noncovalent Ti···F interactions: the hydrogen-bonding-type coupling mechanism **I** (Figure 4.53) is unable to explain why the $^2J_{CF} = 7$ Hz splitting is so much higher than the $^1J_{HF}$, which has not been observed at all. So, for the *o*-fluorinated ion pair **7^Ti a**, direct interaction of two of its *o*-F substituents with the Ti center is suggested (Figure 4.54).

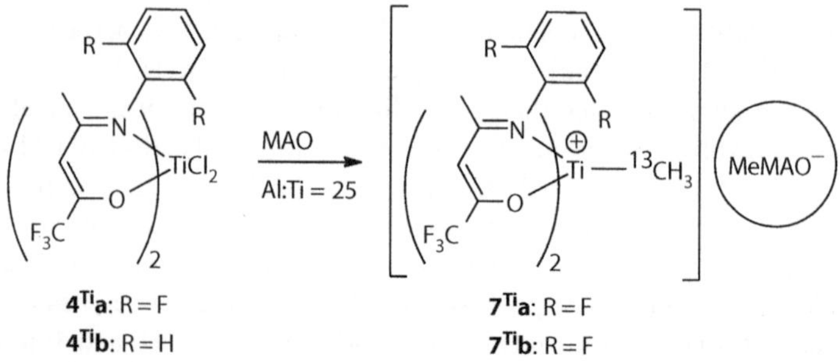

SCHEME 4.11 Activation of bis(enolatoimine) titanium catalysts with ^{13}C-MAO.

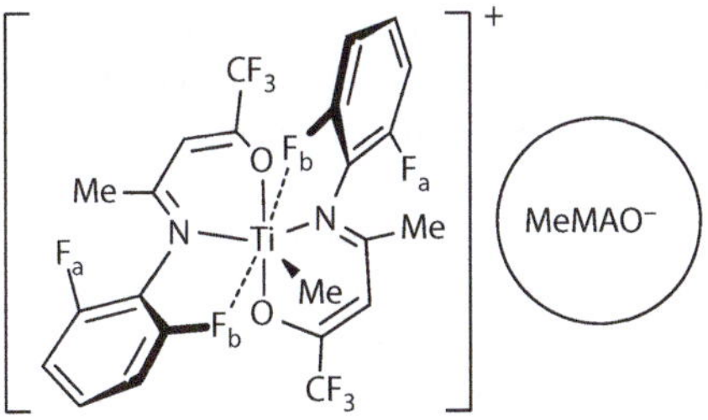

NMR and EPR Spectroscopy of the Mechanisms of Metallocene

FIGURE 4.54 Proposed weak noncovalent Ti···F interactions in the intermediate **7[Ti]a**.

It is very important to establish whether the interaction of this type also exists in the chain-propagating species containing the growing polymeryl chain P, keeping the polymerization catalysis living by suppressing the chain transfer to AlMe$_3$ and β-hydride transfer processes. It has been found that fluorinated and nonfluorinated intermediates [L$_a$TiP]$^+$[MeMAO]$^-$ (**8[Ti]a**) and [L$_b$TiP]$^+$[MeMAO]$^-$ (**8[Ti]b**) (Figure 4.55) display remarkably different NMR spectroscopic properties, that shed light on the distinct coordination geometries of these species [174]. As expected, both the α and β protons (Ti–^{13}CHH– and Ti–^{13}CH$_2$–^{13}CHH–) of **8[Ti]a,b** are diastereotopic and exhibit individual chemical shifts (documented by a pairs of cross-peaks in the ^{1}H–^{13}C HSQC spectra). The differences between the chemical shifts within the pairs of α and β protons (the diastereotopic splitting) are denoted as Δδ$_α$ and Δδ$_β$, respectively. Interestingly, for **8[Ti]a**, Δδ$_α$ > Δδ$_β$ (0.32 vs. 0.12 ppm), while for **8[Ti]b**, Δδ$_α$ < Δδ$_β$ (0.09 vs. 0.23 ppm). It is well known that the diastereotopic splitting in prochiral groups decreases with increasing distance of the prochiral sensor group from the perturbing group [175]. The higher diastereotopic splitting observed for the β-methylene protons of the nonfluorinated complex **8[Ti]b**, compared with that of fluorinated complex **8[Ti]a** (0.23 vs. 0.12 ppm), suggests that in **8[Ti]b**, the β-methylene protons are closer to and more rigidly situated with respect to the perturbing titanium center and aryl rings of the ligand. While in **8[Ti]a** (as well as in **7[Ti]a**), the o-F substituents can be assumed to make the titanium center formally seven-coordinate

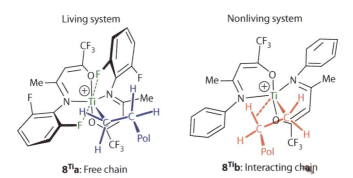

FIGURE 4.55 Proposed structures of propagating species **8[Ti]a** and **8[Ti]b**. (Reproduced with permission from Möller, H. M. et al. 2012. *Chem. Eur. J.* 18: 848–856.)

202   Applications of EPR and NMR Spectroscopy in Homogeneous Catalysis

and would thus shield it from close contacts with the β-methylene group. We would expect that, in the absence of o-F in $8^{Ti}b$, the polymeryl chain will bend and approach the electron-deficient titanium center to establish a weak β-agostic interaction. The structures proposed for the propagating species $8^{Ti}a$ and $8^{Ti}b$ (symbolically denoted as "free chain" and "interacting chain") are presented in Figure 4.55. We conclude that noncovalent aryl o-F⋯Ti interactions are likely to operate in the propagating species $8^{Ti}a$ and suppress the chain termination through β-hydrogen transfer—by preventing close contacts of the titanium center with the β-methylene group of the polymeryl chain. Likewise, polymeryl chain transfer to $AlMe_3$ is probably suppressed by the o-F-induced coordinative saturation of the titanium center in the propagating intermediate $8^{Ti}a$. Similar mechanism may be the case for structurally related o-fluorinated bis(phenoxy-imine) and bis(phenoxy-ketimine)titanium living catalysts $1^{Ti}–3^{Ti}$ [174,176].

4.4 SELECTIVE ETHYLENE TRIMERIZATION BY TITANIUM COMPLEX BEARING PHENOXY IMINE LIGAND

1-Hexene is industrially used as comonomer in the production of linear low-density polyethylene (LLDPE). In industry, 1-hexene is usually obtained by nonselective (statistical) oligomerization of ethylene [177]. The only commercial process capable of selectively producing 1-hexene utilizes chromium-based catalyst [178]. Catalyst systems capable of selective 1-hexene production would be of great industrial and academic interest (see the review by McGuinness [179]). Apart from chromium-based trimerization catalysts, systems relying on titanium complexes have attracted significant interest. Hessen and coworkers discovered a highly active and selective titanium-based catalyst system for this transformation [180]. One of the major recent developments has been the emergence of complex $(FI)TiCl_3$ (9^{Ti}) (FI = phenoxy imine ligand with additional O-donor, Figure 4.56). When activated with MAO, 9^{Ti} produced 1-hexene with exceptionally high activity (up to 132 kg of 1-hexene [g of Ti]$^{-1}$ h^{-1} bar^{-1}) [181].

A serious drawback of titanium-based trimerization catalysts is the formation of a small amount (2%–5%) of high-molecular-weight polyethylene (HMWPE), which leads to reactor fouling. The nature of the activator has a dramatic effect on the

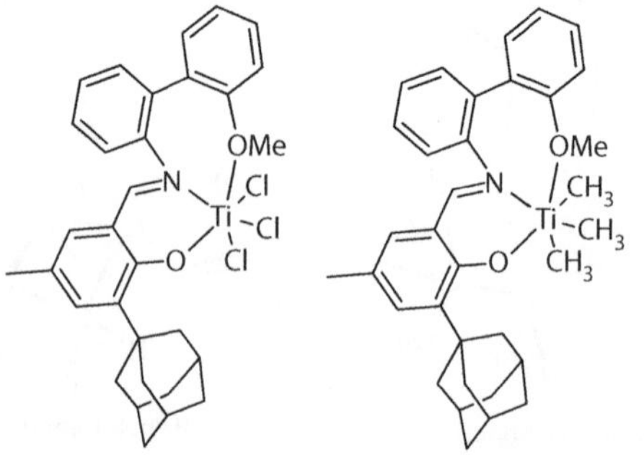

FIGURE 4.56   Structures of complexes 9^{Ti} and 9^{Ti}-Me.

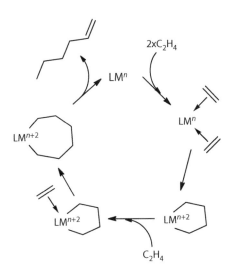

SCHEME 4.12 Proposed mechanism of selective ethylene trimerization on Ti- and Cr-based catalysts. M = Ti or Cr, L = ligand environment, n = 2 for Ti and 1 or 2 for Cr.

1-hexene/PE ratio [182]. For the improvement of the existing trimerization titanium catalysts, a clear understanding of the mechanisms of ethylene trimerization and polymerization is required. The hypothetical mechanism of ethylene trimerization by the catalyst system 9^{Ti}/MAO [181] is similar to the mechanisms previously proposed for chromium [177] and titanium [178] systems (Scheme 4.12).

Bercaw and coworkers showed that the activation of complex 9^{Ti}-Me (Figure 4.56) with 1 equivalent of $B(C_6F_5)_3$ affords ion pair [(Fl)TiMe$_2$]$^+$[MeB(C$_6$F$_5$)$_3$]$^-$ (10^{Ti}), which is an effective precatalyst for selective ethylene trimerization. Treatment of 10^{Ti} with a 1:1 C_2H_4:C_2D_4 mixture at 25°C produced only C_6H_{12}, $C_6H_8D_4$, $C_6H_4D_8$, and C_6D_{12} isotopologs of 1-hexene, which is in accord with the metallacyclic mechanism. Mechanistic studies indicate that the catalyst activation involves the generation of an active Ti^{II} species via olefin insertion into the Ti^{IV}-Me bond, followed by β-H elimination (releasing α-olefin) and reductive elimination of methane [183]. However, Ti^{II} species were not observed spectroscopically in the catalyst system 9^{Ti}-Me/B(C$_6$F$_5$)$_3$.

Our NMR and EPR spectroscopic investigation of the catalyst systems 9^{Ti}/AlMe$_3$/[Ph$_3$C][B(C$_6$F$_5$)$_4$], 9^{Ti}/MAO, and 9^{Ti}/MMAO demonstrated that the ion pairs [(Fl)TiMe$_2$]$^+$[A]$^-$ ([A]$^-$ = [B(C$_6$F$_5$)$_4$]$^-$, [MeMAO]$^-$, and [MeMMAO]$^-$) prevail in the reaction solution only in the initial stage of reaction. Then, depending on the nature of the activator, Ti(III) or Ti(II) species can be observed in the reaction solution [184]. If the activator contains AliBu$_3$ (MMAO, AliBu$_3$/MAO, and AliBu$_3$/[Ph$_3$C][B(C$_6$F$_5$)$_4$]), the concentration of Ti(III) is higher than the concentration of Ti(II). If the activator contains AlMe$_3$ (MAO, AlMe$_3$/[Ph$_3$C][B(C$_6$F$_5$)$_4$]), the concentration of Ti(II) is higher than the concentration of Ti(III).

The EPR spectrum (25°C, toluene) of the sample 9^{Ti}/AliBu$_3$/[Ph$_3$C][B(C$_6$F$_5$)$_4$] displays three resonances (Figure 4.57). The multiplet at g = 2.004 is from the

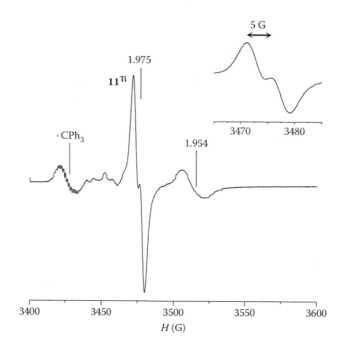

FIGURE 4.57 EPR spectrum (25°C, toluene) of the sample $9^{Ti}/AliBu_3/[Ph_3C]^+[B(C_6F_5)]_4^-$ ([Ti]:[Al]:[B] = 1:10:1.2, [Ti] = 5 × 10^{-3} M). Inset shows the expanded resonance of [(FI)Ti^{III}(μ-H)(μ-Cl)AliBu$_2$]$^+$[B(C$_6$F$_5$)$_4$]$^-$ (**11^{Ti}**).

·CPh$_3$ radical, while the resonance at $g = 1.954$ belongs to the [(FI)Ti^{III}(iBu)$_2$] complex. The intense signal at $g = 1.975$ with resolved hfs ($a_H = 5$ G) from one proton (Figure 4.57, insert) was assigned to a cationic hydride complex **11^{Ti}**. In accordance with its cationic structure, complex **11^{Ti}** was observed only in the presence of ion-forming agents [Ph$_3$C][B(C$_6$F$_5$)$_4$], MAO, and MMAO [184]. For **11^{Ti}**, we proposed the alkylaluminum-complexed structure [(FI)Ti^{III}(μ-H)(μ-Cl)AliBu$_2$]$^+$[B(C$_6$F$_5$)$_4$]$^-$. Similar structures were previously reported for neutral and cationic Zr^{IV} and Zr^{III} hydride species [76,85,185].

In the systems $9^{Ti}/AliBu_3/[Ph_3C][B(C_6F_5)_4]$, $9^{Ti}/AliBu_3/MAO$, and $9^{Ti}/MMAO$, the total concentration of Ti^{III} exceeded 50% of the titanium content, whereas in the systems $9^{Ti}/AlMe_3/MAO$ and $9^{Ti}/AlMe_3/[Ph_3C][B(C_6F_5)_4]$, only minor part of titanium (less than 10%) was present in the reaction solution in the form of Ti^{III} species (neutral complex LTi^{III}Me$_2$), the rest mainly retaining the Ti^{IV} oxidation state.

To test the possibility of the ^{1}H NMR spectroscopic detection of Ti^{II} species in the catalyst systems studied, we started from the reaction of **9^{Ti}** with various trialkylaluminum compounds AlR$_3$ (R = Me, Et, iBu). The ^{1}H NMR spectrum of the sample $9^{Ti}/AlMe_3$ ([Al]:[Ti] = 5:1) displayed broad paramagnetically shifted resonances in the range of δ −20 to +20, which was ascribed to high-spin ($S = 1$) Ti^{II} complex LTi^{II}Cl (**12^{Ti}**) (Figure 4.58a). When the [Al]/[Ti] ratio is increased to 15:1, resonances of new complex with the proposed structure LTi^{II}Me (**13^{Ti}**) grew up at the expense

NMR and EPR Spectroscopy of the Mechanisms of Metallocene

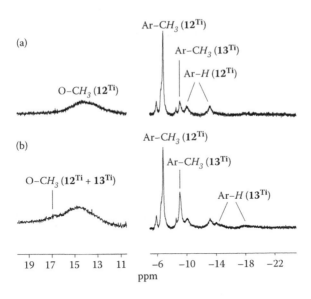

FIGURE 4.58 ^{1}H NMR spectra (25°C, toluene-d_8) of the samples **9^{Ti}**/AlMe$_3$ ([Al]/[Ti] = 5, [Ti] = 5 × 10^{-3} M) (a); **9^{Ti}**/AlMe$_3$ ([Al]/[Ti] = 15, [Ti] = 5 × 10^{-3} M) (b).

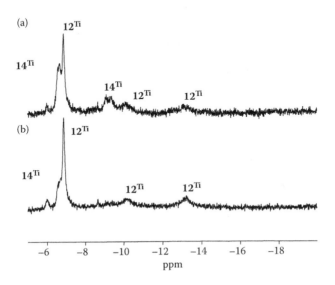

FIGURE 4.59 ^{1}H NMR spectra (25°C, toluene-d_8) of the sample **9^{Ti}**/AlMe$_3$/[Ph$_3$C]$^+$[B(C$_6$F$_5$)$_4$]$^-$ (([Ti]:[Al]:[B] = 1:10:1.3, [Ti] = 5 × 10^{-3} M): 5 min after combining the reagents at 25°C (a) and 30 min after onset of the reaction (b).

206   Applications of EPR and NMR Spectroscopy in Homogeneous Catalysis

of those of 12^{Ti} (Figure 4.58b). A similar picture was documented for the system 9^{Ti}/AlEt$_3$, where complexes with the proposed structures LTi^{II}Cl and LTi^{II}Et were observed. In contrast to the samples 9^{Ti}/AlMe$_3$ and 9^{Ti}/AlEt$_3$, the formation of Ti^{II} species was not detected in the reaction of 9^{Ti} with AliBu$_3$. The stronger ability of AlEt$_3$ to "overreduce" Ti^{IV} into Ti^{II} in comparison to that of AliBu$_3$ was previously reported for the MgCl$_2$/ethylbenzoate/TiCl$_4$-AlR$_3$ catalyst systems [186].

It was proposed previously that ion pairs [LTi^{II}]$^+$[MeMAO]$^-$ could be responsible for the trimerization of ethylene [179,183]. We have undertaken the search for species of this type in the catalyst systems 9^{Ti}/AlMe$_3$/[Ph$_3$C][B(C$_6$F$_5$)$_4$] and 9^{Ti}/MAO [184]. It was found that after mixing 9^{Ti} and AlMe$_3$/[Ph$_3$C][B(C$_6$F$_5$)$_4$] for 5 min at 25°C, the ^{1}H NMR spectrum of the sample mixture exhibited resonances of LTi^{II}Cl and of a new complex 14^{Ti} (Figure 4.59a). The concentration of 14^{Ti} decreased with time (Figure 4.59b). It is tempting to assign 14^{Ti} to an ion pair of type [LTi^{II}]$^+$[B(C$_6$F$_5$)$_4$]$^-$. Catalytic studies witnessed that catalyst system 9^{Ti}/MMAO produced a much higher amount of PE and a lower amount of 1-hexene than the system 9^{Ti}/MAO [184]. As was noted above, the major part of titanium exists in the catalyst system 9^{Ti}/MMAO in the form of Ti^{III} species. It is likely to propose that Ti^{III} species are responsible for the generation of PE by the catalyst system 9^{Ti}/MMAO. Further studies are needed to verify the key role of [LTi^{II}]$^+$[B(C$_6$F$_5$)$_4$]$^-$ species in the selective trimerization of ethylene and the role of Ti^{III} species in the formation of HMWPE.

REFERENCES

1. Ziegler, K., Holzkamp, E., Breil, H., Martin. 1955. Das Mülheimer Normaldruck-Polyäthylene-Verfahren. *Angew. Chem.* 67: 541–547.
2. Natta, G. 1955. Une nouvelle classe de polymers d'α-olefines ayant une régularité de structure exceptionnelle. *J. Polym. Sci.* 16: 143–154.
3. Tullo, A. H. 2010. Metallocenes rise again. *Chem. Eng. News* 88: 10–16.
4. Baier, M. C., Zuideveld, M. A., Mecking, S. 2014. Post-metallocenes in the industrial production of polyolefins. *Angew. Chem. Int. Ed.* 53: 9722–9744.
5. Sinn, H., Kaminsky, W., Vollmer, H. J., Woldt, R. 1980. "Lebender Polymere" bei Ziegler-Katalyzatoren extremer produktivität. *Angew. Chem.* 92: 396–402.
6. Chen, E. Y. X., Marks, T. J. 2000. Cocatalysts for metal-catalyzed olefin polymerization: Activators, activation processes, and structure-activity relationships. *Chem. Rev.* 100: 1391–1434.
7. Resconi, L., Cavallo, L., Fait, A., Piemontesi, F. 2000. Selectivity in propene polymerization with metallocene catalysts. *Chem. Rev.* 100: 1253–1346.
8. Gibson, V. C., Spitzmesser, S. K. 2003. Advances in non-metallocene olefin polymerization catalysis. *Chem. Rev.* 103: 283–315.
9. Macchioni, A. 2005. Ion pairing in transition-metal organometallic chemistry. *Chem. Rev.* 105: 2039–2079.
10. Gibson, V. C., Redshaw, C., Solan, G. A. 2007. Bis(imino)pyridines: Surprisingly reactive ligands and a gateway to new families of catalysts. *Chem. Rev.* 107: 1745–1776.
11. Bochmann, M. 2010. The chemistry of catalyst activation: The case of group 4 polymerization catalysts. *Organometallics* 29: 4711–4740.
12. Makio, H., Terao, H., Iwashita, A., Fujita, T. 2011. FI catalysts for olefin polymerization—A comprehensive treatment. *Chem. Rev.* 111: 2363–2449.
13. Bryliakov, K. P., Talsi, E. P. 2012. Frontiers of mechanistic studies of coordination polymerization and oligomerization of α-olefins. *Coord. Chem. Rev.* 256: 2994–3007.

NMR and EPR Spectroscopy of the Mechanisms of Metallocene 207

14. Ewen, J. A. 1984. Mechanisms of stereochemical control in propylene polymerizations with soluble group 4B metallocene/methylalumoxane catalysts. *J. Am. Chem. Soc.* 106: 6355–6364.

15. Kaminsky, W., Külper, K., Brintzinger, H. H., Wild, F. R. W. P. 1985. Polymerization of propene and butene with a chiral zirconocene and methylalumoxane as cocatalyst. *Angew. Chem. Int. Ed.* 24: 507–508.

16. Kaminsky, W., Sinn, H. 2013. Methylaluminoxane: Key compound for new polymerization catalysts. In: *Polyolefins: 50 Years after Ziegler and Natta II*, ed. Kaminsky, W., pp. 1–28. *Advances in Polymer Science Series*, volume 258. Springer-Verlag, Berlin–Heidelberg, vol. 1.

17. Andresen, A., Cordes, H. G., Herwig, J., Kaminsky, W., Mercki, A., Mottweiler, R., Pein, J., Sinn, H., Vollmer, H. J. 1976. Halogen-free soluble Ziegler catalysts for the polymerization of ethylene. Control of molecular weight by choice of temperature. *Angew. Chem. Int. Ed.* 15: 630–632.

18. Benn, R., Rufińska, A., Lehmkull, H., Janssen, E., Krüger, C. 1983. ^{27}Al-NMR spectroscopy: A probe for three-, four-, five-, and sixfold coordinated Al atoms in organoaluminum compounds. *Angew. Chem. Int. Ed.* 22: 779–780.

19. Sugano, T., Matsubara, K., Fujita, T., Takahashi, T. 1993. Characterization of alumoxanes by ^{27}Al–NMR spectra. *J. Mol. Catal.* 82: 93–101.

20. Nekhaeva, L. A., Bondarenko, G. N., Rykov, S. V., Nekhaev, A. I., Krentzel, B. A., Mar'in, V. P., Vyshinskaya, L. I., Khrapova, I. M., Polonskii, A. V., Korenev, N. N. 1991. IR and NMR studies on zirconocene dichloride/methylalumoxane systems—Catalysts for olefin polymerization. *J. Organomet. Chem.* 406: 139–146.

21. Benn, R., Rufińska, A. 1986. High-resolution metal NMR spectroscopy of organometallic compounds [new analytical methods (30)]. *Angew. Chem. Int. Ed.* 25: 861–881.

22. Babushkin, D. E., Semikolenova, N. V., Panchenko, V. N., Sobolev, A. P., Zakharov, V. A., Talsi, E. P. 1997. Multinuclear NMR investigation of methylaluminoxane. *Macromol. Chem. Phys.* 198: 3845–3854.

23. Mason, M. R., Smith, J. M., Bott, S. G., Barron, A. R. 1993. Hydrolysis of tri-tert-butylaluminum: The first structural characterization of alkylalumoxanes [(R$_2$Al)$_2$O]$_n$ and (RAlO)$_n$. *J. Am. Chem. Soc.* 115: 4971–4984.

24. Hansen, E. W., Blom, R., Kvernberg, P. O. 2001. Diffusion of methylaluminoxane (MAO) in toluene probed by ^{1}H NMR spin-lattice relaxation time. *Macromol. Chem. Phys.* 202: 2880–2889.

25. Rocchigiani, L., Busico, V., Pastore, A., Macchioni, A. 2013. Probing the interactions between all components of the catalytic pool for homogeneous olefin polymerization by diffusion NMR spectroscopy. *Dalton Trans.* 42: 9104–9111.

26. Giotto, F., Pateraki, C., Tanskanen, J., Severn, J. R., Luehmann, N., Kusmin, A., Stellbrink, J., Linnolahti, M., Bochmann, M. 2013. Probing the structure of methylalumoxane (MAO) by a combined chemical, spectroscopic, neutron scattering, and computational approach. *Organometallics* 32: 3354–3362.

27. Lozos, G. P., Hoffman, B. M. 1974. Electron paramagnetic resonance of a nitroxide adsorbed on silica, silica-alumina, alumina and decationated zeolites. *J. Phys. Chem.* 78: 2110–2116.

28. Bryliakov, K. P., Semikolenova, N. V., Panchenko, V. N., Zakharov, V. A., Brintzinger, H. H., Talsi, E. P. 2006. Activation of *rac*-Me$_2$Si(Ind)$_2$ZrCl$_2$ by methylalumoxane modified by aluminum alkyls: An EPR spin-probe, ^{1}H NMR, and polymerization study. *Macromol. Chem. Phys.* 207: 327–335.

29. Trefz, T. K., Henderson, M. A., Wang, M. Y., Collins, S., McIndoe, J. S. 2013. Mass spectrometric characterization of methylaluminoxane. *Organometallics* 32: 3149–3152.

30. Dyachkowski, F. S., Shilova, A. K., Shilov, A. E. 1967. The role of free ions in reactions of olefins with soluble complex catalysts. *J. Polym. Sci. C* 16: 2333–2339.

208   Applications of EPR and NMR Spectroscopy in Homogeneous Catalysis

31. Jordan, R. F. 1991. Chemistry of cationic dicyclopentadienyl group 4 metal-alkyl complexes. *Adv. Organomet. Chem.* 32: 325–387.
32. Marks, T. J. 1992. Surface-bound metal hydrocarbyls. Organometallic connections between heterogeneous and homogeneous catalysis. *Acc. Chem. Res.* 25: 57–65.
33. Bochmann, M., Jaggar, A. J., Nicholls, J. C. 1990. Base-free cationic 14-electron titanium and zirconium alkyls: *In situ* generation, solution structures, and olefin polymerization activity. *Angew. Chem. Int. Ed.* 7: 780–782.
34. Trefz, T. K., Henderson, M. A., Linnolahti, M., Collins, S., McIndoe, J. S. 2015. Mass spectrometric characterization of methylaluminoxane-activated metallocene complexes. *Chem. Eur. J.* 21: 2980–2991.
35. Zijlstra, H. S., Harder, S. 2015. Methylalumoxane-history, production, properties, and application. *Eur. J. Inorg. Chem.* 19–43.
36. Bochmann, M., Lancaster, S. J. 1994. Monomer-dimer equilibria in homo- and heterodinuclear cationic alkylzirconium complexes and their role in polymerization catalysis. *Angew. Chem. Int. Ed.* 33: 1634–1637.
37. Tritto, I., Donetti, R., Sacchi, M. C., Locatelli, R., Zannoni, G. 1997. Dimethylzirconocene-methylaluminoxane catalyst for olefin polymerization: NMR study of reaction equilibria. *Macromolecules* 30: 1248–1252.
38. Babushkin, D. E., Semikolenova, N. V., Zakharov, V. A., Talsi, E. P. 2000. Mechanism of dimethylzirconocene activation with methylalumoxane: NMR monitoring of intermediates at high Al/Zr ratios. *Macromol. Chem. Phys.* 201: 558–567.
39. Bochmann, M., Sarsfield, M. J. 1998. Reaction of AlR_3 with $[CPh_3][B(C_6F_5)_4]$: Fasile degradation of $[B(C_6F_5)_4]^-$ by transient "$[AlR_2]$". *Organometallics* 17: 5908–5912.
40. Babushkin, D. E., Brintzinger, H. H. 2002. Activation of dimethyl zirconocene by methylaluminoxane (MAO)—Size estimate for Me-MAO⁻ anions by pulsed field-gradient NMR. *J. Am. Chem. Soc.* 124: 12869–12873.
41. Bryliakov, K. P., Semikolenova, N. V., Yudaev, D. V., Zakharov, V. A., Brintzinger, H. H., Ystenes, M., Rytter, E., Talsi, E. P. 2003. ^1H, ^{13}C-NMR and ethylene polymerization studies of zirconocene/MAO catalysts: Effect of the ligand structure on the formation of active intermediates and polymerization kinetics. *J. Organomet. Chem.* 683: 92–102.
42. Theurkauff, G., Bader, M., Marquet, N., Bondon, A., Roisnel, T., Guegan, J.-P., Amar, A., Boueckkine, A., Carpentier, J.-F., Kirillov, E. 2016. Solution dynamics, trimethylaluminium adducts, and implications in propylene polymerization. *Organometallics* 35: 258–276.
43. Theurkauff, G., Bondon, A., Dorcet, V., Carpentier, J.-F., Kirillov, E. 2015. Heterobi- and-trimetallic ion pairs of zirconocene-based isoselective olefin polymerization catalysts with $AlMe_3$. *Angew. Chem. Int. Ed.* 54: 6343–6346.
44. Bryliakov, K. P., Talsi, E. P., Bochmann, M. 2004. ^1H and ^{13}C NMR spectroscopic study of titanium(IV) species formed by activation of Cp_2TiCl_2 and $[(Me_4C_5)SiMe_2N^tBu]$ $TiCl_2$ with methylaluminoxane (MAO). *Organometallics* 23: 149–152.
45. Bryliakov, K. P., Babushkin, D. E., Talsi, E. P., Voskoboynikov, A. Z., Gritzo, H., Schröder, L., Damrau, H. R. H., Wieser, U., Schaper, F., Brintzinger, H. H. 2005. *Ansa*-titanocene catalysts for α-olefin polymerization. Synthesis, structures, and reactions with methylaluminoxane and boron-based activators. *Organometallics* 24: 894–904.
46. Holton, J., Lappert, M. F., Ballard, D. G. H., Pearce, R., Atwood, J. L., Hunter, W. E. 1979. Alkyl-bridged complexes of the d- and f-block elements. Part 1. Di-μ-alkyl-bis(η-cyclopentadienyl)metal(III) dialkylaluminium(III) complexes and the crystal and molecular structure of the ytterbium methyl species. *J. Chem. Soc. Dalton Trans.* 45–53.
47. Bryliakov, K. P., Semikolenova, N. V., Zakharov, V. A., Talsi, E. P. 2003. ^{13}C-NMR study of Ti(IV) species formed by Cp^*TiMe_3 and Cp^*TiCl_3 activation with methylaluminoxane (MAO). *J. Organomet. Chem.* 683: 23–28.

48. Chen, Y. X., Stern, C. L., Marks, T. J. 1997. Very large counteranion modulation of cationic metallocene polymerization activity and stereoregulation by a sterically congested (perfluoroaryl)fluoroaluminate. *J. Am. Chem. Soc.* 119: 2582–2583.
49. Forlini, F., Fan, Z. Q., Tritto, I., Locatelli, P., Sacchi, M. C. 1997. Metallocene-catalyzed propene/1-hexene copolymerization: Influence of amount and bulkiness of cocatalyst and of solvent polarity. *Macromol. Chem. Phys.* 198: 2397–2408.
50. Kleinschmidt, R., van der Leek, Y., Reffke, M., Fink, G. 1999. Kinetics and mechanistic insight into propylene polymerization with different metallocenes and various aluminum alkyls as cocatalysts. *J. Mol. Catal. A* 148: 29–41.
51. Seraidaris, T., Löfgren, B., Mäkelä-Vaarne, N., Lehmus, P., Stehing, U. 2004. Activation and polymerization of *ansa*-zirconocene chloroamido complexes. *Macromol. Chem. Phys.* 205: 1064–1069.
52. Rishina, L. A., Galashina, N. M., Nedorezova, P. M., Klyamkina, A. N., Aladyshev, A. M., Tsvetkova, V. I., Baranov, O. A., Optov, V. A., Kissin, Y. V. 2004. Copolymerization of propylene and 1-hexene in the presence of homogeneous metallocene catalysts. *Polym. Sci. Ser. A* 46: 911–920.
53. Bhrian, N. N., Brintzinger, H. H., Ruchatz, D., Fink, G. 2005. Polymeryl exchange between *ansa*-zirconocene catalysts for norbornene-ethene copolymerization and aluminum and zinc alkyls. *Macromolecules* 38: 2056–2063.
54. Babushkin, D. E., Brintzinger, H. H. 2007. Modification of methylaluminoxane-activated *ansa*-zirconocene catalysts with triisobutylaluminum—Transformations of reactive cations studied by NMR spectroscopy. *Chem. Eur. J.* 13: 5294–5299.
55. Coates, G. W., Waymouth, R. M. 1995. Oscillating stereocontrol: A strategy for the synthesis of thermoplastic elastomeric polypropylene. *Science* 267: 217–219.
56. Lin, S., Waymouth, R. M. 2002. 2-Arylindene metallocenes: Conformationally dynamic catalysts to control the structure and properties of polypropylene. *Acc. Chem. Res.* 35: 765–773.
57. Busico, V., Castelli, V. V. A., Aprea, P., Cipullo, R., Serge, A., Talarico, G., Vacatello, M. 2003. "Oscillating" metallocene catalysts: What stops the oscillation? *J. Am. Chem. Soc.* 125: 5451–5460.
58. Wilmes, G. M., Polse, J. L., Waymouth, R. M. 2002. Influence of cocatalyst on the stereoselectivity of unbridged 2-phenylindenyl metallocene catalysts. *Macromolecules* 35: 6766–6772.
59. Lyakin, O. Y., Bryliakov, K. P., Semikolenova, N. V., Lebedev, A. Y., Voskoboynikov, A. Z., Zakharov, V. A., Talsi, E. P. 2007. ^{1}H and ^{13}C NMR studies of cationic intermediates formed upon activation of "oscillating" catalyst (2-PhInd)$_2$ZrCl$_2$ with MAO, MMAO, and AlMe$_3$/[Ph$_3$C]$^+$[B(C$_6$F$_5$)$_4$]$^-$. *Organometallics* 26: 1536–1540.
60. Chien, J. C. W., Song, W., Rausch, M. D. 1994. Polymerization of propylene by zirconocenium catalysts with different counter-ions. *J. Polym. Sci. A Polym. Chem.* 32: 2387–2393.
61. Yano, A., Sone, M., Yamada, S., Hasegawa, S., Akimoto, A. 1999. Homo- and copolymerization of ethylene at high temperature with cationic zirconocene catalysts. *Macromol. Chem. Phys.* 200: 917–923.
62. Panin, A. N., Sukhova, T. A., Bravaya, N. M. 2001. Triisobutylaluminum as cocatalyst for zirconocenes. I. Sterically opened zirconocene/triisobutylaluminum/perfluorophenylborate as highly effective ternary catalytic system for synthesis of low molecular weight polyethylenes. *J. Polym. Sci. A Polym. Chem.* 39: 1901–1914.
63. Panin, A. N., Dzhabieva, Z. M., Nedorezova, P. N., Tsvetkova, V. I., Saratovskikh, S. L., Babkina, O. N., Bravaya, N. M. 2001. Triisobutylaluminum as cocatalyst for zirconocenes. II Triisobutylaluminum as a component of a cocatalyst system and as an effective cocatalyst for olefin polymerization derived from dimethylated zirconocenes. *J. Polym. Sci. A Polym. Chem.* 39: 1915–1930.

210   Applications of EPR and NMR Spectroscopy in Homogeneous Catalysis

64. Song, F., Hannant, M. D., Cannon, R. D., Bochmann, M. 2004. Zirconocene-catalyzed propene polymerization: Kinetics, mechanism, and the role of the anion. *Macromol. Symp.* 213: 173–185.

65. Götz, C., Rau, A., Luft, G. 2002. Ternary metallocene catalyst systems based on metallocene dichlorides and $AliBu_3/[PhNMe_2H][B(C_6F_5)_4]$. NMR investigations of the influence of Al/Zr ratios on alkylation and on formation of the precursors of the active metallocene species. *J. Mol. Catal. A Chem.* 184: 95–110.

66. Bryliakov, K. P., Talsi, E. P., Semikolenova, N. V., Zakharov, V. A., Brand, J., Alonso-Moreno, C., Bochmann, M. 2007. Formation and structures of cationic zirconium complexes in ternary systems $rac\text{-}(SBI)ZrX_2/AliBu_3/[Ph_3C][B(C_6F_5)_4]$ (X = Cl, Me). *J. Organomet. Chem.* 692: 859–868.

67. Wu, F., Dash, A. K., Jordan, R. F. 2004. Structures and reactivity of Zr(IV) chlorobenzene complexes. *J. Am. Chem. Soc.* 126: 15360–15361.

68. Rieger, B., Troll, C., Preuschen, J. 2002. Ultrahigh molecular weight polypropylene elastomers by high activity "dual-side" hafnocene catalysts. *Macromolecules* 35: 5742–5743.

69. Hild, S., Cobzaru, C., Troll, C., Rieger, B. 2006. Elastomeric poly(propylene) from "dual-side" metallocenes: Reversible chain transfer and its influence on polymer microstructure. *Macromol. Chem. Phys.* 207: 665–683.

70. Cobzaru, C., Rieger, B. 2006. Control of ultrahigh molecular weight polypropene microstructures via asymmetric "dual-side" catalysts. *Macromol. Symp.* 236: 151–155.

71. Boussie, T. R., Diamond, G. M., Goh, C., Hall, K. A., LaPointe, A. M., Leclerc, M. K., Murphy, V. et al. 2006. Nonconventional catalysts for isotactic propene polymerization in solution developed by using high-throughput-screening technologies. *Angew. Chem. Int. Ed.* 45: 3278–3283.

72. Froese, R. D. J., Hustad, P. D., Kuhlman, R. L., Wenzel, T. T. 2007. Mechanism of activation of a hafnium pyridylamide olefin polymerization catalyst: Ligand modification by monomer. *J. Am. Chem. Soc.* 129: 7831–7840.

73. Arriola, D. J., Canakhan, E. M., Hustad, P. D., Kuhlman, R. L., Wenzel, T. T. 2006. Catalytic production of olefin block copolymers via chain shuttling polymerization. *Science* 312: 714–718.

74. Bryliakov, K. P., Talsi, E. P., Voskoboinikov, A. Z., Lancaster, S. J., Bochmann, M. 2008. Formation and structures of hafnocene complexes in MAO- and $AliBu_3/[Ph_3C][B(C_6F_5)_4]$–activated systems. *Organometallics* 27: 6333–6342.

75. Busico, V., Cipullo, R., Pellecchia, R., Talarico, G., Razavi, A. 2009. Hafnocenes and MAO: Beware of trimethylaluminum! *Macromolecules* 42: 1789–1791.

76. Baldwin, S. M., Bercaw, J. E., Henling, L. M., Day, M. W., Brintzinger, H. H. 2011. Cationic alkylaluminum-complexed zirconocene hydrides: NMR-spectroscopic identification, crystallographic structure determination, and interconversion with other zirconocene cations. *J. Am. Chem. Soc.* 133: 1805–1813.

77. Tritto, I., Donetti, R., Sacchi, M. C., Locatelli, P., Zannoni, G. 1999. Evidence of zirconium-polymeryl ion pairs from ^{13}C NMR *in situ* $^{13}C_2H_4$ polymerization with $Cp_2Zr(^{13}CH_3)_2$-based catalysts. *Macromolecules* 32: 264–269.

78. Landis, C. R., Rosaaen, K. A., Sillars, D. R. 2003. Direct observation of insertion events at $rac(C_2H_4(1\text{-}indenyl)_2Zr(MeB(C_6F_5)_3)$-polymeryl intermediates: Distinction between continuous and intermittent propagation modes. *J. Am. Chem. Soc.* 125: 1710–1711.

79. Sillars, D. R., Landis, C. R. 2003. Catalytic propene polymerization: Determination of propagation, termination, and epimerization kinetics by direct NMR observation of the $(EBI)Zr(MeB(C_6F_5)_3)propenyl$ catalyst species. *J. Am. Chem. Soc.* 125: 9894–9895.

80. Landis, C. R., Christianson, M. D. 2006. Metallocene-catalyzed alkene polymerization and the observation of Zr-allyls. *Proc. Natl. Acad. Sci. U. S. A.* 103: 15349–15354.

81. Vatanamu, M. 2015. Observation of zirconium allyl species formed during zirconocene-catalyzed α-olefin polymerizations. *J. Catal.* 323: 112–120.

NMR and EPR Spectroscopy of the Mechanisms of Metallocene   211

82. Babushkin, D. E., Panchenko, V. N., Brintzinger, H. H. 2014. Zirconium allyl complexes as participants in zirconocene-catalyzed α-olefin polymerizations. *Angew. Chem. Int. Ed.* 53: 9645–9649.

83. Panchenko, V. N., Babushkin, D. E., Brintzinger, H. H. 2015. Zirconium-allyl complexes as resting states in zirconocene-catalyzed α-olefin polymerization. *Macromol. Rapid. Commun.* 36: 249–253.

84. Babushkin, D. E., Brintzinger, H. H. 2010. Reactive intermediates formed during olefin polymerization by methylalumoxane-activated *ansa*-zirconocene catalysts: Identification of a chain carrying intermediate by NMR methods. *J. Am. Chem. Soc.* 132: 452–453.

85. Lenton, T. N., Bercaw, J. E., Panchenko, V. N., Zakharov, V. A., Babushkin, D. E., Soshnikov, I. E., Talsi, E. P., Brintzinger, H. H. 2013. Formation of trivalent zirconocene complexes from *ansa*-zirconocene-based olefin-polymerization precatalysts: An EPR- and NMR-spectroscopic study. *J. Am. Chem. Soc.* 135: 10710–10719.

86. Small, B. L., Brookhart, M., Bennett, A. M. A. 1998. Highly active iron and cobalt catalysts for the polymerization of ethylene. *J. Am. Chem. Soc.* 120: 4049–4050.

87. Britovsek, G. J. P., Gibson, V. C., Kimberley, B. S., Maddox, P. J., McTavish, S. J., Solan, G. A., White, A. J. P., Williams, D. J. 1998. Novel olefin polymerization catalysts based on iron and cobalt. *Chem. Commun.* 849–850.

88. Britovsek, G. J. P., Bruce, M., Gibson, V. C., Kimberley, B. S., Maddox, P. J., Mastroianni, S., McTavish, S. J. et al. 1999. Iron and cobalt ethylene polymerization catalysts bearing 2,6-bis(imino)pyridyl ligands: Synthesis, structures, and polymerization studies. *J. Am. Chem. Soc.* 121: 8728–8740.

89. Britovsek, G. J. P., Gibson, V. C., Wass, D. F. 1999. The search for new-generation polymerization catalysts: Life beyond metallocenes. *Angew. Chem. Int. Ed.* 38: 428–447.

90. Gibson, V. C., Spitzmesser, S. K. 2003. Advances in non-metallocene olefin polymerization catalysis. *Chem. Rev.* 103: 283–315.

91. Kawakami, T., Ito, S., Nozaki, K. 2015. Iron-catalyzed homo- and copolymerization of propylene: Steric influence of bis(imino)pyridine ligands. *Dalton Trans.* 44: 20745–20752.

92. Smit, T. M., Tomov, A. K., Britovsek, G. J. P., Gibson, V. C., White, A. J. P., Williams, D. J. 2012. The effect of imine-carbon substituents in bis(imino)pyridine-based ethylene polymerization catalysts across the transition series. *Catal. Sci. Technol.* 2: 643–655.

93. Flisak, Z., Sun, W. H. 2015. Progression of diiminopyridines: From single application to catalytic versatility. *ACS Catal.* 5: 4713–4724.

94. Britovsek, G. J. P., Mastroianni, S., Solan, G. A., Baugh, S. P. D., Redshaw, C., Gibson, V. C., White, A. J. P., Williams, D. J., Elsegood, M. R. J. 2000. Oligomerization of ethylene by bis(imino)pyridyl iron and cobalt complexes. *Chem. Eur. J.* 6: 2221–2231.

95. Barabanov, A. A., Bukatov, G. D., Zakharov, V. A., Semikolenova, N. V., Mikenas, T. B., Echevskaja, L. G., Matsko, M. A. 2006. Kinetic study of ethylene polymerization over supported bis(imino)pyridine iron(II) catalysts. *Macromol. Chem. Phys.* 207: 1368–1375.

96. Kissin, Y. V., Qian, G., Xie, G., Chen, Y. 2006. Multi-center nature of ethylene polymerization catalysts based on 2,6-bis(imino)pyridyl complexes of iron and cobalt. *J. Polym. Sci. A Polym. Chem.* 44: 6159–6170.

97. Talsi, E. P., Babushkin, D. E., Semikolenova, N. V., Zudin, V. N., Panchenko, V. N., Zakharov, V. A. 2001. Polymerization of ethylene catalyzed by iron complexes bearing 2.6-bis(imine)pyridyl ligand: ^{1}H and ^{2}H NMR monitoring of ferrous species formed via catalyst activation with $AlMe_3$, MAO, $AlMe_3/B(C_6F_5)_3$ and $AlMe_3/[Ph_3C][B(C_6F_5)_4]$. *Macromol. Chem. Phys.* 202: 2046–2051.

98. Bryliakov, K. P., Semikolenova, N. V., Zakharov, V. A., Talsi, E. P. 2004. Active intermediates of ethylene polymerization over 2,6-bis(imino)pyridyl iron complex activated with aluminum trialkyls and methylaluminoxane. *Organometallics* 23: 5375–5378.

99. Bryliakov, K. P., Semikolenova, N. V., Zakharov, V. A., Talsi, E. P. 2009. Formation and nature of the active sites in bis(imino)pyridine iron-based polymerization catalysts. *Organometallics* 28: 3225–3232.

100. Lichtenberg, C., Vicin, L., Adelhardt, M., Sutter, J., Meyer, K., de Bruin, B., Grutzmacher, H. 2015. Low-valent iron(I) amido olefin complexes as promotors for dehydrogenation reactions. *Angew. Chem. Int. Ed.* 54: 5766–5771.

101. Bouwkamp, M. W., Bart, S. C., Hawrelak, E. J., Trovitch, R. J., Lobkowsky, E., Chiric, P. J. 2005. Square planar bis(imino)pyridine iron halide and alkyl complexes. *Chem. Commun.* 3406–3408.

102. Bart, S. C., Chlopek, K., Bill, E., Bouwkamp, M. W., Lobkowsky, E., Neese, F., Wieghardt, K., Chirik, P. J. 2006. Electronic structure of bis(imino)pyridine iron dichloride, monochloride, and neutral ligand complexes: A combined structural, spectroscopic, and computational study. *J. Am. Chem. Soc.* 128: 13901–13912.

103. Fernández, I., Trovitch, R. J., Lobkowski, E., Chiric, P. J. 2008. Synthesis of bis(imino) pyridine iron di- and monoalkyl complexes: Stability differences between $FeCH_2SiMe_3$ and $FeCH_2CMe_3$ derivatives. *Organometallics* 27: 109–118.

104. Trovitch, R. J., Lobkowsky, E., Chiric, P. 2008. Bis(imino)pyridine iron alkyls containing β-hydrogens: Synthesis, evaluation of kinetic stability, and decomposition pathways involving chelate participation. *J. Am. Chem. Soc.* 130: 11631–11640.

105. Tondreau, A. M., Milsmann, C., Patrick, A. D., Hoyt, H. M., Lobkowsky, E., Wieghardt, K., Chirik, P. J. 2010. Synthesis and electronic structure of cationic, neutral, and anionic bis(imino)pyridine iron alkyl complexes: Evaluation of redox activity in single-component ethylene polymerization catalysts. *J. Am. Chem. Soc.* 132: 15046–15059.

106. Scott, J., Gambarotta, S., Korobkov, I., Knijnenburg, Q., de Bruin, B., Budzelaar, P. M. H. 2005. Formation of a paramagnetic Al complex and extrusion of Fe during the reaction of (diiminepyridine)Fe with AlR_3 (R = Me, Et). *J. Am. Chem. Soc.* 127: 17204–17206.

107. Gibson, V. C., Humphries, M. J., Tellmann, K. P., Wass, D. F., White, A. J. P., Williams, D. J. 2001. The nature of the active species in bis(imino)pyridyl cobalt ethylene polymerization catalysts. *Chem. Commun.* 2252–2253.

108. Humphries, M. J., Tellmann, K. P., Gibson, V. C., White, A. J. P., Williams, D. J. 2005. Investigations into the mechanism of activation and initiation of ethylene polymerization by bis(imino)pyridine cobalt catalysts: Synthesis, structures, and deuterium labeling studies. *Organometallics* 24: 2039–2050.

109. Kooistra, T. M., Knijnenburg, Q., Smits, J. M. M., Horton, A. D., Budzelaar, P. H. M., Gal, A. W. 2001. Olefin polymerization with [{bis(imino)pyridyl}Co^{II}Cl$_2$]: Generation of the active species involves Co^I. *Angew. Chem. Int. Ed.* 40: 4719–4722.

110. Semikolenova, N. V., Zakharov, V. A., Talsi, E. P., Babushkin, D. E., Sobolev, A. P., Echevskaja, L. G., Khusniyarov, M. M. 2002. Study of the ethylene polymerization over homogeneous and supported catalysts based on 2,6-bis(imino)pyridyl complexes of Fe(II) and Co(II). *J. Mol. Catal. A Chem.* 182: 283–294.

111. Soshnikov, I. E., Semikolenova, N. V., Bushmelev, A. N., Bryliakov, K. P., Lyakin, O. Y., Redshaw, C., Zakharov, V. A., Talsi, E. P. 2009. Investigating the nature of the active species in bis(imino)pyridine cobalt ethylene polymerization catalysts. *Organometallics* 28: 6003–6013.

112. Hagen, H., Boersma, J., van Koten, G. 2002. Homogeneous vanadium-based catalysts for the Ziegler–Natta polymerization of α-olefins. *Chem. Soc. Rev.* 31: 357–364.

113. Gambarotta, S. 2003. Vanadium-based Ziegler–Natta: Challenges, promises, problems. *Coord. Chem. Rev.* 237: 229–243.

NMR and EPR Spectroscopy of the Mechanisms of Metallocene   213

114. Redshaw, C. 2010. Vanadium procatalysts bearing chelating aryloxides: Structure-activity trends in ethylene polymerization. *Dalton Trans.* 39: 5595–5604.
115. Nomura, K., Zhang, S. 2011. Design of vanadium complex catalysts for precise olefin polymerization. *Chem. Rev.* 111: 2342–2362.
116. Wu, J.-Q., Li, Y.-S. 2011. Well-defined vanadium complexes as the catalysts for olefin polymerization. *Coord. Chem. Rev.* 255: 2303–2314.
117. Small, B. L. 2015. Discovery and development of pyridine-bis(imine) and related catalysts for olefin polymerization and oligomerization. *Acc. Chem. Res.* 48: 2599–2611.
118. Zambelli, A., Allegra, G. 1980. Reaction mechanism for syndiotactic specific polymerization of propene. *Macromolecules* 23: 42–49.
119. Igarashi, A., Zhang, S., Nomura, K. 2012. Ethylene dimerization/polymerization catalyzed by (adamantylimido)vanadium(V) complexes containing (2-anilidomethyl)pyridine ligands: Factors affecting the ethylene reactivity. *Organometallics* 31: 3575–3581.
120. Reardon, D., Conan, F., Gambarotta, S., Yap, G., Wang, Q. 1999. Life and death of an active ethylene polymerization catalyst. Ligand involvement in catalyst activation and deactivation. Isolation and characterization of two unprecedented neutral and anionic vanadium(I) alkyls. *J. Am. Chem. Soc.* 121: 9318–9325.
121. Milone, S., Cavallo, G., Tedesco, C., Grassi, A. 2002. Synthesis of α-diimine V(III) complexes and their role as ethylene polymerization catalysts. *J. Chem. Soc. Dalton Trans.* 1839–1846.
122. Romero, J., Carillo-Hermosilla, F., Antiñolo, A., Otero, A. 2006. Homogeneous and supported bis(imino)pyridyl vanadium(III) catalysts. *J. Mol. Catal. A Chem.* 304: 180–186.
123. Soshnikov, I. E., Semikolenova, N, V., Antonov, A. A., Bryliakov, K. P., Zakharov, V. A., Talsi, E. P. 2014. ^1H and ^2H NMR spectroscopic characterization of heterobinuclear ion pairs formed upon the activation of bis(imino)pyridine vanadium(III) precatalysts with $AlMe_3/[Ph_3C]^+[B(C_6F_5)_4]^-$ and MAO. *Organometallics* 33: 2583–2587.
124. Soshnikov, I. E., Semikolenova, N. V., Bryliakov, K. P., Zakharov, V. A., Talsi, E. P. 2016. ^1H and ^2H NMR spectroscopic study of the ion pairs formed upon the activation of vanadium(III) alpha-diimine precatalyst with $AlMe_3/[Ph_3C][B(C_6F_5)_4]$ and MAO. *J Mol. Catal. A: Chem.* 423: 333–338.
125. Soshnikov, I. E., Semikolenova, N. V., Bryliakov, K. P., Zakharov, V. A., Redshaw, C., Talsi, E. P. 2009. An EPR study of vanadium species formed upon interaction of N and C-capped tris(phenolate)complexes with $AlEt_3$ and $AlEt_2Cl$. *J. Mol. Catal. A Chem.* 303: 23–29.
126. Soshnikov, I. E., Semikolenova, N. V., Bryliakov, K. P., Shubin, A. A., Zakharov, V. A., Redshaw, C., Talsi, E. P. 2009. An EPR study of the V(IV) species formed upon activation of a vanadyl phenoxyimine polymerization catalyst with AlR_3 and AlR_2Cl (R = Me, Et). *Macromol. Chem. Phys.* 210: 542–548.
127. Soshnikov, I. E., Semikolenova, N. V., Shubin, A. A., Bryliakov, K. P., Zakharov, V. A., Redshaw, C., Talsi, E. P. 2009. EPR monitoring of vanadium(IV) species formed upon activation of vanadium(V) polyphenolate precatalysts with AlR_2Cl and $AlR_2Cl/$ethyltrichloroacetate (R = Me, Et). *Organometallics* 28: 6714–6720.
128. Ittel, S. D., Johnson, L. K., Brookhart, M. 2000. Late-metal catalysts for ethylene homo- and copolymerization. *Chem. Rev.* 100: 1169–1203.
129. Nakamura, A., Ito, S., Nozaki, K. 2009. Coordination-insertion copolymerization of fundamental polar monomers. *Chem. Rev.* 109: 5215–5244.
130. Wiedemann, T., Voit, G., Tchernook, A., Roesle, P., Göttker-Schnettmann, I., Mecking, S. 2013. Monofunctional hyperbranched ethylene oligomers. *J. Am. Chem. Soc.* 136: 2078–2085.
131. Yonkin, T. R., Connor, E. F., Henderson, J. I., Friedrich, S. K., Grubbs, R. H., Bunsleben, D. A. 2000. Neutral, single-component nickel(II) polyolefin catalysts that tolerate heteroatoms. *Science* 287: 460–462.

214   Applications of EPR and NMR Spectroscopy in Homogeneous Catalysis

132. Zuideveld, M. A., Wehrmann, P., Röhr, C., Mecking, S. 2004. Remote substituents controlling catalytic polymerization by very active and robust neutral Ni(II) complexes. *Angew. Chem. Int. Ed.* 43: 869–873.
133. Osichow, A., Rabe, C., Vogtt, K., Narayanan, T., Harnau, L., Drechsler, M., Ballauff, M., Mecking, S. 2013. Ideal polyethylene nanocrystals. *J. Am. Chem. Soc.* 135: 11645–11650.
134. Jenkins, J. C., Brookhart, M. 2004. A mechanistic investigation of the polymerization of ethylene catalyzed by neutral Ni(II) complexes derived from bulky anilinotropone ligands. *J. Am. Chem. Soc.* 126: 5827–5842.
135. Waltman, A. W., Yonkin, T. R., Grubbs, R. H. 2004. Insights into the deactivation of neutral nickel ethylene polymerization catalysts in the presence of functionalized olefins. *Organometallics* 23: 5121–5123.
136. Berkefeld, A., Mecking, S. 2009. Deactivation pathways of neutral Ni(II) polymerization catalysts. *J. Am. Chem. Soc.* 131: 1565–1574.
137. Berkefeld, A., Drexler, M., Möller, H. M., Mecking, S. 2009. Mechanistic insights on the copolymerization of polar vinyl monomers with neutral Ni(II) catalysts. *J. Am. Chem. Soc.* 131: 12613–12622.
138. Weberski, M. P., Jr., Chen, C., Delferro, M., Marks, T. J. 2012. Ligand steric and fluoroalkyl substituent effects on enchainment cooperativity and stability in bimetallic nickel(II) polymerization catalysts. *Chem. Eur. J.* 18: 10715–10732.
139. Ölscher, F., Göttker-Schnetmann, I., Monteil, V., Mecking, S. 2015. Role of radical species in salicylaldiminato Ni(II) mediated polymer chain growth: A case study for the migratory insertion polymerization of ethylene in the presence of methyl methacrylate. *J. Am. Chem. Soc.* 137: 14819–14828.
140. Soshnikov, I. E., Semikolenova, N. V., Zakharov, V. A., Möller, H. M., Ölscher, F., Osichow, A., Göttker-Schnettmann, I., Mecking, S., Talsi, E. P., Bryliakov, K. P. 2013. Formation and evolution of chain-propagating species upon ethylene polymerization with neutral salicylaldiminato nickel(II) catalysts. *Chem. Eur. J.* 19: 11409–11417.
141. Gibbs, S. J., Johnson, C. S., Jr. 1991. A PFG NMR experiment for accurate diffusion and flow studies in the presence of eddy currents. *J. Magn. Reson.* 93: 395–402.
142. Wu, D. H., Chen, A. D., Johnson, C. S., Jr. 1995. An improved diffusion-ordered spectroscopy experiment incorporating bipolar-gradient pulses. *J. Magn. Res. A* 115: 260–264.
143. Beck, S., Geyer, A., Brintzinger, H. H. 1999. Diffusion coefficients of zirconoceneborate ion pairs studied by pulsed field-gradient NMR—Evidence for ion quadruples in benzene solutions. *Chem. Commun.* 2477–2478.
144. Zuccaccia, C., Stahl, N. G., Macchioni, A., Chen, M.-C., Roberts, J. A., Marks, T. J. 2004. NOE and PGSE NMR spectroscopic studies of solution structure and aggregation in metallocenium ion-pairs. *J. Am. Chem. Soc.* 126: 1448–1464.
145. Alonso-Moreno, C., Lancaster, S. J., Zuccaccia, C., Macchioni, A., Bochmann, M. 2007. Evidence for mixed-ion clusters in metallocene catalysts: Influence on ligand exchange dynamics and catalyst activity. *J. Am. Chem. Soc.* 129: 9282–9283.
146. Holm, R. H., Hawkins, C. J. 1973. Stereochemistry and structural and electronic equilibria, Chapter 7. In *NMR of Paramagnetic Molecules: Principles and Applications*, eds. La Mar, G. N., Horrocks, W. DeW., R. H. Holm, Jr. pp. 243–332. Academic Press, New York.
147. Sacchi, M. C., Sonzogni, M., Losio, S., Forlini, F., Locatelli, P., Tritto, I., Licchelli, M. 2001. Vinylic polymerization of norbornene by late transition metal-based catalysis. *Macromol. Chem. Phys.* 202: 2052–2058.
148. Huang, Y., Chen, J., Chi, L., Wei, C., Zhang, Z., Li, Z., Li, A., Zhang, L. 2009. Vinyl polymerization of norbornene with bis(imino)pyridyl nickel(II) complexes. *J. Appl. Polym. Sci.* 112: 1486–1495.

NMR and EPR Spectroscopy of the Mechanisms of Metallocene 215

149. Antonov, A. A., Semikolenova, N. V., Zakharov, V. A., Zhang, W., Wang, Y., Sun, W.-H., Talsi, E. P., Bryliakov, K. P. 2012. Vinyl polymerization of norbornene on nickel complexes with bis(imino)pyridine ligands containing electron-withdrawing groups. *Organometallics* 31: 1143–1149.

150. Antonov, A. A., Samsonenko, D. G., Talsi, E. P., Bryliakov, K. P. 2013. Formation of cationic intermediates upon the activation of bis(imino)pyridine nickel catalysts. *Organometallics* 32: 2187–2191.

151. Leatherman, M. D., Svejda, S. A., Johnson, L. K., Brookhart, M. 2003. Mechanistic studies of nickel(II) alkyl agostic cations and alkyl ethylene complexes: Investigations of chain propagation and isomerization in (α-diimine)Ni(II)-catalyzed ethylene polymerization. *J. Am. Chem. Soc.* 125: 3068–3081.

152. Shannon, R. D. 1976. Cf. the effective ionic radii of high-spin Fe^{2+} (0.78 Å) and of low-spin Ni^{2+} (0.69 Å). *Acta Crystallogr. Sect. A* 32: 751–767.

153. Dong, Z., Ye, Z. 2012. Hyperbranched polyethylenes by chain walking polymerization: Synthesis, properties, functionalization, and applications. *Polym. Chem.* 3: 286–301.

154. Wang, S., Sun, W.-H., Redshaw, C. 2014. Recent progress on nickel-based systems for ethylene oligo-/polymerization catalysis. *J. Organomet. Chem.* 751: 717–741.

155. Yu, J., Zeng, Y., Huang, W., Hao, X., Sun, W.-H. 2011. N-(5,6,7-trihydroquinolin-8-ylidene)arylaminonickel dichlorides as highly active single-site pro-catalysts in ethylene polymerization. *Dalton Trans.* 40: 8436–8443.

156. Soshnikov, I. E., Semikolenova, N. V., Bryliakov, K. P., Zakharov, V. A., Sun, W.-H., Talsi, E. P. 2015. NMR and EPR spectroscopic identification of intermediates formed upon activation of 8-mesitylimino-5,6,7-trihydroquinolyl nickel dichloride with AlR_2Cl (R = Me, Et). *Organometallics* 34: 3222–3227.

157. Coates, G. W., Hustad, P. D., Reinartz, S. 2002. Catalysts for the living insertion polymerization of alkenes: Access to new polyolefin architectures using Ziegler–Natta chemistry. *Angew. Chem. Int. Ed.* 41: 2236–2257.

158. Makio, H., Kashiwa, N., Fujita, T. 2002. FI catalysts: A new family of high performance catalysts for olefin polymerization. *Adv. Synth. Catal.* 344: 477–493.

159. Domski, G. J., Rose, J. M., Coates, G. W., Bolig, A. D., Brookhart, M. 2007. Living alkene polymerization: New methods for the precision synthesis of polyolefins. *Prog. Polym. Sci.* 32: 30–92.

160. Tian, J., Hustad, P. D., Coates, G. W. 2001. A new catalyst for highly syndiospecific living olefin polymerization: Homopolymers and block copolymers from ethylene and propylene. *J. Am. Chem. Soc.* 123: 5134–5135.

161. Saito, J., Mitani, M., Mohri, J., Ishii, S., Yoshida, Y., Matsugi, T., Kojoh, S., Kashiwa, N., Fujita, T. 2001. Highly syndiospecific living polymerization of propylene using a titanium complex having two phenoxy-imine chelate ligands. *Chem. Lett.* 30: 576–577.

162. Saito, J., Mitani, M., Mohri, J., Yoshida, Y., Matsui, S., Ishii, S., Kojoh, S., Kashiwa, N., Fujita, T. 2001. Living polymerization of ethylene with a titanium complex containing two phenoxy-imine chelate ligands. *Angew. Chem. Int. Ed.* 40: 2918–2920.

163. Mitani, M., Mohri, J., Yoshida, Y., Saito, J., Ishii, S., Tsuru, K., Matsui, S. et al. 2002. Living polymerization of ethylene catalyzed by titanium complexes having fluorine-containing phenoxy-imine chelate ligands. *J. Am. Chem. Soc.* 124: 3327–3336.

164. Mason, A. F., Coates, G. F. 2004. New phenoxyketimine titanium complexes: Combining isotacticity and living behavior in propylene polymerization. *J. Am. Chem. Soc.* 126: 16326–16327.

165. Edson, J. B., Wang, Z., Kramer, E. J., Coates, G. W. 2008. Fluorinated bis(phenoxyketimine)titanium complexes for the living, isoselective polymerization of propylene: Multiblock isotactic polypropylene copolymers via sequential monomer addition. *J. Am. Chem. Soc.* 130: 4968–4977.

216 Applications of EPR and NMR Spectroscopy in Homogeneous Catalysis

166. Yu, S., Mecking, S. 2008. Extremely narrow-dispersed high molecular weight polyethylene from living polymerization at elevated temperatures with o-F substituted Ti enolatoimines. *J. Am. Chem. Soc.* 130: 13204–13205.
167. Li, X.-F., Dai, K., Ye, W.-P., Pan, L., Li, Y.-S. 2004. New titanium complexes with two β-enaminoketonato chelate ligands: Synthesis, structures, and olefin polymerization activities. *Organometallics* 23: 1223–1230.
168. Talarico, G., Busico, V., Cavallo, L. 2004. "Living" propene polymerization with bis(phenoxyimine) group 4 metal catalysts: New strategies and old concepts. *Organometallics* 23: 5889–5993.
169. Villani, V., Giammarino, G. 2011. Fluorine interactions in a post-metallocene titanium catalyst: An ab initio study. *Macromol. Theory Simul.* 20: 171–173.
170. Kui, S. C. F., Zhu, N., Chan, M. C. W. 2003. Observation of intramolecular C-H$\cdots$F-C contacts in non-metallocene polyolefin catalysts: Model for weak attractive interactions between polymer chain and noninnocent ligand. *Angew. Chem. Int. Ed.* 42: 1628–1632.
171. Chan, M. C. W., Kui, S. C. F., Cole, J. M., McIntyre, G. J., Matsui, S., Zhu, N., Tam, K.-H. 2006. Neutron and X-ray diffraction and spectroscopic investigations of intramolecular [C-H$\cdots$F-C] contacts in post-metallocene polyolefin catalysts: Modelling weak attractive polymer-ligand interactions. *Chem. Eur. J.* 12: 2607–2619.
172. So, L.-C., Liu, C.-C., Chan, M. C. W., Lo, J. C. Y., Sze, K.-H., Zhu, N. 2012. Scalar coupling across [C-H$\cdots$F-C] interactions in (σ-aryl)-chelating post-metallocenes. *Chem. Eur. J.* 18: 565–573.
173. Bryliakov, K. P., Talsi, E. P., Möller, H. M., Baier, M. C., Mecking, S. 2010. Noncovalent interactions in o-fluorinated post-titanocene living ethylene polymerization catalyst. *Organometallics* 29: 4428–4430.
174. Möller, H. M., Baier, M. C., Mecking, S., Talsi, E. P., Bryliakov, K. P. 2012. The origin of living polymerization with an o-fluorinated catalyst: NMR spectroscopic characterization of chain-carrying species. *Chem. Eur. J.* 18: 848–856.
175. Jennings, W. B. 1975. Chemical shift nonequivalence in prochiral groups. *Chem. Rev.* 75: 307–322.
176. Talsi, E. P., Bryliakov, K. P. 2013. On the origin of living polymerization over o-fluorinated post-titanocene catalysts. *Top. Catal.* 56: 914–922.
177. Breuil, P.-A. R., Magna, L., Olivier-Bourbigou, H. 2015. Role of homogeneous catalysis in oligomerization of olefins: Focus on selected examples based on group 4 to group 10 transition metal complexes. *Catal. Lett.* 145: 173–192.
178. Dixon, J. T., Green, M. J., Hess, F. M., Morgan, D. H. 2004. Advances in selective ethylene trimerization—A critical overview. *J. Organomet. Chem.* 689: 3641–3668.
179. McGuinness, D. S. 2011. Olefin oligomerization via metallocycle: Dimerization, trimerization, tetramerization, and beyond. *Chem. Rev.* 111: 2321–2342.
180. Deckers, P. J. W., Hessen, B., Teuben, J. H. 2001. Switching a catalyst system from ethene polymerization to ethene trimerization with a hemilabile ancillary ligand. *Angew. Chem. Int. Ed.* 40: 2516–2519.
181. Suzuki, Y., Kinoshita, S., Shibahara, A., Ishii, S., Kawamura, K., Inoue, Y., Fujita, T. 2010. Trimerization of ethylene to 1-hexene with titanium complexes bearing phenoxy-imine ligands with pendant donors combined with MAO. *Organometallics* 29: 2394–2396.
182. Hagen, H., Kretschmer, W. P., van Buren, F. R., Hessen, B., van Oeffelen, D. A. 2006. Selective ethylene trimerization: A study into the mechanism and the reduction of PE formation. *J. Mol. Catal. A Chem.* 248: 237–247.
183. Sattler, A., Labinger, J. A., Bercaw, J. E. 2013. Highly selective olefin trimerization catalysis by a borane-activated titanium trimethyl complex. *Organometallics* 32: 6899–6902.

NMR and EPR Spectroscopy of the Mechanisms of Metallocene

184. Soshnikov, I. E., Semikolenova, N. V., Ma, J., Zhao, K.-Q., Zakharov, V. A., Bryliakov, K. P., Redshaw, C., Talsi, E. P. 2014. Selective ethylene trimerization by titanium complexes bearing phenoxy-imine ligands: NMR and EPR spectroscopic studies of the reaction intermediates. *Organometallics* 33: 1431–1439.
185. Baldwin, S. M., Bercaw, J. E., Brintzinger, H. H. 2008. Alkylaluminum-complexed zirconocene hydrides: Identification of hydride-bridged species by NMR spectroscopy. *J. Am. Chem. Soc.* 130: 17423–17433.
186. Al-Arifi, A. S. N. 2004. Propylene polymerization using $MgCl_2$/ethylbenzoate/$TiCl_4$ catalyst: Determination of titanium oxidation states. *J. Appl. Polym. Sci.* 93: 56–62.

Index

A

Acetonitrile dissociation, 50
1-Acetyl-1-cyclohexene, 104–105
Acquisition, 12, 29
ADC, *see* Analog-to-digital converter
AlEt$_2$Cl
  cationic intermediates formation upon activation of Ni(II) catalysts with, 194–198
  reaction of calix[4]arene vanadium(V) complex with, 176–178
AlEt$_3$, reaction of calix[4]arene vanadium(V) complex with, 181
Alkylaluminum-complexed
  structure [(FI)Ti^{III}(μ-H)(μ-Cl)AliBu$_2$]$^+$ [B(C$_6$F$_5$)$_4$]$^-$, 204
  zirconocene trihydride cation [(SBI)Zr(μ-H)$_3$(AliBu)$_2$]$^+$ (VII), 155
Alkylperoxo complexes, 67, 68, 75; *see also* Superoxo complexes
  molybdenum, 68–72
  titanium, 72–76
  vanadium, 76–80
Alkylperoxo intermediate, 66
AlMe$_2$Cl
  cationic intermediates formation upon activation of Ni(II) catalysts with, 194–198
  reaction of Calix[4]arene Vanadium(V) Complex with, 178–181
AlMe$_3$
  activation of L^{2iPr}Co^{II}Cl$_2$ with, 169–171
  activation of L^{2iPr}FeCl$_2$ with, 161–166
AlMe$_3$/MAO combination, 160
^{27}Al NMR spectra of DMAO solution, 129
α-diimine vanadium(III) ethylene polymerization catalysts, 171
  L^{2Me}VCl$_3$/AlMe$_3$/[Ph$_3$C][B(C$_6$F$_5$)$_4$] system, 172–175
  L^{2Me}VCl$_3$/MAO system, 175–176
Aminopyridine ligands, iron complexes with, 103
Aminopyridine ligands, substituted, 104
  asymmetric epoxidation of chalcone, 112
  carboxylic acids, 109
  CH3COOH, 105–106
  elusive Fe^V=O intermediates, EPR spectroscopic detection of, 104
  EPR data for oxoferryl intermediates, 110
  EPR spectra, CH$_2$Cl$_2$/CH$_3$CN, 106

  EPR spectroscopic data for S = 1/2 iron species, 106
  iron complexes, 107–108, 109–110
  iron–oxygen intermediates, 108–109
  NMR spectroscopic characterization of metal complexes, 113
  olefin epoxidation, 104–105, 108
  transition metal complexes, 111–112
Analog-to-digital converter (ADC), 12
Antiferromagnetic dimmers
  experimental temperature dependence of chemical shifts, 46
  μ-O-bridged diferric complexes, 47
  temperature dependence of paramagnetic shift of, 45
A-tensors, 8, 9, 58
Axially anisotropic spectrum, 9

B

(benacen)Co^{III}(py)(O$_2$$^{·-}$), 56, 58
Biomimetic catalysis, 55
Bis-superoxo complex((IPr)$_2$Pd(η^1-O$_2$$^{·-}$)$_2$), 60, 61
Bis(3-(salicylidenamino)propyl)methylamine, 56
1,3-Bis(diisopropyl)phenylimidazol-2-ylidene, 61
Bis(enolatoimine) titanium catalysts with ^{13}C-MAO, 200
Bis(imino)pyridine cobalt ethylene polymerization catalysts, 166
  activation of L^{2iPr}Co^{II}Cl$_2$ with AlMe$_3$, 169–171
  activation of L^{2iPr}Co^{II}Cl$_2$ with MAO, 166–169
Bis(imino)pyridine cobalt systems, 190–191
Bis(imino)pyridine iron(II) precatalysts, 160
Bis(imino)pyridine iron ethylene polymerization catalysts, 159
  activation of L^{2iPr}FeCl$_2$ with AlMe$_3$, 161–166
  activation of L^{2iPr}FeCl$_2$ with MAO, 160–161
Bis(imino)pyridine iron systems, 190–191
Bis(imino)pyridine nickel catalysts 5^{Ni}–9^{Ni}, 190–191
Bis(imino)pyridine Nickel Catalysts activation, cationic intermediates formation upon, 190–194
4,4′-Bis(n-hexyl)-2,2′-bipyridine, 71
Bis[2-iPr$_2$P(4-fluoro-phenyl)amido] (^{F}PNP), 60, 61

219

220 Index

Bisperoxo intermediates, 80–81
Bloch equations, 16–21
Bohr magneton, 1
Boltzmann distribution, 2
[(BPMEN)Fe^{II}(CH$_3$CN)$_2$](ClO$_4$)$_2$, 96
[(BPMEN)Fe^V=O(OAc)]$^{2+}$, 102
[(BQEN)Fe^{II}(OTf)$_2$], 96
[Bu$_4$N]$_4$[HPTi(O$_2$)W$_{11}$O$_{39}$], 87
[Bu$_4$N]$_5$[PTi(O$_2$)W$_{11}$O$_{39}$], 87–88
[Bu$_4$N]$_8$[(PTiW$_{11}$O$_{39}$)$_2$O], 87–88

C

Calix[4]arene vanadium(V) complex,
  with AlEt$_2$Cl, reaction of, 176–178
  with AlEt$_3$, reaction of, 181
  with AlMe$_2$Cl, reaction of, 178–181
  ethylene polymerization precatalyst based
    on, 176
  reaction with AlEt$_2$Cl, 176–178
  reaction with AlEt$_3$, 181
  reaction with AlMe$_2$Cl, 178–181
CAO, see Copper-containing amine oxidases
Carboxylic acids, 99–100, 109
Catalyst deactivation, 188–190
Catalyst systems, 202
  9$_{Ti}$/AlMe$_3$/[Ph$_3$C][B(C$_6$F$_5$)$_4$], 206
  9^{Ti}/MMAO, 206
Cationic group 4 metalalkyl complexes, 132
Cationic intermediates formation
  upon activation of Bis(imino)pyridine Nickel
    Catalysts, 190–194
  upon activation of Ni(II) catalysts with
    AlMe$_2$Cl and AlEt$_2$Cl, 194–198
^{13}C-enriched MAO, 133
^{13}C{^{1}H} NMR spectra, 33, 87
CH$_3$CN/H$_2$O mixtures, 84
Chemical kinetics, 55
Chemical shift, 6, 12–14
Chemical shift anisotropy (CSA), 15
Chiral complex, 99–100
Chiral hydroxamic acid, 78, 79
3-Chloro-acetylacetonate (Cl-acac), 108
Chromium-salen-catalyzed epoxidation, 93
Chromium(III)-salen catalysts, 89
^{13}C J-modulated NMR spectrum, 34
^{13}C NMR spectra, 27, 73, 134, 136, 137, 148
  ^{1}H COSY vs. ^{1}H TOCSY spectra, 37–39
  one-dimensional ^{1}H spectra and, 29–37
  of sample compound, 29
  two-dimensional ^{1}H spectra and, 29–37
Co(III) acetate in solution, 113–116
Co(III), superoxo complexes of, 56
  EPR parameters, 58
  EPR spectrum of Co^{III}(acacen)(Py)(η^1-
    O$_2$$^{\cdot-}$), 59
  oxidation of substituted phenols, 57

Cobalt bis(imino)pyridine precatalysts, structures
  of, 167
Co^{II}(acacen)(py), 56, 57
Co^{III}(acacen)(py)(O$_2$$^{\cdot-}$), 56, 58, 59
Co^{III}(Salpr)(O$_2$$^{\cdot-}$), 56
COLOC, see COrrelation through LOng-range
  Coupling
Computer simulation, 58
"Contact ion pair" Iva, 149
Continuous-wave (CW), 3
  spectrometers, 3–4
  spectroscopy, 19
Copper-containing amine oxidases (CAO), 63
Copper(II), superoxo complexes of, 63–64
Copper enzymes, 63
"Correlations", 27
COrrelation through LOng-range Coupling
  (COLOC), 37
COSY spectrum, 32, 37
Cp$_2$TiCl$_2$ with MAO, ion pairs forming upon
  interaction of, 143–148
^{13}C resonances, 33
[Cr^{III}(salen)]$^+$ complexes, 88–89
Cross-peak, 32
[CrV = O (salen)]$^+$, 88–91
Cryospray-assisted variable-temperature mass
  spectrometry (Cryospray-assisted
  VT-MS), 100
CSA, see Chemical shift anisotropy
^{13}C spectroscopy, 153
[Cu^{II}(HB(3-tBu-5-iPrpz)$_3$)(η^2-O$_2$$^{\cdot-}$)], 63, 64
[Cu^{II}(L$_1$)(η^1-O$_2$$^{\cdot-}$)], 63, 64
[Cu^{II}(L$_2$)(η^1-O$_2$$^{\cdot-}$)], 63, 64
[Cu^{II}(L$_3$)(η^1-O$_2$$^{\cdot-}$)], 63, 64
[Cu^{II}(L$_4$)(η^1-O$_2$$^{\cdot-}$)], 64
Curie law, 43, 46
CW, see Continuous-wave
Cyclic voltammetry study, 88
Cyclohexene, 97
Cyclohexene-1-carbonitrile, 104–105
Cyclohexane-(1,2,3,4,5,6-hexa)ol,
  see myo-inositol (MI)
Cyclohexene oxide, 103
2-Cyclohexen-1-one and 2-cyclohexen-
  1-ol, 103

D

Delta scale, 12
Density functional theory (DFT), 61
DEPT pulse sequence, 33
Deuteration, selective, 40
DFT, see Density functional theory
DG, see D-Glucuronate
Diastereomeric alkylperoxo vanadium(V)
  complexes, 79, 80
Diastereotopic protons, 37

Index

221

1,4-Diazabutadiene, 71
9,10-Dihydroanthracene, 62
Diisopropyl tartrate (DIPT), 72
Dimethyl sulfoxide (DMSO), 60
Dinuclear complex, 68
Dinuclear η2-alkylperoxo complex, 72, 73
Diperoxo molybdenum complex $Mo(O_2)_2 \cdot HMPA$, 80–81
Dipolar contributions, 40
Dipolar interactions, 44
Dipole–dipole
  coupling, 15
  interaction, 8, 14, 15
  relaxation, 15
Diprotonated complex $[H_2PTi(O_2)W_{11}O_{39}]^{3-}$, 88
DIPT, *see* Diisopropyl tartrate
DMSO, *see* Dimethyl sulfoxide
Dopamine-β-monooxygenase (DβM), 63
Downfield, 13
DβM, *see* Dopamine-β-monooxygenase

E

Electric field gradients, 15–16
Electromagnetic irradiation, 1
Electron angular momentum, 6
Electron magnetic moments in magnetic field, 1–4
Electron paramagnetic resonance spectroscopy (EPR spectroscopy), 1, 132, 203–204
  characterizing, 4–8
  characterizing EPR spectra, 4–8
  electron and nuclear magnetic moments in magnetic field, 1–4
  "EPR silent", 47
  of frozen solutions, 8–12
  instrumentation for CW, 4
  spectroscopists, 2
  spin-probe study of MAO, 131
Electron spin, 2, 7, 45
  angular momentum, 6–7
  density, 163
  magnetic moment, 1
  Zeeman splitting, 2
Electrospray ionization mass spectrometry (ESI-MS), 72–73
Elusive $Fe^V=O$ intermediates, EPR spectroscopic detection of, 104
  asymmetric epoxidation of chalcone, 112
  carboxylic acids, 109
  CH_3COOH, 105–106
  EPR data for oxoferryl intermediates, 110
  EPR spectra, CH_2Cl_2/CH_3CN, 106
  EPR spectroscopic data, 106
  iron complexes, 107–108, 109–110
  iron–oxygen intermediates, 108–109

NMR spectroscopic characterization of metal complexes, 113
  transition metal complexes, 111–112
Epoxidation, 88
  olefin, 104–105, 108
EPR spectroscopy, *see* Electron paramagnetic resonance spectroscopy
Equilibrium magnetization, 2
ESI-MS, *see* Electrospray ionization mass spectrometry
ETA, *see* Ethyl trichloroacetate
η2-(*tert*-butylperoxo)titanatrane dimer, 68, 70, 72
Ethylene (C_2H_4), 60
Ethylene polymerization, 37, 169, 172; *see also* Living polymerization
  activity, 181
  calix[4]arene vanadium(V) complex reaction with $AlEt_2Cl$, 176–178
  calix[4]arene vanadium(V) complex reaction with $AlEt_3$, 181
  calix[4]arene vanadium(V) complex reaction with $AlMe_2Cl$, 178–181
  catalyzed by $10^{Ni}/AlR_2Cl$, 195
  chain-propagating species forming upon, 182–186
  over $(Cp-R)_2ZrCl_2/MAO$ Catalysts, 139
  precatalyst based on calix[4]arene vanadium(V) complex, 176
Ethylene trimerization,
  EPR spectrum, 204
  ^1H NMR spectra, 205
  ion pairs $[LTi^{II}]^+[MeMAO]^-$, 206
  selective ethylene trimerization on Ti-and Cr-based catalysts, 203
  structures of complexes 9^{Ti} and 9^{Ti}-Me, 202
  by titanium complex bearing phenoxy imine ligand, 202
Ethyl trichloroacetate (ETA), 176
Evans method, 47–51
Extended X-ray absorption fine structure (EXAFS), 61
External magnetic field, 1, 2
Extra absorption peaks, 10

F

$[Fe^{II}(BPMEN)(CH_3CN)_2]$ $(ClO_4)_2$ complex, 48, 49
$Fe^{III}Fe^{III}-O_2{}^{\cdot-}$, Diiron(μ-hydroxo) structures, 66
$Fe^{III}-O_2{}^{\cdot-}$ structure, 66
Fermi contact, 40
  contribution, 43
  interaction, 6
  paramagnetic shift, 51
  shift, 46
Ferric-superoxo complex, 65–66

222 Index

Ferric complexes, 104
 asymmetric epoxidation of chalcone, 112
 carboxylic acids, 109
 CH_3COOH, 105–106
 elusive $Fe^V=O$ intermediates, EPR
  spectroscopic detection of, 104
 EPR data for oxoferryl intermediates, 110
 EPR spectra, CH_2Cl_2/CH_3CN, 106
 EPR spectroscopic data for $S = 1/2$ iron
  species, 106
 iron complexes, 107–108
 iron complexes with aminopyridine ligands,
  109–110
 iron–oxygen intermediates, 108–109
 NMR spectroscopic characterization of metal
  complexes, 113
 olefin epoxidation, 104–105
 olefin epoxidation by H_2O_2, 108
 transition metal complexes, 111–112
FID, *see* Free induction decay
Fluorinated intermediates $[L_aTiP]^+[MeMAO]^-$
  (8Tia), 201
^{19}F NMR spectroscopy, 153
Fourier transform (FT), 4, 24
 NMR spectroscopy, 4
"Free chain" structure, 202
Free induction decay (FID), 4, 12, 22–23
Frequency-domain spectrum, 12
FT, *see* Fourier transform

G

$[\gamma-PV_2W_{10}O_{38}(OH)_2]^{3-}$, 85
Geraniol, 80
g-factor, 1–2
 anisotropy, 7–8
D-Glucuronate (DG), 65
"Green" catalyst systems, 55
Group 4 metallocene catalysts, 159–160
g-tensor, 7, 8, 195
 anisotropy, 44
 rhombic, 62, 65
Gyromagnetic ratio, 1–2

H

Hafnium precatalysts, 155
Hafnocene complexes, 155
Halcon epoxidation, 67
Hammond–Leffler principle, 110
Harmonic oscillations, 23
^1H COSY spectra, 37–39
$^1H,^1H$ COSY, *see* 2D homo-nuclear *COrrelation
  SpectroscopY*
$^1H,^{13}C$-COSY spectra, 37
$^1H,^{13}C$ heteronuclear correlated spectrum, 35
Heme enzymes, 95–96

HETCOR, *see* Heteronuclear H,X-correlation
  spectroscopy (HXCO)
Heteronuclear H,X-correlation spectroscopy
  (HXCO), 35, 37
Heteronuclear multiple bond correlation
  spectroscopy (HMBC
  spectroscopy), 37
Heteronuclear multiple quantum coherence
  spectroscopy (HMQC
  spectroscopy), 37
Heteronuclear single quantum coherence
  spectroscopy (HSQC
  spectroscopy), 37
Hexamethylphosphoric triamide (HMPA),
  80–81, 82
Hexane, 30
1-Hexene, 202
 polymerization, 157
HF, *see* Hyperfine
High-molecular-weight polyethylene
  (HMWPE), 202
HMBC spectroscopy, *see* Heteronuclear multiple
  bond correlation spectroscopy
HMPA, *see* Hexamethylphosphoric triamide
HMQC spectroscopy, *see* Heteronuclear multiple
  quantum coherence spectroscopy
HMWPE, *see* High-molecular-weight
  polyethylene
^1H NMR–characterized oxomanganese(V)
  complexes, 93
^1H NMR monitoring, 161–162
^1H NMR spectra, 73, 137, 151, 160
 of cationic (SBI)Zr complexes, 158
 of Cp_2TiCl_2 + MAO, 144
 of L_2ZrCl_2/MAO systems, 139–140
Homogeneous catalysis, 27
 ^1H and ^{13}C NMR chemical shifts
  diagram, 28
 ^1H spectra of paramagnetic molecules,
  39–51
 measuring ^1H and ^{13}C NMR spectra of sample
  compound, 29–39
Homogeneous catalytic processes, 67
Homogeneous single-site catalysts, 128
Homonuclear, ^{13}C, ^{13}C-correlation
  spectroscopy, 37
Homoprotocatechuate 2, 3-dioxygenase (2,
  3-HPCD), 65, 66
2,3-HPCD, *see* Homoprotocatechuate
  2,3-dioxygenase (2,3-HPCD)
^1H pulsed field-gradient spin echo NMR
  spectroscopy (PFGSE NMR
  spectroscopy), 187
^1H spectra of paramagnetic molecules, 39
 probing structure of unknown Ni(II)
  complex by multinuclear NMR,
  40–42, 43

Index

223

temperature dependence of paramagnetic shift, 42–51

^{1}H spectra of sample compound, 29

^{1}H COSY *vs.* ^{1}H TOCSY spectra, 37–39

one-dimensional ^{13}C NMR spectra and, 29–37

two-dimensional ^{13}C NMR spectra and, 29–37

^{1}H spectroscopy, 153

HSQC spectroscopy, *see* Heteronuclear single quantum coherence spectroscopy

^{1}H TOCSY spectra, 37–39

180° pulse, 22

H,X-COSY, *see* Heteronuclear H,X-correlation spectroscopy (HXCO)

HXCO, *see* Heteronuclear H,X-correlation spectroscopy

Hybrid density functional approach, 80–81

Hydrogen peroxide, 61, 80

Hydroxamic acids, 78, 79

Hyperfine (HF), 5–6

splitting constant, 6

I

ICT, *see* Intervalence charge transfer

INADEQUATE, 37

Instrumentation

for CW EPR spectroscopy, 4, 5

for pulsed FT NMR spectroscopy, 11, 12

Integral intensity, 30

"Interacting chain" structure, 202

Intervalence charge transfer (ICT), 91

Ion pairs formation

in catalyst systems (Cp-R)$_2$ZrCl$_2$/MAO, detection of, 137–139

in catalyst systems metallocene/AliBu$_3$/ [Ph$_3$C][B(C$_6$F$_5$)$_4$], 152–156

in catalyst systems zirconocene/activator/α-olefin, 156–159

in catalyst systems zirconocene/MMAO, observation of, 149–152

upon activation of (C$_5$Me$_5$)TiCl$_3$ and [(Me$_4$C$_5$) SiMe$_2$N^{t}Bu]TiCl$_2$ with MAO, 148–149

upon activation of (Cp-R)$_2$ZrCl$_2$ (R = nBu, tBu) with MAO, detection of, 134–137

upon activation of *Ansa*-Zirconocenes with MAO, 139–143

upon interaction of Cp$_2$TiCl$_2$ and *Rac*-C$_2$H$_4$(Ind)$_2$TiCl$_2$ with MAO, 143–148

upon interaction of Cp$_2$ZrMe$_2$ with MAO, structure of, 132–134

IPr, 61

Iron(III), superoxo complexes of, 65–67

Iron complexes, 96

with aminopyridine ligands, 103, 109–110

Isomeric Zr-allyl complexes, 157

J

Jacobsen's catalyst, 93

J-modulated carbon spectra (JMOD), 32–33

K

Kagan–Modena asymmetric sulfoxidation system, 72

Katsuki–Sharpless, asymmetric epoxidation system, 72

epoxidation, 67

Klystron, 4

Kronecker symbol, 7

L

Lactol, 80

Larmor frequency, 17

Larmor precession, 17

Lewis acidic sites, 130–131, 150

[(L)FeV=O]$^{3+}$ oxocomplexes, 95

asymmetric epoxidation of chalcone, 101

chiral complex, 99–100

cyclohexene, 97

EPR spectroscopic data for $S = 1/2$ iron–oxygen species, 100

EPR spectroscopy, 96–97

ferric-peracetate complexes, 104

heme enzymes, 95–96

iron complexes, 96

iron complexes with aminopyridine ligands, 103

nonheme iron complexes, 96

putative oxoiron(V) intermediates, 102

[LHf(μ-H)$_2$Al(H) iBu)$^+$, 155

[LHf(μ-H)$_2$AliBu$_2$)$^+$, 155

[LHf(μ-Me)$_2$AlMe$_2$]$^+$ [MeMAO]$^-$ (III$_{Hf}$), 154

[LHf(μ-Me)$_2$AlMe$_2$]$^+$[B(C6F5)4]$^-$ (III$'_{Hf}$), 154

Linear low-density polyethylene (LLDPE), 202

L^{2iPr}Co^{II}Cl$_2$

with AlMe$_3$, activation of, 169–171

with MAO, activation of, 166–169

L^{2iPr}FeCl$_2$

with AlMe$_3$, activation of, 161–166

with MAO, activation of, 160–161

Living polymerization; *see also* Ethylene polymerization

β-hydrogen transfer, 202

bis(enolatoimine) titanium catalysts with ^{13}C-MAO, 200

over o-fluorinated post-titanocene catalysts, 198

structure of complexes 1^{Ti}–4^{Ti}, 199

structure of complexes 5^M and 6^M, 200

structures of propagating species 8^{Ti}a and 8^{Ti}b, 201

weak noncovalent Ti···F interactions in intermediate 7^{Ti}a, 201

LLDPE, *see* Linear low-density polyethylene
$L^{2Me}VCl_3/AlMe_3/[Ph_3C][B(C_6F_5)_4]$ System,
    172–175
$L^{2Me}VCl_3/MAO$ System, 175–176
Longitudinal relaxation time, 16, 18
Lorentzian line shape, 20
LS–HS equilibrium, 51
$[LTi^{II}]+[B(C_6F_5)_4]^-$ species, 206
$[(L)Zr(\mu-Me)_2AlMe_2)^+[Me-MAO]^-$, 141–142
$[(L)Zr(\mu-Me)_2AlMe_2]^+[B(C_6F_5)_4)^-$, 141–142
$[(L)ZrMe(NMe_2Ph)]^+[B(C_6F_5)_4]^-$, 141–142
$[(L)ZrMe]^+[B(C_6F_5)_4]^-$, 141–142

M

m-CPBA, *see meta*-chloroperoxybenzoic acid
Macrocyclic tetraamide ligand (TAML), 100
Macroscopic magnetization, 3, 16–17
Magnetic field, 4–5, 12–16, 18, 20–22
Magnetic moment of atom, 6
Magnetic nuclei, 3
Magnetic resonance, 2
Magnetic susceptibility measurement for
    studying spin equilibrium, 47
  $[Fe^{II}(BPMEN)(CH_3CN)_2]$ $(ClO_4)_2$ complex, 49
  LS–HS equilibrium, 51
  NMR timescale process, 50
  preparation of NMR sample, 48
Magnetization transfer, 32
MAO, *see* Methylalumoxane
Mass magnetic susceptibility equation, 47
Me_2AlEt, 148
$[(Me_2PyTACN)Fe^{II}(OTf)_2]$, 96
$[(Me_2PyTACN)Fe^V=O(OH)]^{2+}$, 102
$MeAl(C_6F_5)_3$, 142
$MeB(C_6F_5)_3$, 142
$MeE(C_6F_5)_3$, 141–142
meta-chloroperoxybenzoic acid (*m*-CPBA), 95,
    96–97
Metallocene-based polymers, 127–128
Metallocene catalysts, 127; *see also* Post-
    metallocene catalysts
  on active centers of MAO, 130–132
  ion pairs detection forming in catalyst
    systems (Cp-R)$_2$ZrCl$_2$/MAO,
    137–139
  ion pairs detection forming upon activation
    of (Cp-R)$_2$ZrCl$_2$ (R = *n*Bu, *t*Bu) with
    MAO, 134–137
  ion pairs forming in catalyst systems
    metallocene/Al*i*Bu$_3$/[Ph$_3$C][B(C$_6$F$_5$)$_4$],
    152–156
  ion pairs forming upon activation of (C$_5$Me$_5$)
    TiCl$_3$ and [(Me$_4$C$_5$) SiMe$_2$N*t*Bu]TiCl$_2$
    with MAO, 148–149
  ion pairs forming upon activation of *Ansa*-
    Zirconocenes with MAO, 139–143

ion pairs forming upon interaction of
    Cp$_2$TiCl$_2$ and *Rac*-C$_2$H$_4$(Ind)$_2$TiCl$_2$
    with MAO, 143–148
ion pairs operating in catalyst systems
    zirconocene/activator/α-olefin,
    156–159
metallocene-based polymers, 127–128
observation of ion pairs forming in catalyst
    systems zirconocene/MMAO,
    149–152
size of MAO oligomers, 128–130
structure of ion pairs forming upon
    interaction of Cp$_2$ZrMe$_2$ with MAO,
    132–134
Metal–oxygen intermediates, 55
Methylalumoxane (MAO), 128
  activation of L^{2iPr}Co^{II}Cl$_2$ with, 166–169
  activation of L^{2iPr}FeCl$_2$ with, 160–161
  on active centers of, 130–132
  on active centers of MAO, 130–132
  ion pairs detection forming in catalyst
    systems (Cp-R)$_2$ZrCl$_2$/MAO, 137–139
  ion pairs detection forming upon activation
    of (Cp-R)$_2$ZrCl$_2$ (R = *n*Bu, *t*Bu) with,
    134–137
  ion pairs forming upon activation of (C$_5$Me$_5$)
    TiCl$_3$ and [(Me$_4$C$_5$) SiMe$_2$N*t*Bu]TiCl$_2$
    with, 148–149
  ion pairs forming upon activation of *Ansa*-
    Zirconocenes with, 139–143
  ion pairs forming upon interaction of
    Cp$_2$TiCl$_2$ and *Rac*-C$_2$H$_4$(Ind)$_2$TiCl$_2$
    with MAO, 143–148
  oligomer size, 128–130
  size of MAO oligomers, 128–130
  structure of ion pairs forming upon
    interaction of Cp$_2$ZrMe$_2$ with,
    132–134
MI, *see myo*-inositol
Microwave frequency (MW frequency), 3
MIOX, *see myo*-inositol oxygenase
MMAO, *see* Modified methylalumoxane
$[Mn^V = O(salen)]^+$, 91–95
Mo(VI)-1,2-diol complexes, 68
Modified methylalumoxane (MMAO), 128
  observation of ion pairs forming in catalyst
    systems, 149–152
Molar susceptibility, 48
Molybdenum
  complexes, 81
  diperoxo complex of molybdenum(VI), 81
  peroxo complexes of, 80–83
Molybdenum(VI), 68
  alkylperoxo complexes of, 68
  dinuclear complex, 68–69
  ^{17}O NMR spectra, 69–71
  ^{95}Mo NMR spectrum, 69

Index

225

structure, 72
Mono-intermediates, 80–81
Monomeric compounds
  ^{1}H NMR spectra, 44
  linear temperature dependence, 45
  temperature dependence of paramagnetic
    shift of, 43
Monomer Pd(OAc)$_2$S(η^2-O$_2$$^{\cdot-}$), 61
Mononuclear alkylperoxo complex, 73
Mononuclear species, 64
Mössbauer spectroscopy, 67
Multinuclear NMR
  ^{1}H NMR spectrum, 41
  ^{1}H spectral data of complex, 3, 41
  molecular structure of compound, 43
  ^{31}P and ^{19}F NMR spectra, 42
  probing structure of unknown Ni(II) complex
    by, 40
  synthesis of Ni(TPA) complex, 40
Multiplets, 12
MW frequency, *see* Microwave frequency
myo-inositol (MI), 2, 3, 4, 5, 65
myo-inositol oxygenase (MIOX), 65

N

N4Py, *see* N,N-bis(2-pyridylmethyl)-N-(bis-2-
  pyridylmethyl)amine
4NC, *see* 4-Nitrocatechol
Net magnetization, 2
Neutral Ni^{II}κ^2-(N, O)-salicylaldiminato olefin
  polymerization catalysts, 182
  catalyst deactivation, 188–190
  chain-propagating species, 182–186
  evaluation of size of Ni–Polymeryl species by
    PFG NMR spectroscopy, 187–188
N-Fluoroaryl bis(phenoxyketimine) Ti complex
  3^{Ti}, 198–199
Ni-butyl species, 39
Ni-polymeryl species, 37, 39
  evaluation of size of Ni-polymeryl species,
    187–188
Ni-propyl species, 39
Ni(II)
  catalysts, 194–198
  complex by multinuclear NMR, 40–42, 43
  metal center, 40
  superoxo complexes of, 61–63
Ni(TPA) complex synthesis, 40
Nickel(II) complexes 5^{Ni}–9^{Ni}, 191
[Ni^{II}(13-TMC)(η^1-O$_2$$^{\cdot-}$)]$^+$, 62, 63
[Ni^{II}(14-TMC)(η^1-O$_2$$^{\cdot-}$)]$^+$, 61, 62
[Ni^{II}(dkim)(η^1-O$_2$$^{\cdot-}$)], 62
[Ni^{II}(PhTtAd) (η^2-O$_2$$^{\cdot-}$)], 61, 62
[Ni^{III}(13-TMC)(η^2-O$_2$$^{2-}$)]$^+$, 62, 63
90° pulse, 22
4-Nitrocatechol (4NC), 66

N-methylmorpholine-N-oxide (NMO), 94–95
NMO, *see* N-methylmorpholine-N-oxide
NMR spectroscopy, *see* Nuclear magnetic
  resonance spectroscopy
N,N-bis(2-pyridylmethyl)-N-(bis-2-
  pyridylmethyl)amine (N4Py), 100
N,N′-Ethylenebis(benzoylacetoniminide), 58
N,N,N′,N′-Tetrakis(2-pyridylmethyl)-
  ethylenediamine (TPEN), 100
NOE, *see* Nuclear Overhauser effect
Non-heme iron enzymes, 66
Nonfluorinated intermediates [L$_b$TiP]$^+$[MeMAO]$^-$
  (8Tib), 201
Nonzero electric quadrupole moments, 15
Norbornene polymerization, 190
  on bis(imino)pyridine Ni complexes, 191
Nuclear magnetic moments in magnetic
  field, 1–4
Nuclear magnetic resonance spectroscopy (NMR
  spectroscopy), 1
  Bloch equations, 16–21
  characteristics, 12
  chemical shift, 12–14
  line width, 14–16
  measurements, 16
  "NMR-silent" species, 159
  pulsed FT NMR spectroscopy, 11, 12,
    21–24
  spectroscopic analyses, 27
  spectroscopists, 2
  spin–spin coupling, 14
Nuclear magneton, 1
Nuclear Overhauser effect (NOE), 32
  NOE-enhanced multiplets, 33
  signal, 32
Nuclear spin relaxation, 15

O

Octamethyltrisiloxane (OMTS), 131
o-fluorinated bis-(phenoxyimine) titanium
  catalysts, 199
o-fluorinated post-titanocene catalysts,
  β-hydrogen transfer, 202
  bis(enolatoimine) titanium catalysts with
    ^{13}C-MAO, 200
  living polymerization over, 198
  structure of complexes 1^{Ti}–4^{Ti}, 199
  structure of complexes 5^M and 6^M, 200
  structures of propagating species 8^{Ti}a and
    8^{Ti}b, 201
  weak noncovalent Ti···F interactions in the
    intermediate 7^{Ti}a, 201
  o-F-substituted bis(enolatoimine) titanium
    complex 4Tia (R = F), 199
^{18}O labeling studies, 80
Olefin epoxidation by H$_2$O$_2$, 108

Index

Olefins; 76; *see also* Selective catalytic epoxidation of olefins
active species of selective epoxidation, 95
asymmetric epoxidation of chalcone, 101
chiral complex, 99–100
cyclohexene, 97
EPR spectroscopy, 96–97, 100
ferric-peracetate complexes, 104
heme enzymes, 95–96
iron complexes, 96
iron complexes with aminopyridine ligands, 103
nonheme iron complexes, 96
putative oxoiron(V) intermediates, 102
OMTS, *see* Octamethyltrisiloxane
One-dimensional ^1H and ^{13}C NMR spectra, 29
$^{13}C\{^1H\}$ NMR spectrum, 33
^{13}C J-modulated NMR spectrum, 34
^{13}C NMR experiment with ^1H decoupling, 33
$^1H,^1H$ COSY spectrum, 32
diastereotopic protons, 37
pulse sequence for H, X-COSY experiment ($X = ^{13}C$), 36
pulse sequence for J-modulated, 34
sample compound—dimeric titanium(IV), 30
structure of hydrocarbon framework, 35
two-dimensional ^1H, ^{13}C HXCO spectrum, 36
1^V complex, 177
Orbital angular momentum, 6
operator, 7
Organic compounds, 27
Oscillating field, 3, 4, 18
Oxochromium(V) complex, 89
Oxochromium(V)-salen species, 91
Oxo complexes, 88; *see also* Peroxo complexes
chromium-salen-catalyzed epoxidation, 93
$[Cr^V = O(salen)]^+$, 88–91
EPR spectroscopic detection of elusive $Fe^V=O$ intermediates, 104–113
$[(L)Fe^V=O]^{3+}$, 95–104
$[Mn^V = O(salen)]^+$, 91–95
(salen)manganese(III) complexes, 93
Oxygen transfer, 89

P

Palladium(II) (Pd(II)), 59
superoxo complexes, 59–61
para-substituted 2,6-di-*tert*-butylphenols (*p*-X-DTBPs), 64
Paramagnetic molecules, ^1H spectra of, 39–51
Paramagnetic shifts, 39
temperature dependence of, 42–51
Paramagnetic species, 3
Parts per million (ppm), 13
Pascal's triangle, 14
PCA, *see* Pyrazine-2-carboxylic acid

$Pd_3(OAc)_6(\eta^1-O_2{}^{\cdot-})$, 59, 60
$[((S,S)-PDP)Fe^{II}(CH_3CN)_2](SbF_6)_2$, 96
$[((S,S)-PDP)Fe^V=O(OAc)]^{2+}$, 102
PE, *see* Polyethylene
Peptidylglycine-α-hydroxylating monooxygenase (PHM), 63
Peroxo complexes, 80; *see also* Oxo complexes
of molybdenum, 80–83
of titanium, 87–88
of vanadium, 83–87
Peroxy-*p*-quinolato cobalt(III) complex, 56
Perturbation theory, 7
PFG NMR spectroscopy, *see* Pulsed field-gradient NMR spectroscopy
PFGSE NMR spectroscopy, *see* ^1H pulsed field-gradient spin echo NMR spectroscopy
Ph3PO, *see* Triphenylphosphine oxide
(2-PhInd)$_2$ZrCl$_2$ (1-Cl), 152
PHM, *see* Peptidylglycine-α-hydroxylating monooxygenase
Phosphotungstate, 86
Planar-chiral [2.2] paracyclophane-derived hydroxamate, 79–80
Planck constant, 1
Polyethylene (PE), 159
Polyolefins, 198
Polyoxoanion $[PTiW_{11}O_{39}]^{3-}$, 87
Polypyridyl, 71
Post-metallocene catalysts; *see also* Metallocene catalysts
α-diimine vanadium(III) ethylene polymerization catalysts, 171–176
bis(imino)pyridine cobalt ethylene polymerization catalysts, 166–171
bis(imino)pyridine iron ethylene polymerization catalysts, 159–166
cationic intermediates formation upon activation of Ni(II) catalysts with AlMe$_2$Cl and AlEt$_2$Cl, 194–198
ethylene polymerization precatalyst based on calix[4]arene vanadium(V) complex, 176–181
formation of cationic intermediates upon activation of bis(imino)pyridine nickel catalysts, 190–194
neutral Ni^{II}κ^2-(*N*,*O*)-salicylaldiminato olefin polymerization catalysts, 182–190
Post-metallocenes, 159
ppm, *see* Parts per million
Principal-axis system, 7
"Projections", 31, 35
Proton magnetic moment, 1
Pulsed field-gradient NMR spectroscopy (PFG NMR spectroscopy), 133
evaluation of size of Ni–Polymeryl species by, 187–188
Pulsed Fourier transform (Pulsed FT)

Index

instruments, 21
NMR spectroscopy, 4, 21–24, 29
Pulsed magnetic resonance spectroscopy, 3
Pulsed spectrometers, 3–4
"Pure spin" ground state, 7
p-X-DTBPs, *see para*-substituted
   2,6-di-*tert*-butylphenols
PyO, *see* Pyridine N-oxide
Pyrazine-2-carboxylic acid (PCA), 84
2,6-Pyridinedicarboxylate, 72
Pyridine N-oxide (PyO), 88–89
Pyridyl-amide hafnium complexes, 154

Q

Quadrature detection, 19
Quadrupolar interactions, 15–16
Quadrupole moment, 16

R

Rac-$C_2H_4(Ind)_2TiCl_2$, 143–148
Radio frequency (RF), 3
   pulse, 4, 12, 21, 22–23
Reference frequency, 19
Reflection spectrometers, 4
Relaxation, 17–18
Relaxation delay, 29
Resonance absorption, 3
Resonance Raman spectroscopy, 67
Resonating nucleus, 12
Resonator cavity, 4
RF, *see* Radio frequency
Rhombically anisotropic spectrum, 9
Rotating frame, 21
Rotation matrix formalism, 19

S

S=1/2 diiron(III/III)-superoxo species, 65
Salan ligand, 30
Salen-type ligand, 71
[(salen)Mn^V = O]$^+$ species, 93–94
Salicylaldiminato nickel(II) complexes, 2, 37, 39
SANS, *see* Small-angle neutron scattering
Scalar coupling (J-coupling), *see* Spin–spin
   coupling
Schiff base, 86
Second-rank tensor, 7
Selective catalytic epoxidation of olefins, 104;
   see also Olefins
   asymmetric epoxidation of chalcone, 112
   carboxylic acids, 109
   CH_3COOH, 105–106
   elusive Fe^V=O intermediates, EPR
      spectroscopic detection of, 104
   EPR data for oxoferryl intermediates, 110

EPR spectra, CH_2Cl_2/CH_3CN, 106
EPR spectroscopic data for S = 1/2 iron
   species, 106
iron complexes, 107–108
iron complexes with aminopyridine ligands,
   109–110
iron–oxygen intermediates, 108–109
NMR spectroscopic characterization of metal
   complexes, 113
olefin epoxidation, 104–105
olefin epoxidation by H_2O_2, 108
transition metal complexes, 111–112
Selective ^1H decoupling, 35
shf, *see* Superhyperfine
Shielding constant, 12
Side–on bond dioxygen species, 66
Small-angle neutron scattering (SANS), 130
Sodium amalgam (NaHg), 159
Spectroscopic techniques, 62
"Spin-echo" experiments, 32
Spin-forbidden process, 55
"Spin-only" electron, 7
Spin-probe technique, 131
Spin equilibrium, measuring magnetic
   susceptibility for studying, 47–51
Spin Hamiltonian, 8, 9
Spin–lattice relaxation time, *see* Longitudinal
   relaxation time
Spin–orbit coupling, 7, 8
Spin–spin coupling, 6, 14
Spin–spin relaxation time, *see* Transverse
   relaxation time
Superhyperfine (shf), 58
Superoxide anion ($O_2^{\cdot-}$), 56
Superoxo complexes, 56; *see also* Alkylperoxo
   complexes
   Co(III), 56–59
   Co(II), 63–64
   iron(III), 65–67
   Ni(III), 61–63
   Pd(II), 59–61
Synthesized alkylperoxo complexes, 76–77
Synthetic polymer chemistry, 198

T

TAML, *see* Macrocyclic tetraamide ligand
[(TAML)Fe^V=O]$^-$, 102
$(tBuAlO)_n$ cage structures, 130
Temperature dependence of paramagnetic shift,
   42
   of antiferromagnetic dimmers, 45–47
   measuring magnetic susceptibility for
      studying spin equilibrium, 47–51
   of monomeric compounds, 43–45
(2,2,6,6-Tetramethylpiperidin-1-yl)oxyl
   (TEMPO), 4-5, 131

Index

3,3,5,5-Tetramethyl-pyrroline-*N*-oxide (TMPO), 84
Thermodynamic equilibration, 20–21
8Tia, *see* Fluorinated intermediates [L$_a$TiP]$^+$[MeMAO]$^-$
8Tib, *see* Nonfluorinated intermediates [L$_b$TiP]$^+$[MeMAO]$^-$
4^{Ti}a,b/AlMe$_3$/[Ph$_3$C][B(C$_6$F$_5$)$_4$]/C$_2$H$_4$, catalyst systems, 199–200
4^{Ti}a,b/MAO/C$_2$H$_4$, catalyst systems, 199–200
9^{Ti}/AlEt$_3$ System, 206
9^{Ti}/AlMe$_3$/[Ph$_3$C][B(C$_6$F$_5$)$_4$] Systems, 204
9^{Ti}/AlMe$_3$/MAO Systems, 204
TIBA, *see* Triisobutylaluminum
Ti(OOtBu)(OiPr)$_3$, 73, 74
Ti(OOtBu)$_2$(OiPr)$_2$, 73, 74
Ti(OOtBu)$_3$(OiPr), 73, 74
Ti(OOtBu)$_4$, 73, 74
Titanium, 154
  ^{13}C and ^{1}H NMR chemical shifts, 75–76
  ^{13}C NMR spectrum, 75
  alkylperoxo complexes, 72
  EPR spectrum, 204
  ESI-MS, 72–73
  ethylene trimerization by titanium complex bearing phenoxy imine ligand, 202
  ^{1}H NMR spectra, 205
  ion pairs [LTi^{II}]$^+$[MeMAO]$^-$, 206
  olefins, 76
  peroxo complexes, 87–88
  selective ethylene trimerization on Ti-and Cr-based catalysts, 203
  structures, 73, 202
Titanium silicalites (TS-1), 67
[(TM-4-PyP)Mn^V=O]$^{5+}$, 93
TMA, *see* Trimethylaluminum
TMBQ, *see* 2,3,5-Trimethyl-*p*-benzoquinone (TMBQ)
[(TMC)Fe^V=O(NC(O)CH$_3$)]$^+$, 102
[(TMG$_3$tren)Cu^{II}(η^1-O$_2$$^{•-}$)]$^+$, 63, 64
TMP, *see* 2,3,6-Trimethylphenol (TMP)
TMPO, *see* 3,3,5,5-Tetramethyl-pyrroline-*N*-oxide (TMPO)
TOCSY, *see* TOtal Correlation SpectroscopY
Torque, 17
TOtal Correlation SpectroscopY (TOCSY), 37
[(TPA)Fe^{II}(CH$_3$CN)$_2$](ClO$_4$)$_2$, 96
[(TPA)Fe^V=O(OAc)]$^{2+}$, 102
TPA ligand, 40
TPEN, *see* N,N,N′,N′-Tetrakis(2-pyridylmethyl)-ethylenediamine
Transition metal–catalyzed oxidations, intermediates of; *see also* Oxo complexes
  alkylperoxo complexes, 67–80
  Co(III) acetate in solution, 113–116

  peroxo complexes, 80–88
  superoxo complexes, 56–67
Transition metal complexes, 29, 47
Transition metal ions, 61
Transverse relaxation time, 18
2,4,6-Tri-*tert*-butyl phenol, 56
Triisobutylaluminum (TIBA), 148
1,4,7-Trimethyl-1,4,7-triazacyclononane (Me$_3$tacn), 108
2,3,5-Trimethyl-*p*-benzoquinone (TMBQ), 86
Trimethylaluminum (TMA), 128
2,3,6-Trimethylphenol (TMP), 86
Triphenylmethane, 62
Triphenylphosphine oxide (Ph3PO), 88–89
Triplet oxygen (^{3}O$_2$), 55
Triplet structure, 5–6
2,4,6-Trisubstituted phenol, 56
TS-1, *see* Titanium silicalites
2D heteronuclear correlation spectroscopy, 35
2D homo-nuclear *COrrelation SpectroscopY* (^{1}H,^{1}H COSY), 30, 31, 37
Two-dimensional ^{1}H and ^{13}C NMR spectra, 29
  ^{13}C{^{1}H} NMR spectrum, 33
  ^{13}C *J*-modulated NMR spectrum, 34
  ^{13}C NMR experiment with ^{1}H decoupling, 33
  ^{1}H, ^{1}H COSY spectrum, 32
  2D ^{1}H, ^{1}H COSY experiment, 31
  diastereotopic protons, 37
  pulse sequence for H, X-COSY experiment, 36
  pulse sequence for *J*-modulated, 34
  sample compound—dimeric titanium(IV), 30
  structure of hydrocarbon framework, 35
  two-dimensional ^{1}H, ^{13}C HXCO spectrum, 36
Two-dimensional NMR techniques (2D NMR techniques), 27

V

Vanadium
  chemo-and stereoselective oxidizing catalyst systems, 84
  [γ-PV$_2$W$_{10}$O$_{38}$(OH)$_2$]$^{3-}$, 85
  mono-and bis(peroxo) complexes, 83–84
  monoperoxo complexes of, 84
  oxidation of hydrocarbons, 85
  peroxo complexes of, 83
  phosphotungstate, 86
  Schiff base, 86
  ^{51}V and ^{13}C{^{1}H} NMR spectroscopy, 87
Vanadium(V), 72
  alkylperoxo complexes of, 76
  diastereomeric vanadium(V) *tert*-butyl peroxo complexes, 79–80
  η^2-alkylperoxo complex, 76
  structure and reactivity of, 78–79

Index

synthesized alkylperoxo complexes, 76–77
Visualize scalar (J), 27
^{51}V NMR spectroscopy, 80, 87
VO(acac)$_2$(OOtBu), 77
VO(O$_2$)(Pic)(H$_2$O)(CH$_3$CN), 84
VO(O$_2$)(Pic)(H$_2$O)$_2$, 84
VO(OOtBu)(OnBu)$_2$, 77
VO(OOtBu)(OtBu)$_2$, 78
VO(OOtBu)$_2$(OnBu), 77
VO(OOtBu)$_2$(OtBu), 78
VO(OOtBu)$_3$ complex, 77
VO(OtBu)$_3$, 78

W

Waveguides, 4

X

X-band CW EPR spectroscopy, 4
X-band EPR spectrometers, 3, 4
X-ray adsorption spectroscopy (XAS), 61

Z

Ziegler–Natta catalysts, 127
Ziegler–Natta polymerization catalysts, 127
Zirconium complexes, 154
Zirconium precatalysts, 151
Zirconocene/MMAO catalyst systems, 149–152
Zr-polymeryl species VIII, VIII$_{ethene}$, and VIII$_{propene}$, 156
"Zwitterion-like" intermediates, 133–134